Methods in Molecular Biology™

AF180549

Series Editor
John M. Walker
School of Life Sciences
University of Hertfordshire
Hatfield, Hertfordshire, AL10 9AB, UK

For further volumes:
http://www.springer.com/series/7651

Glycosylation Engineering of Biopharmaceuticals

Methods and Protocols

Edited by

Alain Beck

Antibody Physico-Chemistry Department, Centre d'Immunologie Pierre-Fabre, 5 Avenue Napoléon III, BP 60497, Saint Julien-en-Genevois, France

Humana Press

Editor
Alain Beck, Ph.D.
Antibody Physico-Chemistry Department
Centre d'Immunologie Pierre-Fabre
5 Avenue Napoléon III, BP 60497
Saint Julien-en-Genevois, France

ISSN 1064-3745 ISSN 1940-6029 (electronic)
ISBN 978-1-4939-6266-2 ISBN 978-1-62703-327-5 (eBook)
DOI 10.1007/978-1-62703-327-5
Springer New York Heidelberg Dordrecht London

© Springer Science+Business Media New York 2013
Softcover re-print of the Hardcover 1st edition 2013
This work is subject to copyright. All rights are reserved by the Publisher, whether the whole or part of the material is concerned, specifically the rights of translation, reprinting, reuse of illustrations, recitation, broadcasting, reproduction on microfilms or in any other physical way, and transmission or information storage and retrieval, electronic adaptation, computer software, or by similar or dissimilar methodology now known or hereafter developed. Exempted from this legal reservation are brief excerpts in connection with reviews or scholarly analysis or material supplied specifically for the purpose of being entered and executed on a computer system, for exclusive use by the purchaser of the work. Duplication of this publication or parts thereof is permitted only under the provisions of the Copyright Law of the Publisher's location, in its current version, and permission for use must always be obtained from Springer. Permissions for use may be obtained through RightsLink at the Copyright Clearance Center. Violations are liable to prosecution under the respective Copyright Law.
The use of general descriptive names, registered names, trademarks, service marks, etc. in this publication does not imply, even in the absence of a specific statement, that such names are exempt from the relevant protective laws and regulations and therefore free for general use.
While the advice and information in this book are believed to be true and accurate at the date of publication, neither the authors nor the editors nor the publisher can accept any legal responsibility for any errors or omissions that may be made. The publisher makes no warranty, express or implied, with respect to the material contained herein.

Printed on acid-free paper

Humana Press is a brand of Springer
Springer is part of Springer Science+Business Media (www.springer.com)

Foreword: Glyco-Engineering of Monoclonal Antibodies

Monoclonal antibodies (mAbs) have become increasingly important as innovative therapeutic agents during the past two decades. Interest by biopharmaceutical firms in developing these products has been driven by critical advances in the engineering, design and manufacturing of mAbs, as well as improved understanding of target biology and antibody mechanisms of action. As of 2012, the rate at which new mAb product candidates enter clinical study has risen to ~60 per year, a total of ~350 mAbs are undergoing evaluation in clinical studies, over 30 mAbs have been approved in either the United States or European Union, and annual global sales of mAb therapeutics total nearly $60 billion.

Although canonical antibodies (i.e., unmodified full-length molecules) have a notable record of success, the biopharmaceutical industry, often in collaboration with academic and government organizations, is now dedicating substantial resources to the exploration of new types of mAbs that may fulfill unmet medical need, or prove safer and more efficacious compared with the currently marketed products. In particular, research has focused on improving critical antibody properties (e.g., potency) by controlling the composition of carbohydrates on the molecule. A commonly used approach to glyco-engineering antibodies involves reduction of fucose, which has been shown to enhance antibody-dependent cell-mediated cytotoxicity (ADCC). Two companies, BioWa/Kyowa Hakko Kirin and GlycArt/Roche, have developed proprietary cell lines that yield defucosylated antibodies. To achieve this result, BioWa/Kyowa Kirin Hakko established α-1,6-fucosyltransferase (FUT8) enzyme knockout Chinese hamster ovary cell line (POTELLIGENT® technology), while GlycArt/Roche chose to over-express heterologous β1,4-N-acetylglucosaminyltransferase III in antibody-producing cells (GlycoMab™ technology). This glyco-engineering approach was validated with the approval in Japan of mogamulizumab (POTELIGEO®), which was derived from the POTELLIGENT® technology. Mogamulizumab's approval as a treatment for patients with relapsed or refractory CCR4-positive adult T-cell leukemia-lymphoma, which was granted March 30, 2012 by the Japanese Ministry of Health, Labour and Welfare, is a triumph for the glyco-engineering field.

To my knowledge, a total of 15 glyco-engineered mAbs are in the commercial clinical pipeline as of mid-2012. Six POTELLIGENT®-derived mAbs are in clinical studies, with three (benralizumab, MEDI-551, BIW-8962) in Phase 2 studies and three (KHK2898, KHK2804, KHK2866) in Phase 1 studies. Benralizumab targets interleukin-5 receptor alpha chain, and it is undergoing evaluation as a treatment for asthma and for chronic obstructive pulmonary disease. MEDI-551, which targets CD19 on B cells, is currently in Phase 2 studies in adults with diffuse large B cell lymphoma or adults with chronic lymphocytic leukemia, Phase 1/2 studies of patients with scleroderma or relapsing-remitting multiple sclerosis, and Phase 1 studies in adults with advanced B cell malignancies. Anti-GM2 ganglioside BIW-8962 was undergoing evaluation as monotherapy in a Phase1/2 study of patients with previously treated multiple myeloma, but the study was terminated due to lack of efficacy. KHK2898, KHK2804, and KHK2866 target CD98, a tumor glycan, and heparin binding epidermal growth factor-like growth factor, respectively; these mAbs are undergoing evaluation in Phase 1 studies as therapies for patients with advanced solid tumors.

Three GlycoMab™-derived mAbs (obinutuzumab, GA201, RO5479599) are in clinical study. Anti-CD20 obinutuzumab, which is also known as GA101, is undergoing evaluation in four Phase 3 studies, four Phase 2 studies, and two Phase 1 studies, all of which include patients with hematological malignancies such as non-Hodgkin's lymphoma and chronic lymphocytic leukemia. The anti-epidermal growth factor receptor (EGFR) GA201 is currently being evaluated in Phase 2 studies of patients with non-small cell lung cancer or colorectal cancer, and a Phase 1 study of patients with head and neck squamous cell carcinoma. RO5479599, which targets human epidermal growth factor receptor (HER)-3, is in a Phase 1 dose-escalation study in patients with HER3-positive solid tumors.

The remaining six of the 15 glyco-engineered mAbs in the clinical pipeline are either derived from YB2/0 rat cells or Glycotope's GlycoExpress™ technology. The YB2/0 cell line expresses FUT8 at a low level, and thus it produces recombinant mAbs with low fucose content. Three mAbs (ecromeximab, roledumab, ublituximab) in clinical studies are derived from YB2/0 cells. The safety and effectiveness of anti-GD3 ecromeximab, developed by Kyowa Hakko and licensed by Life Science Pharmaceuticals, is being evaluated in a Phase 2 study of patients with metastatic melanoma. LFB is developing two low-fucose mAbs, roledumab, and ublituximab. Anti-rhesus (Rh) D roledumab was evaluated in a Phase 2 study designed to demonstrate the ability of LFB-R593 to effectively eliminate exogenously administered RhD-positive red blood cells from the circulation of an RhD-negative individual, thereby preventing RhD alloimmunization. Anti-CD20 ublituximab, which is licensed to TG Therapeutics, is being evaluated in a Phase 1/2 study of patients with relapsed or refractory B-cell lymphoma who were previously treated with rituximab. The GlycoExpress™ technology utilizes glyco-engineered human cell lines that allow expression of antibodies with varying percentages of sialylation, fucosylation, and galactosylation. Three GlycoExpress™-derived mAbs (GT-MAB2.5GEX, GT-MAB5.2GEX, GT-MAB7.3GEX) are in Phase 1 studies. The safety and tolerability of GT-MAB2.5GEX, which targets MUC1, is being evaluated in a dose escalation study in patients with advanced MUC1-positive solid malignancies. Anti-EGFR GT-MAB5.2GEX and anti-HER2 GT-MAB7.3GEX are undergoing evaluation in Phase 1 studies of patients with EGFR-positive and HER2-positive solid tumors, respectively.

Enhancing ADCC through reduction of fucose in antibodies produced in mammalian cells has proven valuable, but there are numerous other approaches to antibody glyco-engineering that are also being explored. Antibodies with humanized glycoforms can now be produced in non-mammalian cell lines such as yeast (*Pichia pastoris* and *Saccharomyces pombe* and *cerevisiae*), filamentous fungi (*Aspergillus niger and nidulans*), duck embryonic stem cells, and plants. Galactosylation levels, which are correlated with complement-dependent cytotoxicity activity, can be manipulated via the concentrations of components of the cell culture medium. In yet another approach, alterations in antibody sialyation, which have been shown to affect anti-inflammatory properties and ADCC, can be achieved through glyco-engineering. These approaches are all being applied to the development of glyco-engineered biobetter or next-generation mAbs that may enter the clinical pipeline in the near future.

Because of the importance of advances in glyco-engineering to antibody development and the obvious promise of designed mAbs as therapeutics, the publication of *Glycosylation Engineering of Biopharmaceuticals*, edited by Alain Beck, is notable and timely. The current methodologies for therapeutic glycoprotein engineering used to achieve a variety of effects, including strategies to enhance cytotoxicity or inflammatory properties, are explained in a clear and concise manner. Production of glyco-engineered biopharmaceuticals from various cell types, including mammalian, yeast or fungi, baculovirus/insect, prokaryote and plant, and via biochemistry- or chemical-based methods, is discussed in detail. In addition, critical protocols for glyco-analysis, functional assays, and pharmacokinetic, pharmacodynamic and toxicology assessment of glyco-engineered biopharmaceuticals are included. *Glycosylation Engineering of Biopharmaceuticals* thus provides comprehensive coverage of this important topic and will most certainly prove to be a valuable resource for both experts and those new to the field.

Janice M. Reichert, Ph.D.

Preface

Therapeutic properties of monoclonal antibodies (mAbs) and other glycoproteins strongly depend on the composition of their glycans. Most of the currently approved biopharmaceuticals are produced in mammalian cell lines, which yield mixtures of different glycoforms that are close to those of humans but not fully identical. Glyco-engineering is being developed as a method to control the composition of carbohydrates and to enhance the pharmacological properties of mAbs and other proteins. The approval, on April 30, 2012, in Japan, of mogamulizumab, the first glyco-engineered antibody to reach the market, was a landmark in the field of engineered biopharmaceuticals. Mogamulizumab is a humanized mAb with enhanced antibody-dependent cell-mediated cytotoxicity (ADCC) activity linked to optimized a-fucosylated glycoforms and an illustration of the therapeutic importance of a tailored glycosylation. The antibody is indicated for patients with relapsed or refractory CCR4-positive adult T-cell leukemia-lymphoma.

The aim of this present volume of *Methods in Molecular Biology* is to provide readers with production and characterization protocols of glycoproteins and glyco-engineered biopharmaceuticals with a focus on mAbs. The volume is divided in four complementary parts dealing with Glyco-engineering of therapeutic proteins (Part I), Glycoanalytics (Part II), Glycoprotein complexes characterization (Part III), and PK/PD assays for therapeutic antibodies (Part IV).

The first two chapters deal with recombinant glycoproteins produced in Chinese Hamster Ovary (CHO) cells, the most frequently used cell line to produce biopharmaceuticals. J. Holgersson (University of Gothenburg, SE) and colleagues report methods to engineer therapeutic and diagnostic O-glycans on recombinant mucin-type immunoglobulin fusion proteins. C. Ronin et al. (Siamed'Xpress, FR) follow up with protocols to engineer human-like glycosylation of therapeutic glycoproteins based on 6-linked sialylation. The next three chapters discuss the use of non-mammalian cell line to produce glycoproteins including yeasts (*Pichia pastoris* and *Saccharomyces cerevisiae*) and insect cells infected with baculoviruses. First, D. Zha (Merck-GlycoFi, US) describes the production of glyco-engineered Pichia-based expression of Monoclonal Antibodies. Then C. Javaud (Glycode, FR) presents the humanization of N-glycosylation of antibodies produced in *S. cerevisiae*. Finally, M. Cérutti (CNRS, FR) and colleagues report methods to engineer the baculovirus genome to produce galactosylated antibodies.

To assess the structure of glycoprotein and glyco-engineered biopharmaceuticals, state-of-the art orthogonal analytical methods are needed. E. Wagner-Rousset (CIPF, FR), C. Schaeffer-Reiss (CNRS-LSMBO), and colleagues describe nanoLC-chips-MS/MS methods for the characterization on N-glycopeptides generated from trypsin digestion of monoclonal antibodies. Alternatively, M.C-Janin Bussat, L. Tonini, and colleagues (CIPF, FR) propose the use of IdeS proteolytic digestion and electrospray ionization—time-of-flight mass spectrometry for antibody fast differential glycoprofilling of cetuximab Fab and Fc glycans. Then, A. Delobel, G. Van Vyncht, et al. (Quality Assistance, BE) report a panel of analytical methods that are used to characterize therapeutic antibody glycosylation for batch release or comparability support, tacking the case of trastuzumab. To have a complete picture, mass spectrometric analysis of *O*-linked oligosaccharides from various recombinant expression systems are described by J. Holgersson, N.G. Karlsson, and colleagues (University of Gothenburg, SE). In complement to mass spectrometry, glycoprofilling can also be performed by liquid chromatography and by electrophoresis based methods. This is illustrated by T.S. Raju (Janssen R&D, US) who reports the assessment of Fc Glycan heterogeneity of therapeutic recombinant mAbs by Normal Phase—HPLC, and by R.R. Rustandi and colleagues (Merck) who report two different Capillary Electrophoresis systems. Alternatively glycoprofilling may be performed based on lectins as illustrated by L. Landemarre (GLYcodiag, FR) and E. Duverger (Université d'Orléans, FR) for recombinant therapeutic Interleukin-7. Glycosylation also has a deep impact on glycoprotein solubility and limits the propensity to aggregate. This can be assessed either by Hydrophobic Interaction Chromatography, as illustrated by R.R. Rustandi (Merck, US), or by Sedimentation Velocity Analytical Ultracentrifugation as reported by W.B. Stine (Abbott, US).

To go a step forward, glycoprotein complexes can be characterized by emerging mass spectrometry methods. S. Sanglier-Cianferani (CNRS-LSMBO), E. Wagner-Rousset (CIPF, FR), and colleagues report the use of non-covalent mass spectrometry for the characterization of antibody/antigen complexes. In addition, conformational analysis of recombinant mAbs can be performed by Hydrogen/Deuterium Exchange Mass Spectrometry as reported by D. Houde (BiogenIdec, US) and J.R. Engen (Northeastern University, Boston, MA). To gain insights on the interaction of antibodies with their target antigens, epitope and paratope can be mapped by different Mass Spectroscopy methods as described by Victor Obungu and colleagues (Lilly, US).

Last, but not least, PK/PD assays for therapeutic antibodies are mandatory to explore the impact of glycosylation or glyco-engineering on pharmacokinetics and potency. For this purpose, M. Broussas, L. Goetsch, and L. Broyer (CIPF, FR) describe a method for the evaluation of antibody-dependent cell cytotoxicity (ADCC) using Lactate Dehydrogenase Measurement and a method for complement-dependent cytotoxicity (CDC) determination using ATP measurement and C1q/C4b binding. As a surrogate in vitro assay, the capture of the human IgG1 antibodies by protein A for the kinetic study of h-IgG/FcgammaR interaction using SPR-based biosensor technology is reported by T. Champion (CIPF, FR). To evaluate the antibody clearance, a mass spectrometry protocol for the absolute quantification of a mAbs in serum with immuno-purification is described by F. Becher and colleagues (CEA, FR).

I would like to acknowledge Janice M. Reichert for her foreword as well as John M. Walker for his invitation to edit this volume of Methods in Molecular Biology and for his enthusiasm and his support.

The book is dedicated to my wife Nathalie, to my daughters Juliette and Louise, and to my parents Paulette and Norbert. Thanks also to Claire Catry for her help in some logistics aspect concerning this book.

Saint Julien-en-Genevois, France ***Alain Beck***

Contents

Contributors

CARRIE ANDERSON • *Vaccine Analytical Development, Merck Research Laboratories, West Point, PA, USA*

CHRISTELLE ARICO • *Glycode S.A.S, Uzerche, France*

CÉDRIC ATMANENE • *Laboratoire de Spectrométrie de Masse Bio-Organique, IPHC, CNRS, UMR7178, Université de Strasbourg, Strasbourg, France*

ALAIN BECK • *Antibody Physico-Chemistry Department, Centre d'Immunologie Pierre-Fabre, Saint Julien-en-Genevois, France*

FRANÇOIS BECHER • *Service de Pharmacologie et d'Immunoanalyse, CEA, Gif-sur-Yvette, France*

AUDREY BEDNARCZYK • *Laboratoire de Spectrométrie de Masse Bio-Organique, IPHC, CNRS, UMR7178, Université de Strasbourg, Strasbourg, France*

CHRISTINE BONNET • *Glycode S.A.S, Uzerche, France*

MATTHIEU BROUSSAS • *Experimental Oncology Department, Centre d'Immunologie Pierre-Fabre, Saint Julien-en-Genevois, France*

LUCILE BROYER • *Experimental Oncology Department, Centre d'Immunologie Pierre-Fabre, Saint Julien-en-Genevois, France*

VALÉRIE CALABRO • *Siamed'Xpress, University Aix-Marseille, Marseille, France*

FABRICE CANTAIS • *Quality Assistance, Donstiennes, Belgium*

ANICET CATRAIN • *Quality Assistance, Donstiennes, Belgium*

MARTINE CÉRUTTI • *CNRS UPS3044, Saint Christol-Lèz-Alès, France*

PIERRE CÉRUTTI • *CNRS UPS3044, Saint Christol-Lèz-Alès, France*

THIERRY CHAMPION • *Physico-Chemistry Department, Centre d'Immunologie Pierre-Fabre, Saint Julien-en-Genevois, France*

GRACE CHEN • *Vaccine Analytical Development, Merck Research Laboratories, West Point, PA, USA*

SYLVIE CHOBLET • *CNRS UPS3044, Saint Christol-Lèz-Alès, France*

OLIVIER COLAS • *Physico-Chemistry Department, Centre d'Immunologie Pierre-Fabre, Saint Julien-en-Genevois, France*

NATHALIE CORVAÏA • *Centre d'Immunologie Pierre-Fabre, Saint-Julien-en-Genevois, France*

ARNAUD DELOBEL • *Quality Assistance, Donstiennes, Belgium*

ERELL DEREUX • *Quality Assistance, Donstiennes, Belgium*

SANDRINE DONADIO-ANDRÉI • *Siamed'Xpress, University Aix-Marseille, Marseille, France*

MATHIEU DUBOIS • *Service de Pharmacologie et d'Immunoanalyse, CEA, Gif-sur-Yvette, France*

ERIC DUVERGER • *Laboratoire de Neurobiologie, Université d'Orléans, Orléans, France*

NASSIMAL EL MAÏ • *Siamed'Xpress, University Aix-Marseille, Marseille, France*

JOHN R. ENGEN • *Department of Chemistry and Chemical Biology, The Barnett Institute of Chemical and Biological Analysis, Northeastern University, Boston, MA, USA*

ERIC EZAN • *Service de Pharmacologie et d'Immunoanalyse, CEA, Gif-sur-Yvette, France*
FRANÇOIS FENAILLE • *Service de Pharmacologie et d'Immunoanalyse, CEA, Gif-sur-Yvette, France*
STEFAN GAUNITZ • *Division of Clinical Immunology and Transfusion Medicine, Karolinska Institute, Stockholm, Sweden*
VALENTINA GELFANOVA • *Lilly Research Laboratories, Indianapolis, IN, USA*
LILIANE GOETSCH • *Experimental Oncology Department, Centre d'Immunologie Pierre-Fabre, Saint Julien-en-Genevois, France*
ANKI GUSTAFSSON • *Recopharma AB, Stockholm, Sweden*
SHA HA • *Vaccine Analytical Development, Merck Research Laboratories, West Point, PA, USA*
MELISSA HAMM • *Vaccine Analytical Development, Merck Research Laboratories, West Point, PA, USA*
ANNE HARDUIN-LEPERS • *Unité de Glycobiologie Structurale et Fonctionnelle, CNRS UMR8576, Université Lille Nord de France, Lille1 Villeneuve-d'Ascq, France*
CATHERINE A. HAYES • *Medical Biochemistry, University of Gothenburg, Gothenburg, Sweden*
JAN HOLGERSSON • *AbSorber AB, Stockholm, Sweden; Division of Clinical Chemistry and Transfusion Medicine, Sahlgrenska Academy, University of Gothenburg, Gothenburg, Sweden*
DAMIAN HOUDE • *Biogen Idec, Inc., Cambridge, MA, USA*
LIHUA HUANG • *Lilly Research Laboratories, Indianapolis, IN, USA*
CÉLINE HUILLET • *Physico-Chemistry Department, Centre d'Immunologie Pierre-Fabre, Saint Julien-en-Genevois, France*
CHLOÉ ISS • *Siamed'Xpress, University Aix-Marseille, Marseille, France*
MARIE-CLAIRE JANIN-BUSSAT • *Physico-Chemistry Department, Centre d'Immunologie Pierre-Fabre, Saint Julien-en-Genevois, France*
CHRISTOPHE JAVAUD • *Glycode S.A.S, Uzerche, France*
SYLVIE JULIANT • *CNRS UPS3044, Saint Christol-Lèz-Alès, France*
NICLAS G. KARLSSON • *Medical Biochemistry, University of Gothenburg, Gothenburg, Sweden*
DIARMUID T. KENNY • *School of Chemistry and Medical Biochemistry, National University Ireland Galway and University of Gothenburg, Galway, Ireland and Gothenburg, Sweden*
CHRISTINE KLINGUER-HAMOUR • *Physico-Chemistry Department, Centre d'Immunologie Pierre-Fabre, Saint Julien-en-Genevois, France*
LUDOVIC LANDEMARRE • *GLYcoDiag, Orléans, France*
MARYLÈNE LÉVÊQUE • *CNRS UPS3044, Saint Christol-Lèz-Alès, France*
LINDA LINDBERG • *AbSorber AB, Stockholm, Sweden*
JINING LIU • *AbSorber AB, Stockholm, Sweden*
VICTOR H. OBUNGU • *Lilly Research Laboratories, Indianapolis, IN, USA*
ANNICK OZIL • *CNRS UPS3044, Saint Christol-Lèz-Alès, France*
T. SHANTHA RAJU • *Biologics Research, Janssen Research and Development, LLC, Radnor, PA, USA*

Janice M. Reichert • *Reichert Biotechnology Consulting LLC, Framingham, MA, USA*
Catherine Ronin • *Siamed'Xpress, University Aix-Marseille, Marseille, France*
Richard R. Rustandi • *Vaccine Analytical Development, Merck Research Laboratories, West Point, PA, USA*
Sarah Sanglier-Cianférani • *Laboratoire de Spectrométrie de Masse Bio-Organique, IPHC, CNRS, UMR7178, Université de Strasbourg, Strasbourg, France*
Christine Schaeffer-Reiss • *Laboratoire de Spectrométrie de Masse Bio-Organique, IPHC, CNRS, UMR7178, Université de Strasbourg, Strasbourg, France*
Magnus Sjöblom • *Department of Civil, Environmental and Natural Resources Engineering, Luleå University of Technology, Luleå, Sweden*
Marie-Christine Slomianny • *Unité de Glycobiologie Structurale et Fonctionnelle, CNRS UMR8576, Université Lille Nord de France, Lille1 Villeneuve-d'Ascq, France*
W. Blaine Stine Jr. • *Abbott Bioresearch Center, Worcester, MA, USA*
Laure Tonini • *Physico-Chemistry Department, Centre d'Immunologie Pierre-Fabre, Saint Julien-en-Genevois, France*
Alain Van Dorsselaer • *Laboratoire de Spectrométrie de Masse Bio-Organique, IPHC, CNRS, UMR7178, Université de Strasbourg, Strasbourg, France*
Géry Van Vyncht • *Quality Assistance, Donstiennes, Belgium*
Marie-Luce Violet • *CNRS UPS3044, Saint Christol-Lèz-Alès, France*
Elsa Wagner-Rousset • *Physico-Chemistry Department, Centre d'Immunologie Pierre-Fabre, Saint Julien-en-Genevois, France*
Dongxing Zha • *GlycoFi Inc., Merck and Co, Lebanon, NH, USA*

Part I

Glyco-Engineering of Therapeutic Proteins

Chapter 1

Engineering of Therapeutic and Diagnostic *O*-Glycans on Recombinant Mucin-Type Immunoglobulin Fusion Proteins Expressed in CHO Cells

Linda Lindberg, Jining Liu, and Jan Holgersson

Abstract

Metabolic engineering of mammalian cells for optimized glycosylation is usually done to improve activity and the pharmacokinetic features of glycoprotein therapeutics. The field is mainly focused around engineering of *N*-glycans. We have created a platform in which recombinant mucin-type immunoglobulin fusion proteins are used as scaffolds for multivalent expression of *O*-glycans with diagnostic or therapeutic potential. The methods used to make stable CHO cell lines secreting a mucin-type fusion protein with blood group A or B determinants following expression of up to five different cDNAs are described.

Key words Glycoengineering, *O*-Glycans, Mucins, CHO cells, Fusion protein, Glycosyltransferase, ABO blood group system

1 Introduction

Many interactions taking place at the cell surface, e.g., microbial and cell adhesion, are dependent on receptor proteins binding to carbohydrate ligands (1). The overall binding strength of such protein–carbohydrate interactions is enhanced by multivalency, i.e., more than one receptor binding multiple copies of a particular carbohydrate ligand (1–3). Thus, inhibitors of such interactions should present several copies of the carbohydrate determinant. Mucins are proteins characterized by a high degree of *O*-glycan substitution of their mucin domains: domains which are built up of short peptide repeats with abundant serines, threonines, and prolines (4, 5). They are important constituents of the protective mucus layer covering our mucosal surfaces as well as the glycocalyx layer of the cell surface (4, 5).

We have used genetic engineering of mammalian, insect, and yeast cells to direct expression of mucin-type immunoglobulin

Alain Beck (ed.), *Glycosylation Engineering of Biopharmaceuticals: Methods and Protocols*, Methods in Molecular Biology, vol. 988, DOI 10.1007/978-1-62703-327-5_1, © Springer Science+Business Media New York 2013

fusion proteins with tailored *O*-glycan substitution (1). Because the glycosyltransferases involved in the biosynthesis of many *O*-glycan core structures and determinants are known, host cells can be glyco-engineered to express multiple copies of a desired carbohydrate determinant on the mucin scaffold (1). Recombinant mucin-type fusion proteins carrying blood group A and B, sialyl-Le^x, Le^b, and other determinants on different *O*-glycan core structures have been engineered (1, 6–9).

In this chapter we describe the protocol that was used to express up to five different cDNAs stably in CHO cells, which under serum-free conditions secreted a mucin-type immunoglobulin fusion protein (P-selectin glycoprotein ligand-1/mouse IgG_{2b}; PSGL-1/mIgG_{2b}) carrying *O*-glycans with blood group A or B determinants on the type 1 (Galβ1,3GlcNAc), 2 (Galβ1,4GlcNAc), or 3 (Galβ1,3GalNAcα) outer core saccharide chains.

While generating a repertoire of stable CHO transfectants, we have used different transfection techniques, transfected multiple cDNAs simultaneously or sequentially, and transfected adherently growing cells cultured in the presence of serum as well as cells cultured in suspension under serum-free conditions. From our gained experience and continuous method improvements, a protocol has evolved in which CHO cells adapted to serum-free conditions are exposed to low levels of serum at the time of transfection and during initial selection. Subsequent selection and clone expansion are done under serum-free conditions. We will focus on this protocol that has worked well in our lab to generate stable transfectants for production of a recombinant protein with tailored *O*-glycan substitution, but also in parallel describe a protocol where serum-dependent cells are transfected and selected. Benefits and drawbacks of the different protocols will be commented on. In addition, methods routinely used in our lab for screening of clones and confirmation of gene expression stability will be presented.

2 Materials

2.1 Cell Culture

1. Chinese hamster ovary (CHO) cells: CCL-61 (ATCC®, Manassas, USA).
2. Dulbecco's modified Eagle's medium (DMEM) supplemented with 10% fetal bovine serum (FBS) (see Note 1), 2 mM L-glutamine, and 25 μg/mL gentamicin.
3. ProCHO-4 medium (Lonza, Basel, Switzerland) supplemented with 2 mM L-glutamine, 100 μg/mL dextran sulfate (M_r 5000), and 25 μg/mL gentamicin.
4. Trypsin–EDTA: 0.05% Trypsin in 0.53 mM EDTA.
5. Phosphate-buffered saline (PBS).
6. Dimethyl sulfoxide (DMSO).

2.2 Transfection

1. Linearized plasmid DNA.
2. Lipofectamine™ 2000 (Invitrogen, Carlsbad, CA, USA).

2.3 Selection Reagents

1. Puromycin.
2. G-418 Sulfate.
3. Zeocin.
4. Hygromycin B.
5. Blasticidin S.
6. GPT Selection Reagent: mycophenolic acid, xanthine, and hypoxanthine.

2.4 Equipment and Plastics

1. Humidified CO_2 incubator.
2. CO_2 shaker incubator.
3. Laboratory centrifuge.
4. Microscope.
5. Corning® CellBIND® Surface tissue culture vessels (Corning Inc, NY, USA).
6. Tissue culture clusters (96-well, 24-well, and 6-well).
7. Tissue culture dishes (10 cm).
8. Tissue culture T-flasks (25 and 75 cm^2).
9. Erlenmeyer flasks for cell culture.
10. Pipetting aid.
11. Micropipettes.
12. Multichannel pipette (200–300 μL).
13. Sterile pipettes.
14. Sterile pipette tips.
15. Polypropylene microtubes.
16. Sterile centrifuge tubes.
17. Sterile reagent-dispensing reservoirs.
18. Cryovials.

3 Methods

3.1 CHO Cell Culture and Adaptation to Serum-Free Conditions

CHO cells can be directly switched from serum-supplemented medium into serum-free medium, or sequentially adapted to serum-free conditions in several steps. Sequential adaptation is less harsh on cells, and if stable transfectants generated in the presence of serum (see Subheading 3.4.2) are adapted to serum-free conditions, it may be advised to perform sequential adaptation to decrease the risk of gene expression being lost.

1. Propagate cells in DMEM in tissue culture flasks at 37°C in a humidified CO_2 incubator.
2. Use cells at low passage numbers and with viability above 90% for adaptation to serum-free conditions. Passage cells into a 1:1 mixture of DMEM and serum-free ProCHO-4 medium. Passage cells successively into 1:4 and 1:8 mixtures of DMEM and ProCHO-4 medium, and finally into 100% ProCHO-4 medium (see Notes 2–4). Use normal seeding conditions and maintain cells at each serum level for two passages before proceeding.
3. Once cells are adapted to serum-free conditions, maintain cells in a single-cell suspension culture in shaker flasks at 100 rpm at 37°C with 5% CO_2.

3.2 Determination of Optimal Concentrations of Antibiotics for Selection

1. Plate approx. 2×10^5 cells per well in a 6-well CellBIND® tissue culture plate in 2 mL of the same medium as to be used in the selection of stable transfectants (see Subheading 3.5). Use ProCHO-4 medium containing 1–2 % FBS if CHO cells adapted to serum-free conditions are to be selected at low serum levels, or DMEM for adherently growing CHO cells (see Note 5).
2. On the next day, add increasing concentrations of antibiotics into individual wells.
3. Examine viability every day.
4. Replace the medium containing antibiotics every 3 days (or as needed).
5. The lowest concentration of antibiotics that results in the onset of massive cell death in 3–5 days and kills all cells within 2 weeks should be used (see Note 6). The concentrations of different antibiotics used by us are given in Table 1.

3.3 Preparation of Plasmid DNA

The expression vectors used by us to generate stable transfectants are bidirectional having the EF1α promoter or the CMV promoter upstream of a polylinker identical to the one in the CDM8 expression vector, a splice donor and acceptor site, and the bidirectional poly(A) additional signal of SV40; opposite in orientation to this transcription unit, and utilizing the poly(A) signals from the opposite direction, is a second transcription unit consisting of the HSV TK promoter followed by the coding sequence for different resistance genes (see Note 7 and Table 2).

1. Purify the plasmid DNA using a method that yields high-quality DNA free of protein, RNA, and chemical and microbial contamination (see Note 8).
2. Linearize the plasmid DNA with a suitable restriction enzyme, and purify DNA from the enzymatic reaction (see Note 9).

Table 1
Antibiotics used for selection in CHO cells

Antibiotic	Working concentration[a] (μg/mL)
Puromycin	2–6
G418	400–900
Zeocin	50–200
Hygromycin	50–200
GPT selection reagent	
Mycophenolic acid	25
Xanthine	0.25
Hypoxanthine	13.6
Blasticidin	1–4

[a]The lower concentrations refer to selection under serum-free conditions and in combination with multiple antibiotics

Table 2
Expression vectors used to generate stable transfectants in CHO cells

Expression vector	cDNA	Resistance gene	Reference
EF1α/PSGL-1/EK/ mIgG_{2b}/PAC	PSGL-1/mIgG_{2b} fusion gene	Puromycin acetyl transferase (puromycin resistance)	Liu et al. (10)
CMV/FUT1/Zeo	α1,2-Fucosyltransferase-I (encoded by the *H*-gene)	Sh*Ble* (zeocin resistance)	Löfling et al. (9)
CMV/FUT2/Neo	α1,2-Fucosyltransferase-II (encoded by the *Se*-gene)	Neomycin phosphotransferase (G418 resistance)	Löfling et al. (9)
EF1α/β1,6GlcNAcT/ Neo	β1,6-*N*-Acetylglucosaminyltransferase-I (core 2 synthase)	Neomycin phosphotransferase (G418 resistance)	Liu et al. (7)
EF1α/β1,6GlcNAcT/ Hyg	β1,6-*N*-Acetylglucosaminyltransferase-I (core 2 synthase)	Hygromycin phosphotransferase (hygromycin B resistance)	Liu et al. (7)
CMV/β1,3GlcNAcT/ Zeo	β1,3-*N*-Acetylglucosaminyltransferase-VI (core 3 synthase)	Sh*Ble* (zeocin resistance)	Löfling et al. (11)
CMV/GalT5/Gpt	β1,3-Galactosyltransferase-V	Guanosine phosphoribosyl transferase (mycophenolic acid, xanthine, and hypoxanthine resistance)	Löfling et al. (11)
CMV/α1,3GalNAcT/ Bsd	α1,3-*N*-Acetylgalactosyltransferase (encoded by the *A*-gene)	Blasticidin S deaminase (blasticidin resistance)	
CMV/α1,3GalT/Bsd	α1,3-Galactosyltransferase (encoded by the *B*-gene)	Blasticidin S deaminase (blasticidin resistance)	

3.4 Transfection Using Lipofectamine™ 2000

Described below are two protocols that we have used to generate stable CHO transfectants: transfection of cells adapted to serum-free conditions or transfection of adherently growing cells. Benefits and drawbacks of the different protocols are commented on (see Note 10).

3.4.1 Transfection of CHO Cells Adapted to Serum-Free Conditions

Day 1

1. On the day before the transfection, plate cells adapted to serum-free conditions (see Subheading 3.1) in ProCHO-4 medium containing 1–2 % FBS in a CellBIND® tissue culture flask of suitable format (see Notes 4 and 11). Use a seeding density that results in 90–95 % confluent cells at the time of transfection, that is, approx. 1 × 10(6) cells per mL in 10 mL medium for a 75 cm^2 flask (see Note 12). Use cells at low passage numbers and with viability above 90 % for transfection. Incubate cells at 37°C in a humidified CO_2 incubator.

Day 2

2. Prepare DNA–Lipofectamine™ 2000 complexes according to the manufacturer's protocol (see Notes 13–15).
3. Thirty minutes before adding the complexes to cells, replace the medium with DMEM (see Note 16).
4. Approx. 6 h after the transfection, replace the medium with ProCHO-4 medium containing 1–2% FBS. Incubate cells at 37°C in a humidified CO_2 incubator.

Day 3

5. Passage cells at a 1:8 dilution (or higher) into ProCHO-4 medium containing 1–2% FBS and plate cells in CellBIND® tissue culture dishes. Incubate cells at 37°C in a humidified CO_2 incubator.

Day 4

6. Start selection in ProCHO-4 medium containing 1–2% FBS and appropriate selection drugs as described below (see Subheading 3.5).

3.4.2 Transfection of Serum-Dependent CHO Cells

Day 1

1. Plate adherently growing cells maintained in DMEM (see Subheading 3.1, step 1) in a tissue culture flask on the day before transfection. Use a seeding density that results in 90–95% confluent cells at the time of transfection (see Note 12). Incubate cells at 37°C in a humidified CO_2 incubator.

Day 2

2. Prepare DNA–Lipofectamine™ complexes according to the manufacturer's protocol (see Notes 13–15).
3. Thirty minutes before adding the transfection complexes to cells, replace the medium with fresh DMEM.

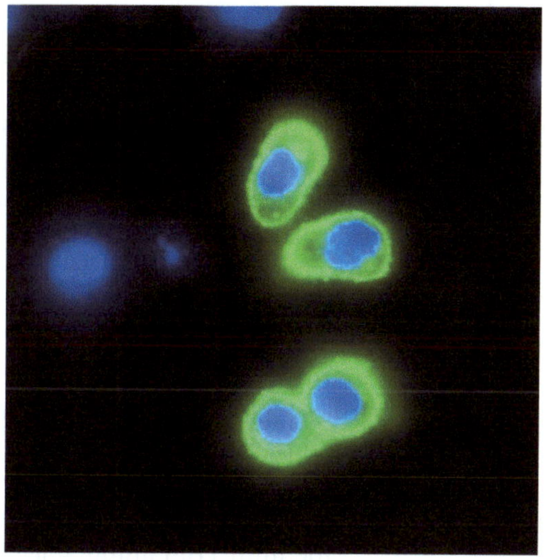

Fig. 1 Indirect immunocytochemical analysis of CHO cells showing transient expression of PSGL-1/mIgG$_{2b}$ 2 days after the transfection. PSGL-1/mIgG$_{2b}$ was stained with FITC-conjugated monoclonal anti-mouse IgG Fc (Sigma). Cell nuclei were stained with 4,6-diamidino-2-phenylindole (DAPI)

Day 3

4. Passage cells at a 1:8 dilution (or higher) into DMEM and plate cells in tissue culture dishes. Incubate cells at 37°C in a humidified CO_2 incubator.

Day 4

5. Start selection in DMEM containing appropriate selection drugs as described below (see Subheading 3.5).

3.4.3 Analysis of Transient Gene Expression (Optional)

Before proceeding into selection of stable transfectants, it is advised to confirm transient gene expression 2–3 days after the transfection (see Subheading 3.8 and Fig. 1).

3.5 Selection of Stable Transfectants

Cells should be maintained in nonselective medium for approx. 48 h after the transfection. Thereafter, they are incubated in cell culture medium containing optimal concentrations of appropriate selection drugs as determined from the establishment of cell death curves (see Subheading 3.2 and Note 6). The concentrations of different antibiotics used by us are given in Table 1.

1. On Day 4 (see Subheading 3.4), replace the medium with ProCHO-4 medium containing 1–2% FBS and selection drugs for selection of cells adapted to serum-free conditions, and with DMEM containing selection drugs for selection of serum-dependent cells.
2. Replace the selective medium every 3 days (or as needed) to eliminate dead cells and debris (see Note 17).

3. Incubate cells at 37°C in a humidified CO_2 incubator for approx. 10–14 days until distinct colonies can be visualized.

3.6 Single-Cell Cloning

Stable transfectants form distinct antibiotic-resistant colonies, which can be isolated by picking using sterile pipette tips. However, when selecting stable transfectants in medium with low serum levels, cells tend to detach, float around, and adhere at new locations with the risk of mixing cells from different clones. In such cases, it is advised to isolate clones by dilution of cells in 96-well plates. It is important to bear in mind that neither of the two single-cell cloning protocols described below guarantee that the colonies are monoclonal. Thus, repeated cloning is advised to increase the likelihood of monoclonality.

3.6.1 Picking of Clones

1. Fill a sterile reagent-dispensing reservoir with cell culture medium containing selection drugs and add 200 μL of the medium to all the wells in a 96-well plate using a multichannel pipette (see Note 18). Use serum-free ProCHO-4 medium (no addition of FBS) for cells preadapted to serum-free conditions and selected at low serum levels, and DMEM for serum-dependent cells.
2. Aspirate the medium from the tissue culture dish, wash the cells once with PBS, and add 10 mL of the same medium as used in step 1.
3. Pick up cell colonies one at a time by using a 200 μL pipette under a microscope. Find the colony under the microscope and loosen the colony by gently scraping with the pipette tip. Suck out 30 μL of medium containing the loosened colony into the tip and transfer into a well of the 96-well plate.
4. Incubate cells in the 96-well plate(s) for approx. 2 weeks until large cell clusters are visible.
5. Replace 100 μL of the medium using a multichannel pipette every 4–7 days (depending on cell density) and maintain cells in medium containing selection drugs.
6. Expand the bulk population of cells remaining after single-cell cloning of individual colonies to establish a backup of frozen vials.

3.6.2 Single-Cell Cloning in 96-Well Plates

1. Retain any viable cells that may have detached from the tissue culture dish by collecting the supernatant, pellet cells by centrifugation, and combine with the cells detached by trypsinization (step 3 below).
2. Wash the cells attached on the tissue culture dish once with PBS, detach cells with trypsin, and resuspend cells in ProCHO-4 medium containing 1–2% FBS.

3. Pellet all cells from steps 1 and 2 by centrifugation and resuspend the cells in serum-free ProCHO-4 medium (no addition of FBS).
4. Count cells using a hemocytometer.
5. Set up four (or more) 96-well plates with an average of 2, 1, 0.5 respective 0.25 cells per well in 200 μL serum-free ProCHO-4 medium (no addition of FBS) (see Notes 18 and 19).
6. Follow steps 4–6 as described above in Subheading 3.6.1.

3.7 Expansion of Clones and Adaptation to Suspension Culture in Shaker Flasks

1. Once large cell clusters have formed in the wells of the 96-well plate(s), it is advised to perform a screening for positive clones (see Subheading 3.8).
2. Transfer positive clones successively to larger tissue culture formats (24-well, 12-well, and 6-well plates, 25 and 75 cm^2 flasks). Use ProCHO-4 medium for cells adapted to serum-free conditions, and DMEM for adherently growing cells. Maintain cells in medium containing selection drugs or alternate between selective and nonselective medium in fairly short cycles. A lower maintenance drug concentration (approx. 1/3 of the concentration used during the initial selection) can be used.
3. Establish frozen cell banks as early as possible.
4. If serum-dependent CHO cells were transfected (see Subheading 3.4.2), perform sequential adaptation of clones to serum-free conditions as described in Subheading 3.1.
5. Once serum-free cultures of individual clones are established, transfer cells to shaker flasks and maintain cells in a single-cell suspension culture at 100 rpm at 37°C with 5% CO_2.
6. Confirm stable gene expression 2–4 weeks after adaptation to shaker flasks before proceeding into large-scale cultivation (see Subheading 3.8).

3.8 Screening for Positive Clones and Confirmation of Stable Gene Expression

Presented below are methods used by us for screening of clones and analysis of gene expression and stability. The mucin-type immunoglobulin fusion protein (PSGL-1/mIgG$_{2b}$) is secreted into the cell culture supernatant from where it can be immunopurified and further analyzed.

1. A sandwich enzyme-linked immunosorbent assay (ELISA) is used to screen for positive clones in the 96-well plate(s) and to assess the expression level of PSGL-1/mIgG$_{2b}$ in supernatants of individual clones (7). Furthermore, the relative density of blood group A or B determinants on PSGL-1/mIgG$_{2b}$ produced in individual clones can be estimated, thus enabling selection of high-producing clones with high density of the desired carbohydrate determinants (see Fig. 2).

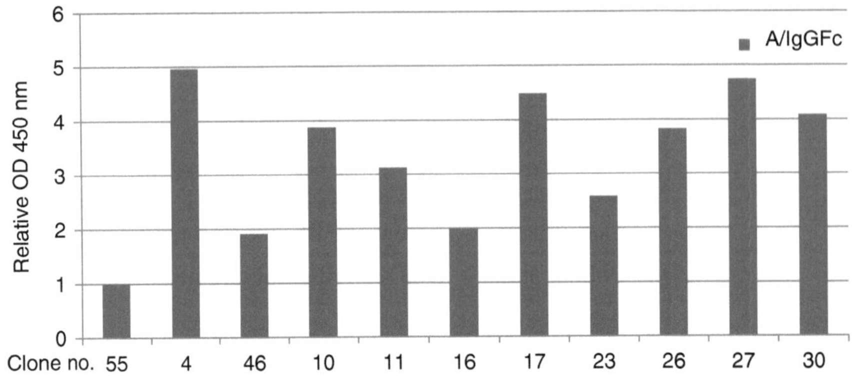

Fig. 2 ELISA analysis showing the relative density of blood group A determinants on PSGL-1/mIgG$_{2b}$ produced in different CHO clones. The blood group A determinant density was determined by comparing the relative OD from two ELISAs (anti-A-reactivity/anti-mIgG Fc reactivity)

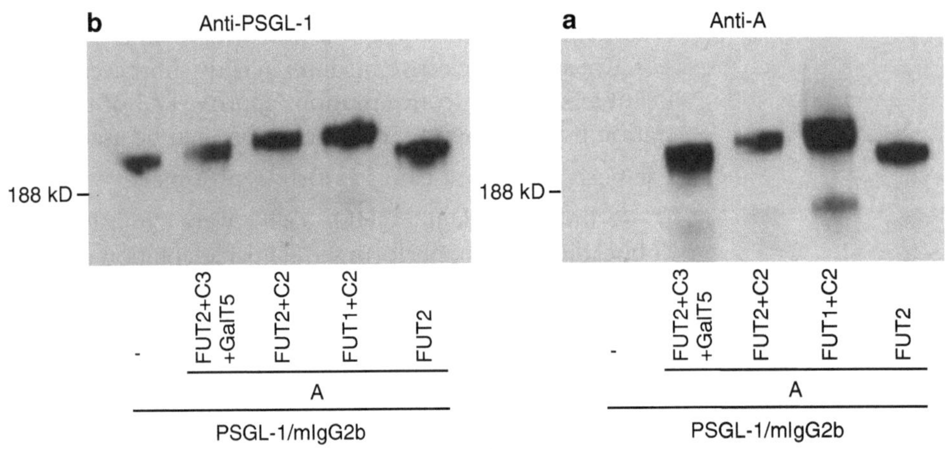

Fig. 3 Western blot analysis of immunopurified PSGL-1/mIgG$_{2b}$ produced in CHO cells transfected with PSGL-1/mIgG$_{2b}$ and different glycosyltransferase cDNAs to direct blood group A determinants on defined core structures. (**a**) Mouse monoclonal anti-PSGL-1 staining. PSGL-1/mIgG$_{2b}$ produced in cells expressing the core 2 (C2) and core 3 (C3) enzymes contained more complex glycans as indicated by the mobility shifts. (**b**) Mouse monoclonal anti-blood group A staining

2. Indirect immunofluorescence is performed according to standard techniques and is used to determine the transfection efficiency but also to confirm that all cells in individual colonies stably express PSGL-1/mIgG$_{2b}$ and blood group A or B determinants on cells.
3. SDS-PAGE and Western blot analysis of purified PSGL-1/mIgG$_{2b}$ is performed according to standard techniques and is used to confirm that the blood group A or B determinants are situated on defined core structures on secreted PSGL-1/mIgG$_{2b}$ (see Fig. 3).

4. MS analysis of saccharides released by β-elimination can be used to verify that correct *O*-glycan structures are carried by the recombinant fusion protein.

4 Notes

1. It is strongly recommended to use FBS originating from countries meeting the United States Department of Agriculture (USDA) requirements and not listed in 9 CFR, Part 94:18.
2. ProCHO-4 can be exchanged for other serum-free media formulations.
3. CHO cells are usually very easy to detach at low serum levels. Normal trypsin concentrations can be used during adaptation, but exposure times should be minimal (approx. 1 min) and cells should be pelleted by centrifugation directly when detached to remove the trypsin. Retain any viable cells that may detach at low serum levels by collecting the supernatant, pellet cells by centrifugation, and combine with the cells detached by trypsinization.
4. Cell clumping often occurs during adaptation to serum-free conditions. Triturate the clumps carefully to break them up when passaging cells. Addition of anionic polymers such as dextran sulfate to the medium will reduce cell clumping and facilitate a single-cell suspension. However, it is important to propagate cells in the absence of charged medium supplements such as dextran sulfate for some passages before the time of transfection as these reagents can interfere with the transfection.
5. The sensitivity to antibiotics can be greatly influenced by cell density. It is thus important to establish a cell death curve using the same seeding conditions as used post transfection. It is also important to bear in mind that cells cultured in serum-free medium are more sensitive to the selection drugs than cells cultured in serum-supplemented medium, as serum proteins may bind a certain amount of drug.
6. If multiple cDNAs in distinct vectors with different selection markers are co-transfected, it is possible to perform the selection with a combination of antibiotics. In such cases, a lower concentration of each of the antibiotics may be required.
7. The transfection efficiency is influenced by many factors including the size of the vector, and the choice of promoter is critical for efficient expression of the transfected gene. The selectable marker can either be present in the same vector as the gene of interest, as in the expression vectors used by us, or in a separate

co-transfected vector. The vectors we have used integrate randomly in the genome. More sophisticated vector systems exist that support site-specific integration in the genome and, thus, also higher expression levels may be achieved in a predictable manner.

8. The quality of the plasmid DNA greatly influences the outcome of transfection experiments. We have used Qiagen Plasmid Purification Kit for purification of plasmid DNA and QIAquick Gel Extraction Kit for DNA cleanup from enzymatic reactions. Other equivalent methods can be used.

9. It is advised to use linear DNA for stable transfections, even though the transfection efficiency might be lower compared to when circular DNA is used. The reason is that circular DNA will recombine into the genome at random positions within the plasmid. Therefore, the expression cassette for the gene of interest might be destroyed, leading to antibiotic resistant clones that do not express the gene of interest. By linearizing the plasmid, the position within the plasmid at which the recombination occurs can be determined, and hence the expression cassette is more likely to be conserved.

10. When generating stable transfectants in the presence of serum, we have encountered problems such as lost or reduced gene expression upon subsequent adaptation of clones to serum-free conditions. In addition, generating stable transfectants in the presence of serum is time consuming as cells subsequently need to be adapted to serum-free conditions for large-scale cultivation and, hence, reanalyzed for stable gene expression. On the other hand, when performing transfections in 100 % serum-free suspension culture, we experienced low transfection efficiency and, hence, fewer clones for screening were generated. Difficulties in the selection of suspension cells were also encountered.

    The protocol described in Subheading 3.4.1 is a compromise in which cells adapted to serum-free conditions are transfected and initially selected in the presence of low levels of serum. This results in high transfection efficiency and enables convenient selection of cells adhered to the surface. The short time the cells are exposed to low levels of serum does not seem to influence the stability of gene expression upon subsequent clone expansion under serum-free conditions. Potential clones not stable under serum-free conditions can be identified early and discarded.

11. A low level of serum is added to the medium to permit cell adhesion to the surface. The use of Corning® Cell BIND® Surface is recommended to facilitate adhesion of cells. If cells resist attachment to the surface, the concentration of serum

can be slightly increased when plating cells. It is important that cells from a single-cell suspension are used for transfection, as cell clumps are more difficult to attach and result in lower transfection efficiency.

12. The uptake efficacy of transfection complexes is influenced by cell health and cell density. Too confluent cells will impair DNA uptake as actively dividing cells take up exogenous DNA better than quiescent cells. On the other hand, too few cells may result in cytotoxicity and low transfection efficiency.
13. The optimal amount of plasmid DNA to be used in the transfection will vary widely depending on the type of vector, transfection reagent, target cell line, and number of cells, and hence needs to be optimized for each new application. We generally use a DNA (μg):Lipofectamine™ (μL) ratio of 1:2.5 when transfecting CHO cells with the expression vectors given in Table 2. For transfection of 1–1.5 × 10(7) cells adapted to serum-free conditions we have used 24 μg plasmid DNA. The DNA–Lipofectamine™ 2000 complexes were prepared in DMEM with no addition of FBS or antibiotics.
14. Transfection reagents other than Lipofectamine™ 2000 can be used but might need optimization. We have obtained successful results using Lipofectamine™ 2000, polyethylenimine (PEI), and nucleofection when generating stable CHO transfectants.
15. If multiple cDNAs are to be expressed stably, cells can be either co-transfected with distinct plasmid DNAs or sequentially transfected with individual plasmids. Sequential transfection requires careful analysis of gene expression stability after each transfection and, hence, is time consuming. Moreover, if the expression of any gene is lost or markedly reduced it will have profound implications for the final product, as expression of all glycosyltransferases is required to generate the desired *O*-glycan substitution on the fusion protein. Co-transfection of plasmid DNAs may increase the likelihood that the transfected genes are integrated at the same site in the host cell genome, and hence acquire similar stability and expression profiles.

    We have performed co-transfection of up to five distinct plasmid DNAs using the described protocols, both of which generated CHO transfectants with stable expression of all transfected genes and which secreted recombinant mucin-type immunoglobulin fusion proteins with desired *O*-glycan substitution. However, we have not as yet evaluated a high enough number of selected clones for a time long enough to be able to draw any conclusions with regard to long-term gene expression stability.

16. Some serum-free media formulations may inhibit cationic lipid-mediated transfection. Therefore, replacing ProCHO-4 medium with DMEM temporarily at the time of transfection leads to higher transfection efficiency when using Lipofectamine™ 2000.
17. The attachment of cells to the surface is influenced by the addition of some types of antibiotics. If cells detach during selection, sediment cells by centrifugation and replate (in a new CellBIND® tissue culture dish). It may also be advisable to increase the serum level slightly during the initial selection if cells appear too loosely attached to the surface.
18. It is advised to use 96-well plates with U-bottom when expanding clones in serum-free medium.
19. It is advised to omit selection drugs in the serum-free medium when diluting single cells in 96-well plates, as cells are likely to be more sensitive at this step. After a week, we usually add the same concentration of selection drugs as was used during the initial selection.

Acknowledgements

This work was supported by AbSorber AB and in part by grants to J.H. from the Swedish Research Council/Medicine (K2011-65X-3031-01-6) and the County Council of Västra Götaland (ALF).

Conflicts of interest. J.H. is founder, board member, and part-time medical director of AbSorber AB. He is also a shareholder in Allenex AB, the main owner of AbSorber AB.

References

1. Gustafsson A, Holgersson J (2006) A new generation of carbohydrate-based therapeutics: recombinant mucin-type fusion proteins as versatile inhibitors of protein-carbohydrate interactions. Expert Opin Drug Discov 1: 161–178
2. Coombs PJ, Harrison R, Pemberton S, Quintero-Martinez A, Parry S, Haslam SM et al (2010) Identification of novel contributions to high-affinity glycoprotein-receptor interactions using engineered ligands. J Mol Biol 396:685–696
3. Mammen M, Choi S, Whitesides G (1998) Polyvalent interactions in biological systems: implications for design and use of multivalent ligands and inhibitors. Angew Chem Int Ed 37:2754–2794
4. Kufe DW (2009) Mucins in cancer: function, prognosis and therapy. Nat Rev Cancer 9: 874–885
5. Linden SK, Sutton P, Karlsson NG, Korolik V, McGuckin MA (2008) Mucins in the mucosal barrier to infection. Mucosal Immunol 1: 183–197
6. Holgersson J, Löfling J (2006) Glycosyltransferases involved in type 1 chain and Lewis antigen biosynthesis exhibit glycan and core chain specificity. Glycobiology 16: 584–593
7. Liu J, Gustafsson A, Breimer ME, Kussak A, Holgersson J (2005) Anti-pig antibody adsorption efficacy of a-Gal carrying recombinant P-selectin glycoprotein ligand-1/immunoglobulin chimeras increases with core 2 b1,

6-N-acetylglucosaminyltransferase expression. Glycobiology 15:571–583

8. Löfling J, Holgersson J (2009) Core saccharide dependence of sialyl Lewis X biosynthesis. Glycoconj J 26:33–40
9. Löfling JC, Hauzenberger E, Holgersson J (2002) Absorption of anti-blood group A antibodies on P-selectin glycoprotein ligand-1/immunoglobulin chimeras carrying blood group A determinants: core saccharide chain specificity of the Se and H gene encoded alpha1,2 fucosyltransferases in different host cells. Glycobiology 12:173–182
10. Liu J, Qian Y, Holgersson J (1997) Removal of xenoreactive human anti-pig antibodies by absorption on recombinant mucin-containing glycoproteins carrying the Gal alpha1, 3Gal epitope. Transplantation 63(11): 1673–1682
11. Löfling J, Diswall M, Eriksson S, Borén T, Breimer ME, Holgersson J (2008) Studies of Lewis antigens and H. pylori adhesion in CHO cell lines engineered to express Lewis b determinants. Glycobiology. 18(7):494–501. doi: 10.1093/glycob/cwn030. Epub 2008 Apr 9

Chapter 2

Engineering a Human-Like Glycosylation to Produce Therapeutic Glycoproteins Based on 6-Linked Sialylation in CHO Cells

Nassimal El Maï, Sandrine Donadio-Andréi, Chloé Iss, Valérie Calabro, and Catherine Ronin

Abstract

When recombinant glycoproteins for therapeutic use are to be produced on an industrial scale, there is a crucial need for technologies that can engineer fast-growing stable cells secreting the protein drug at a high rate and with a defined and safe glycosylation profile. Current cell lines approved for drug production are essentially from rodent origin. Their glycosylation machinery often adds undesired carbohydrate determinants which may alter protein folding, induce immunogenicity, and reduce circulatory life span of the drug. Notably, sialic acid as *N*-acetylneuraminic acid is not efficiently added in most mammalian cells and the 6-linkage is missing in rodent cells. Engineering cells with the various enzymatic activities required for sialic acid transfer has not yet succeeded in providing a human-like pattern of glycoforms to protein drugs. To date, there is a need for engineering animal cells and get highly sialylated products that resemble as closely as possible to human proteins. We have designed ST6Gal minigenes to optimize the ST6GalI sialyltransferase activity and used them to engineer ST6(+)CHO cells. When stably transfected in cells expressing a protein of interest or not, these constructs have proven to equip cell clones with efficient transfer activity of 6-linked sialic acid. In this chapter, we describe a methodology for generating healthy stable adherent clones with hypersialylation activity and high secretion rate.

Key words Sialyltransferases, Glycoengineering, Stable clones, CHO cells, Sialylation

1 Introduction

Most marketed therapeutic proteins, including the wide family of monoclonal antibodies as well as many new candidates under development, are glycoproteins. These glycoproteins are predominantly produced in mammalian (rodent) expression systems such as Chinese hamster ovary (CHO), mouse myeloma NS0, or SP2/0 cell lines because their glycosylation machinery is of the complex type like that of human cells. However terminal glycosylation was found to be significantly different and increase the immunogenicity of the drug (1).

Alain Beck (ed.), *Glycosylation Engineering of Biopharmaceuticals: Methods and Protocols*, Methods in Molecular Biology, vol. 988, DOI 10.1007/978-1-62703-327-5_2, © Springer Science+Business Media New York 2013

Rodent-type glycans substantially differ from human structures in both N-linked and O-linked carbohydrate chains. Sialic acid is not the preferred terminal sugar in mammals which rather terminate their glycans in α-linked galactose or poly *N*-acetyllactosamine. Also among sialic acids, the presence of *N*-glycolylneuraminic acid (NeuGc) predominates over *N*-acetylneuraminic acid (NeuAc) in rodents. NeuGc is not found in human glycoproteins because of a mutation in the CMP-neuraminic acid hydroxylase gene (2). Altogether, such terminal glycosylation was found to substantially generate carbohydrate-based immunogenicity and promote clearance. More particularly, antibodies expressed in NS0 or SP2/0 cells were found to generate strong reactivity in patients administered with these proteins (3). Neutralization of the drugs rapidly occurs through anti-α-Gal antibodies since they represent 1% of our circulating immunoglobulins. Anti-NeuGc antibodies have also been found in some patients (4, 5), and are considered as being linked to chronic inflammation (6, 7).

CHO cells have been approved for drug production very early and very luckily, they fail to express the α3Gal epitope. They express an α2,3-sialyltransferase (ST3) (8) which adds α2,3-linked terminal sialic acid (NeuGc/NeuAc) to *N*-glycans. However, during a fermentation process, this enzyme often fails to fully cover all the branches of *N*-glycans, leaving a variable amount of free β-galactose residues detrimental to prolonged circulatory half-life. Indeed, clearance of asialoglycoproteins rapidly occurs through the liver asialoglycoprotein receptors, a galactose lectin receptor capable of removing from serum with a remarkably high capacity (mg/min) (9), as previously described for erythropoietin (10). Various genetic engineering methods have therefore been developed to enhance glycoprotein sialylation in rodent cells (11). These methods have included various approaches such as increasing the intracellular pool of sialic acids, increasing the availability of nucleotide precursor CMP-NeuAc, increasing the number of sialyltransferase acceptor sites, overexpression of sialyltransferases (STs), and decreasing soluble sialidase activities (12–16). All these methods were found to increase the cellular content in sialic acid and in sialylated endogeneous components but have limited efficacy on the therapeutic proteins. CHO cells also lack the expression of an α2,6-sialyltransferase (ST6) able to transfer sialic acid in the 6-position to N-linked glycans of proteins. Such linkage is the preferred terminal glycosylation of human proteins, especially in blood, and is thus often regarded as the representative human-type glycosylation because yeast, plant, insect, or bird expression systems are not able to carry out 6-linked sialylation. Expressing ST6 activity is thus essential to produce glycoproteins with human-like glycoform pattern.

Since the first transfection of the rat ST6GalI gene into CHO cells (8), several groups have attempted to coexpress this enzyme with various proteins of human origin. Recombinant human tPA was expressed in CHO cells engineered with rat ST6GalI (17),

IFN-γ in CHO clones containing mouse ST3 and rat ST6 (15) or rat ST6GalI alone (18), and more recently, TSH in CHO cells containing rat ST6cDNA (19). All these studies demonstrated a substantial improvement in glycoengineered cells with a 1.5-fold increase in sialic acid content and an average distribution of 60% α2,3 to 40% α2,6-linked sialic acid but no substantial increase in the sialylation of the therapeutic protein of interest. Some studies even failed to detect significant difference in overall sialylation following ST6 transfection (20, 21). It was therefore concluded that the wild-type ST6 enzyme was ineffective to compete with the endogenous ST3 enzyme, either because it did not localize in the same subcellular compartments or because it could not be active enough in the cellular context. Some of us could further demonstrate that efficacy on protein acceptors is restricted by a peptide region in the stem domain of the human hST6GalI enzyme (22).

The gene family of human sialyltransferases (STs) is composed of 20 members and distributes in four subfamilies and according to their specificity: ST3Gal, ST6Gal, ST6GalNAc, and ST8Sia. All STs are type II proteins, resident of the *trans*-Golgi/TGN intracellular compartments. They are all constituted of a short cytoplasmic N-terminal tail (1–31 aa), a transmembrane fragment (16–20 aa) followed by a stem region of variable length (20–300 aa), and a C-terminal catalytic domain (CD) facing the Golgi lumen (23). Unlike other glycosyltransferase families, mammalian and human STs display high sequence homology in their catalytic domain, with four consensus sequences designated as sialylmotifs: L (for long) (24, 25), S (for short) (26, 27) VS (for very short) (28), and sialylmotif-3 (third position in the sequence) (29) (Fig. 1). Site-directed mutagenesis demonstrated that the L-sialylmotif contributes to the binding of donor substrate CMP-NeuAc (24), and the motifs III- and VS contribute to the binding of the acceptor substrate (29). The S-sialyl motif participates in the binding of both the donor and acceptor substrates (30). Alternatively, the cytosolic tail, transmembrane region, and a significant portion of the stem, often designed as the CTS region, govern Golgi localization (31). Accordingly, innovative design of the CTS region has been carried out to provide a panel of a nonnatural membrane anchors to the catalytic domain of the human ST6GalI (cf. Footnote). The catalytic domain of the transferase has been optimized to present a 30-fold increase in transfer efficacy and broadly transfer sialic acid to acceptor substrates of different degree of branching compared to the full-length enzyme. Figure 1 describes the design of the minigenes.

All the engineered transferases were found to be located in the Golgi apparatus as shown in Fig. 2.

It could further be observed that the glycoengineered cells are healthy, good producers and possess high hypersialylation activity. We describe below the transfection procedure and selection of stable clones using such minigenes and an adherent CHO cell line.

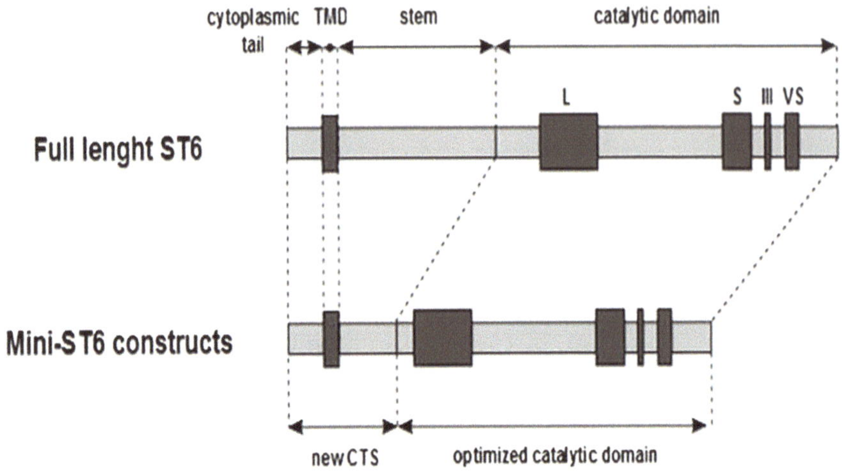

Fig. 1 Design of engineered hST6GalI sialyltransferases

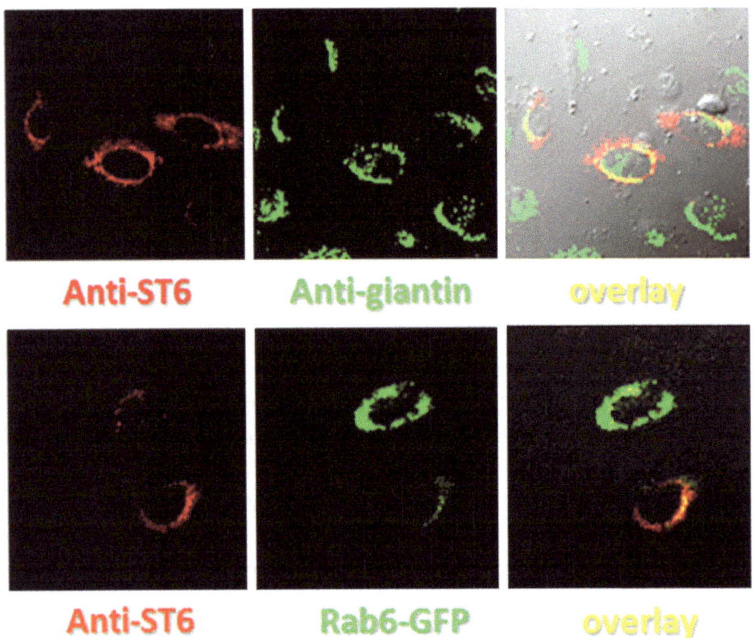

Fig. 2 Localization of engineered sialyltransferases in the Golgi compartments using specific markers and anti-ST6 antibodies. Anti-giantin labeled the *cis* Golgi cisternae in *green*, Rab6 GFP is a marker of *trans* Golgi/ TGN (also in *green*), and staining with ST6 antibodies is in *red*. Overlay shows the colocalization in *yellow*. These experiments are from a transient transfection of CHO cells with minigenes

2 Materials

2.1 Cell Culture

1. CHO cells expressing or not the protein of interest: This cell line was originally derived as a subclone from the parental CHO cell line initiated from a biopsy of an ovary of an adult Chinese hamster. Various CHO cell lines may be used alternatively with similar efficiency.

2. pcDNA3.1(+) vector which encodes a sialyltransferase minigene, ampicillin resistance for selection and maintenance in bacteria, and geneticin (G418) for selection in mammalian cells.
3. Ham's medium with L-glutamine and fetal bovine serum (FBS) Ham's medium is supplemented with 10% FBS with or without 100 U/mL penicillin/100 μg/mL streptomycin. Transfection is performed in the absence of FBS.
4. A commercial transfection reagent.
5. Geneticin selective antibiotic.
6. DPBS (1×) with Ca^{2+} and Mg^{2+}.
7. Clean forceps with fine tips, autoclaved.
8. Trypsin/EDTA.
9. 6-well culture plates.
10. CO_2 incubator (37°C): It maintains optimal temperature, humidity, and conditions of CO_2 and oxygen content.
11. 2-, 5-, and 10-mL serological pipets.

2.2 Selection of Transfected Cells

1. Optic inverted microscope.
2. Geneticin selective antibiotic liquid10131-027.
3. 6-well culture plates.
4. DPBS (1×) with Ca^{2+} and Mg^{2+}.
5. Clean forceps with fine tips, autoclaved.
6. Trypsin/EDTA.
7. 2-, 5-, and 10 mL serological pipets.

2.3 Immunostaining

1. Fluorescence microscope: Filters should be selected to handle double or triple fluorochromes.
2. Fluorescent *Sambucus Nigra* agglutinin and prolong gold DAPI. The SNA lectin is specific for 6-linked sialic acid while DAPI is a specific labeling of the nucleus in cells. The procedure herein describes the use of SNA-FITC but any other fluorophore can be used.
3. DPBS (1×) with Ca^{2+} and Mg^{2+}.
4. Cover glass and microslides.
5. 24-well plates.
6. PBS–PFA 4%: 4 g of paraformaldehyde are dissolved in 100 mL of DPBS and warmed until a colorless solution is obtained. Do not heat above 60°C. Store at −20°C. Wear a mask when weighing paraformaldehyde.
7. PBS/BSA 1%: Dissolve 1 g bovine serum albumin (BSA) in 100 mL of DPBS (1×) and store at 4°C.

2.4 Selection of Stable Cell Clones

1. 96-well culture plates.
2. Automatic pipettor and sterile tips.
3. Ham's medium with L-glutamine and FBS Ham's medium is supplemented with 5% FBS with or without 100 U/mL penicillin/100 µg/mL streptomycin.
4. Geneticin.

3 Methods

3.1 Transfection of CHO Cells

1. Seed 5×10^5 cells/well in a 6-well plate in the presence of Ham's medium + 5% FBS and incubate for 24 h at 37°C under 5% CO_2.
2. Wash the cells three times with 1 mL of DPBS and then add 2 mL of Ham's medium FBS free.
3. Transfect CHO cells with pcDNA3.1(+) plasmids using the transfectant reagent at various ratios (3:2, 3:8, and 4:5 (µL/µg)) and incubate overnight (see Note 1).
4. Remove the medium, wash once with DPBS, then add 2 mL of new Ham's medium + 5% FBS, and incubate for 24 h to allow cell recovery after transfection.
5. Remove the medium and place in Ham's medium + 5% FBS containing G418 at 100 µg/mL. This step will select cells that have stably incorporated the plasmid into their genomic DNA. It is recommended to treat cells with moderate concentration of antibiotic because it reduces cellular growth. Clones can also be rapidly selected after transfection but the most interesting cells are often the ones which grow slowly at this stage.
6. Following addition of the selection medium, cells still grow for 1–2 days but most cells start to die within a week. There is a lot of debris in the culture at this point. Change the medium regularly over this period.
7. It is recommended to include a negative control prepared with non-transfected cells to estimate cell viability at each concentration of antibiotic (see Note 2).

3.2 Selection of Transfected Cells

1. Observe cells every day with an optic inverted microscope. Once confluence is almost reached, split cells to 1:5 dilution in new selective Ham's medium + 5% SVF + 100 µg/mL G418. Splitting of cells should be done three times a week according to cell growth. It is especially important to well separate cells during splitting. Allow the trypsin to act under gentle agitation to keep the cells separated and avoid aggregates.
2. Repeated splitting allows the removal of non-transfected cells and favor increase of resistant cells.

3. After 3 weeks, increase the G418 concentration of the medium to 150 μg/mL. After time, a few cells grow in low concentration of antibiotic and the purpose of this step is to prevent their growth. Take care not to pipet directly on the cells; otherwise the concentration of the G418 will be too strong on these cells and not on others. When a massive cell death is achieved, wash off the bottom of the dish, leaving colonies of resistant cells.

3.3 Immunostaining

1. Pick 2.5×10^4 cells of each well and plate them on sterile cover glass in a 24-well plate.
2. After overnight incubation, the cells are attached to the cover glass and rinsed with DPBS (1×) with Ca^{2+} and Mg^{2+} three times.
3. Each cover glass is treated with 0.5 mL of PBS/PFA 4% for 30 min at RT and then wash three times for 2 min with DPBS (1×) with Ca^{2+} and Mg^{2+}.
4. Add 0.5 mL of PBS/BSA 1% and saturate each cover glass for 30 min at RT. Do not wash the cells after saturation.
5. Prepare the staining solution by adding 5 μL of SNA-FITC to 995 μL of DPBS (1×) with Ca^{2+} and Mg^{2+}. Cover each cover glass with a 50 μL drop of the staining solution so that cells will be directly in contact with the fluorescent lectin. Stain for 1 h at 4°C in the dark. Then place cells upward to wash each cover glass three times with DPBS (1×) with Ca^{2+} and Mg^{2+}. Only cells which show sialyltransferase activity are labeled as shown in Fig. 3.

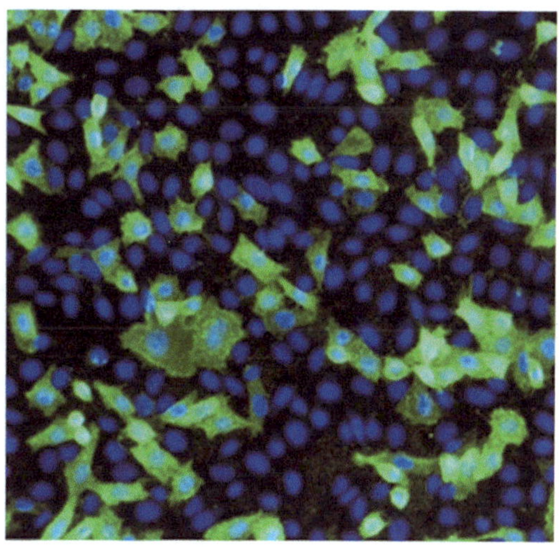

Fig. 3 Double staining of CHO cells transfected with a sialyltransferase minigene and selected over several weeks with a G418 concentration of 100 μg/mL. FITC labeling is in *green* and DAPI in *blue*

6. Dry-clean microslides with a paper soaked in alcohol, put down a drop of prolong gold-DAPI (liquid mountant with DAPI nuclear stain) on the microslide and put one drop of prolong gold on cells, and let it dry at RT in the dark. Store at −20°C in slidebox until use (see Note 3).
7. Count positive stained cells with regard to the total number of cells (based on nuclear staining). Some negative cells still remain at this stage even though cells are under selection. Estimating the number of stained cells allows the identification of wells suitable for further cloning. Save the wells that are best enriched in labeled cells.

3.4 Selection of Stable Cell Clones

1. Resistant cells are harvested and counted. Seed a single cell in each well of 96-well plates in the selection medium and incubate overnight at 37°C, 5% CO_2. The remaining cells are picked and plated in a new 6-well plate. At this stage, they can be frozen and stored at −80°C for further use. On the 96-well plate, the cell number may vary among wells. The wells with a single cell are carefully identified. It takes 1–2 weeks for cells to reach suitable density. At this point, the plates should be carefully inspected under microscope (see Note 4).
2. Transfer every single group of cells from 96-well to 24-well plates. Each well is matched with the earlier staining. Use immunostaining to determine the wells which contain only positive cells and then transfer each positive clone to a new well. All cells should be labeled as in Fig. 4.

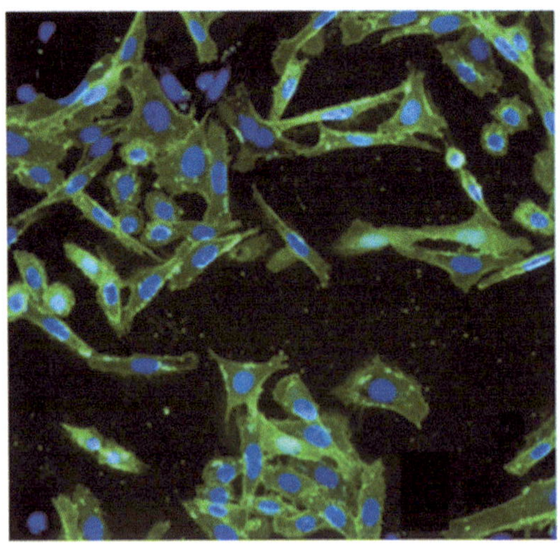

Fig. 4 Double staining of a stable clone expressing a 6-sialylation: 1 at this stage, 100% of the CHO cells are stained by FITC (*green*) and DAPI (*blue*) fluorochromes and are resistant to G418 at the final stage of the limit dilution process

3. Stable clones selected after dilution are propagated in selection medium. When cell number reaches almost 80% confluence in T25 flask, aliquots can be frozen at −80°C. Each selected clone can be propagated at a reduced amount of G418 for maintenance. Keep track of passage numbers to ensure stability of each clone which may vary and even die after a few passages. Periodically check the quality of the clones by immunostaining.

4 Notes

1. The transfection reagent must be adapted to the cell line used. Many different preparations are commercially available and must be tested to optimally allow 10–30% of transfected cells.
2. The wild-type cell line should be initially sensitive to G418 before starting the glycoengineering process. It is recommended to carry out a negative control at each raise in antibiotic concentration.
3. Immunostaining is stable over a week or so. However, it is recommended to analyze the staining as soon as possible to get the best contrast for stained cells.
4. Never mix positive cells from different clones because the sialyltransferase activities may vary from one clone to another. Further selection should determine the best clone to be used in subsequent experiments.

Acknowledgment

The "Preparation and Uses of Gene Sequences encoding Chimerical Glycosyltransferases with Optimized Glycosylation Activity" has been patented as *EP2019864* (C. Ronin and G. Guiraudie-Capraz, inv).

References

1. Wacker C, Berger CN, Girard P, Meier R (2011) Glycosylation profiles of therapeutic antibody pharmaceuticals. Eur J Pharm Biopharm 79:503–507
2. Chou HH, Takematsu H, Diaz S, Iber J, Nickerson E, Wright KL et al (1998) A mutation in human CMP-sialic acid hydroxylase occurred after the Homo-Pan divergence. Proc Natl Acad Sci U S A 95:11751–11756
3. Chung CH, Mirakhur B, Chan E, Le QT, Berlin J, Morse M et al (2008) Cetuximab-induced anaphylaxis and IgE specific for galactose-alpha-1,3-galactose. N Engl J Med 358: 1109–1117
4. Noguchi A, Mukuria CJ, Suzuki E, Naiki M (1995) Immunogenicity of N-glycolylneuraminic acid-containing carbohydrate chains of recombinant human erythropoietin expressed in Chinese hamster ovary cells. J Biochem 117:59–62
5. Zhu A, Hurst R (2002) Anti-N-glycolylneuraminic acid antibodies identified

in healthy human serum. Xenotransplantation 9:376–381
6. Byres E, Paton AW, Paton JC, Lofling JC, Smith DF, Wilce MC et al (2008) Incorporation of a non-human glycan mediates human susceptibility to a bacterial toxin. Nature 456:648–652
7. Hedlund M, Padler-Karavani V, Varki NM, Varki A (2008) Evidence for a human-specific mechanism for diet and antibody-mediated inflammation in carcinoma progression. Proc Natl Acad Sci U S A 105:18936–18941
8. Lee EU, Roth J, Paulson JC (1989) Alteration of terminal glycosylation sequences on N-linked oligosaccharides of Chinese hamster ovary cells by expression of beta-galactoside alpha 2,6-sialyltransferase. J Biol Chem 264:13848–13855
9. Morell AG, Gregoriadis G, Scheinberg IH, Hickman J, Ashwell G (1971) The role of sialic acid in determining the survival of glycoproteins in the circulation. J Biol Chem 246:1461–1467
10. Briggs DW, Fisher JW, George WJ (1974) Hepatic clearance of intact and desialylated erythropoietin. Am J Physiol 227:1385–1388
11. Bork K, Horstkorte R, Weidemann W (2009) Increasing the sialylation of therapeutic glycoproteins: the potential of the sialic acid biosynthetic pathway. J Pharm Sci 98:3499–3508
12. Bork K, Reutter W, Weidemann W, Horstkorte R (2007) Enhanced sialylation of EPO by overexpression of UDP-GlcNAc 2-epimerase/ManAc kinase containing a sialuria mutation in CHO cells. FEBS Lett 581:4195–4198
13. Jeong YT, Choi O, Lim HR, Son YD, Kim HJ, Kim JH (2008) Enhanced sialylation of recombinant erythropoietin in CHO cells by human glycosyltransferase expression. J Microbiol Biotechnol 18:1945–1952
14. Son YD, Jeong YT, Park SY, Kim JH (2011) Enhanced sialylation of recombinant human erythropoietin in Chinese hamster ovary cells by combinatorial engineering of selected genes. Glycobiology 21:1019–1028
15. Fukuta K, Yokomatsu T, Abe R, Asanagi M, Makino T (2000) Genetic engineering of CHO cells producing human interferon-gamma by transfection of sialyltransferases. Glycoconj J 17:895–904
16. Ferrari J, Gunson J, Lofgren J, Krummen L, Warner TG (1998) Chinese hamster ovary cells with constitutively expressed sialidase antisense RNA produce recombinant DNase in batch culture with increased sialic acid. Biotechnol Bioeng 60:589–595
17. Minch SL, Kallio PT, Bailey JE (1995) Tissue plasminogen activator coexpressed in Chinese hamster ovary cells with alpha(2,6)-sialyltransferase contains NeuAc alpha(2,6)Gal beta(1,4)Glc-N-AcR linkages. Biotechnol Prog 11: 348–351
18. Bragonzi A, Distefano G, Buckberry LD, Acerbis G, Foglieni C, Lamotte D et al (2000) A new Chinese hamster ovary cell line expressing alpha2,6-sialyltransferase used as universal host for the production of human-like sialylated recombinant glycoproteins. Biochim Biophys Acta 1474:273–282
19. Damiani R, Oliveira JE, Vorauer-Uhl K, Peroni CN, Vianna EG, Bartolini P et al (2009) Stable expression of a human-like sialylated recombinant thyrotropin in a Chinese hamster ovary cell line expressing alpha2,6-sialyltransferase. Protein Expr Purif 67:7–14
20. Zhang X, Lok SH, Kon OL (1998) Stable expression of human alpha-2,6-sialyltransferase in Chinese hamster ovary cells: functional consequences for human erythropoietin expression and bioactivity. Biochim Biophys Acta 1425:441–452
21. Jassal R, Jenkins N, Charlwood J, Camilleri P, Jefferis R, Lund J (2001) Sialylation of human IgG-Fc carbohydrate by transfected rat alpha2,6-sialyltransferase. Biochem Biophys Res Commun 286:243–249
22. Legaigneur P, Breton C, El Battari A, Guillemot JC, Auge C, Malissard M et al (2001) Exploring the acceptor substrate recognition of the human beta-galactoside alpha 2,6-sialyltransferase. J Biol Chem 276:21608–21617
23. Paulson JC, Colley KJ (1989) Glycosyltransferases. Structure, localization, and control of cell type-specific glycosylation. J Biol Chem 264:17615–17618
24. Datta AK, Paulson JC (1995) The sialyltransferase "sialylmotif" participates in binding the donor substrate CMP-NeuAc. J Biol Chem 270:1497–1500
25. Datta AK, Paulson JC (1997) Sialylmotifs of sialyltransferases. Indian J Biochem Biophys 34:157–165
26. Datta AK, Chammas R, Paulson JC (2001) Conserved cysteines in the sialyltransferase sialylmotifs form an essential disulfide bond. J Biol Chem 276:15200–15207
27. Angata K, Yen TY, El Battari A, Macher BA, Fukuda M (2001) Unique disulfide bond structures found in ST8Sia IV polysialyltransferase are required for its activity. J Biol Chem 276:15369–15377
28. Tsuji S, Datta AK, Paulson JC (1996) Systematic nomenclature for sialyltransferases. Glycobiology 6:v–vii
29. Jeanneau C, Chazalet V, Auge C, Soumpasis DM, Harduin-Lepers A, Delannoy P et al

(2004) Structure-function analysis of the human sialyltransferase ST3Gal I: role of n-glycosylation and a novel conserved sialyl-motif. J Biol Chem 279:13461–13468
30. Datta AK, Sinha A, Paulson JC (1998) Mutation of the sialyltransferase S-sialylmotif alters the kinetics of the donor and acceptor substrates. J Biol Chem 273:9608–9614
31. Fenteany FH, Colley KJ (2005) Multiple signals are required for alpha2,6-sialyltransferase (ST6Gal I) oligomerization and Golgi localization. J Biol Chem 280:5423–5429

Chapter 3

Glycoengineered *Pichia*-Based Expression of Monoclonal Antibodies

Dongxing Zha

Abstract

Currently, mammalian cells are the most commonly used hosts for the production of therapeutic monoclonal antibodies (mAbs). These hosts not only secrete mAbs with properly assembled two heavy and two light chains but also deliver mAbs with a glycosylation profile that is compatible with administration into humans. GlycoFi, a wholly owned subsidiary of Merck & Co., Inc., humanized the *Pichia* glycosylation pathway which allows it to express glycoproteins with a human-like glycan profile. This offers an alternative mAb production platform similar to mammalian hosts and in some cases it even provides more homogenous product and better efficacy, such as enhanced effector function. This chapter describes a protocol for using glycoengineered *Pichia* to produce full-length mAbs. It covers a broad spectrum of mAb expression technologies in yeast including expression vector construction, yeast transformation, high-throughput strain selection to fermentation, and antibody purification.

Key words Glycoengineered *Pichia*, Monoclonal antibody, Glycosylation profile, High-throughput screening, FACS, Fermentation and purification

1 Introduction

Monoclonal antibodies (mAbs) have been used as treatments for a wide range of diseases, including oncology, immunological and infectious diseases, and organ transplantation. At present, at least 28 therapeutic mAbs are marketed and hundreds more are in clinical development (1). Mammalian cells like NS0, SP20, and in particular Chinese hamster ovary (CHO) cells are commonly used as the host platform for therapeutic mAb production. However, alternative mAb platforms including plant, insect cells, and microbial expression systems are being explored (2–6). Recent studies by Grohs et al. and Komarova et al. have demonstrated that a plant-produced antibody (trastuzumab) was similar to a CHO-produced counterpart (4, 5). Moreover, yeasts are regarded as compelling alternatives to mammalian cell culture for the production of recombinant antibodies and antibody fragments. There are multiple

Alain Beck (ed.), *Glycosylation Engineering of Biopharmaceuticals: Methods and Protocols*, Methods in Molecular Biology, vol. 988,
DOI 10.1007/978-1-62703-327-5_3, © Springer Science+Business Media New York 2013

advantages of using yeast expression systems for therapeutic glycoprotein production, including ease of genetic manipulation, stable expression, rapid cell growth, high yield of secreted protein, scalable fermentation processes, and no risk of human pathogenic virus contamination. There is one example of *Pichia*-expressed full-length antibody in clinical trial. ALD518, a humanized mAb that targets interleukin-6 (IL-6) which contains a mutation at N297 that abolishes glycosylation as a benign blocker, has successfully completed Phase 2 (2, 6). However, glycosylation plays important roles in mAb pharmacokinetics and efficacy, so it is essential to eliminate fungal type glycosylation in therapeutic antibodies from *Pichia*. By humanization of the *Pichia* glycosylation pathway, secreted mAbs have glycosylation profiles that are similar to proteins found in humans. We have been able to demonstrate that an afucosylated anti-CD20 antibody from glycoengineered *Pichia* has similar antigen binding but enhanced Fc gamma IIIa receptor binding and better ex vivo B cell depletion activity when compared with CHO-produced Rituximab (3). More recently, we showed that glycoengineered *Pichia*-produced anti-Her2 is comparable to CHO-produced counterpart trastuzumab in preclinical studies of efficacy. *Pichia*-produced anti-Her2 and CHO-produced trastuzumab displayed similar in vitro biological functions, in vivo antitumor efficacy, and pharmacokinetics in both mice and nonhuman primates (7). Furthermore, a robust and scalable process for producing more than 1 g/l of functional mAbs with uniform N-linked glycans was developed (8). The methods described here cover a broad spectrum of techniques for producing full-length mAbs in glycoengineered *Pichia*. It includes expression cassette construction, glycoengineered *Pichia* transformation, and clone selection using high-throughput screening to identify high-expression cell lines. In addition, protocols of mAb production strain fermentation and antibody purification from the supernatant are also part of this chapter.

2 Materials

2.1 Construction of Expression Cassette, Glycoengineered Pichia Transformation, and Antibody Strain Selection

1. Restriction enzymes *Eco*RI, *Fse*I, *Pst*I, and *Swa*I (New England BioLabs).
2. One Shot® TOP10/P3 Chemically Competent *Escherichia coli* (Invitrogen).
3. S.O.C. medium at room temperature (Invitrogen).
4. LB plates contain 1% Bacto-tryptone, 0.5% yeast extract, 1% NaCl, and 2% agar and then adjust pH to 7.5 with NaOH. Add ampicillin (Sigma) to final concentration of 50–100 μg/ml before pouring plates.

Table 1
Sequences of primers used in this study

Primer name	Primer sequence
5′ AOX1/up	GCGACTGGTTCCAATTGACAAGCTTTTGATTTTAACG
Hc/lp	ACCAGGGGACAAAGACAAGGACTTTTGAG
Lc/lp	ACACTCTCCTCTGTTGAAGGACTTAG

5. 96-well deep-well titer block (USA Scientific, Inc.).
6. Yeast extract peptone dextrose (YPD) plates contain 1% yeast extract (Becton Dickinson), 2% peptone (Becton Dickinson), 2% dextrose (glucose) (Becton Dickinson, Franklin Lakes, NJ), and 2% agar (Becton Dickinson, Franklin Lakes, NJ).
7. 1 M D-sorbitol sterile and ice cold (Becton Dickinson).
8. Zeocin (Invitrogen).
9. Oligonucleotides were synthesized by Integrated DNA Technologies, Inc. (Coralville, Iowa).
10. Primers and their sequences are listed in Table 1.
11. mAb expression pGLY-mAb-vector was constructed according to Zhang et al. (7). The plasmid map is illustrated in Fig. 1. The vector was derived from pUC19 which contains ampicillin-resistant gene for *E. coli* transformation selection. Zeocin-resistant cassette was included in the vector for *Pichia* transformation selection. Restriction sites of *Eco*RI and *Fse*I, or *Pst*I and *Swa*I, are engineered for heavy- and light-chain cloning. Heavy and light chains were codon optimized using *Pichia*-preferred codon usage. Alpha mating factor predomain and Kozak sequence were fused at the 5′ end of heavy and light chain, respectively, and DNA was synthesized by Genscript, Inc.
12. Glycoengineered *Pichia* strain YGLY638 was described by Potgieter et al. (8).
13. YGLY8316 is a *Pichia* host that contains the enzymatic and transport machinery necessary for adding galactose to the nonreducing ends of *N*-acetylglucosamine in bi-antennary N-linked glycans.
14. Transformation recovery media contains YPDS which contains 1% yeast extract, 2% peptone, 2% dextrose (D-glucose), and 1 M sorbitol.
15. Gene Pulser MXcell Electroporation System (Bio-Rad Laboratories).

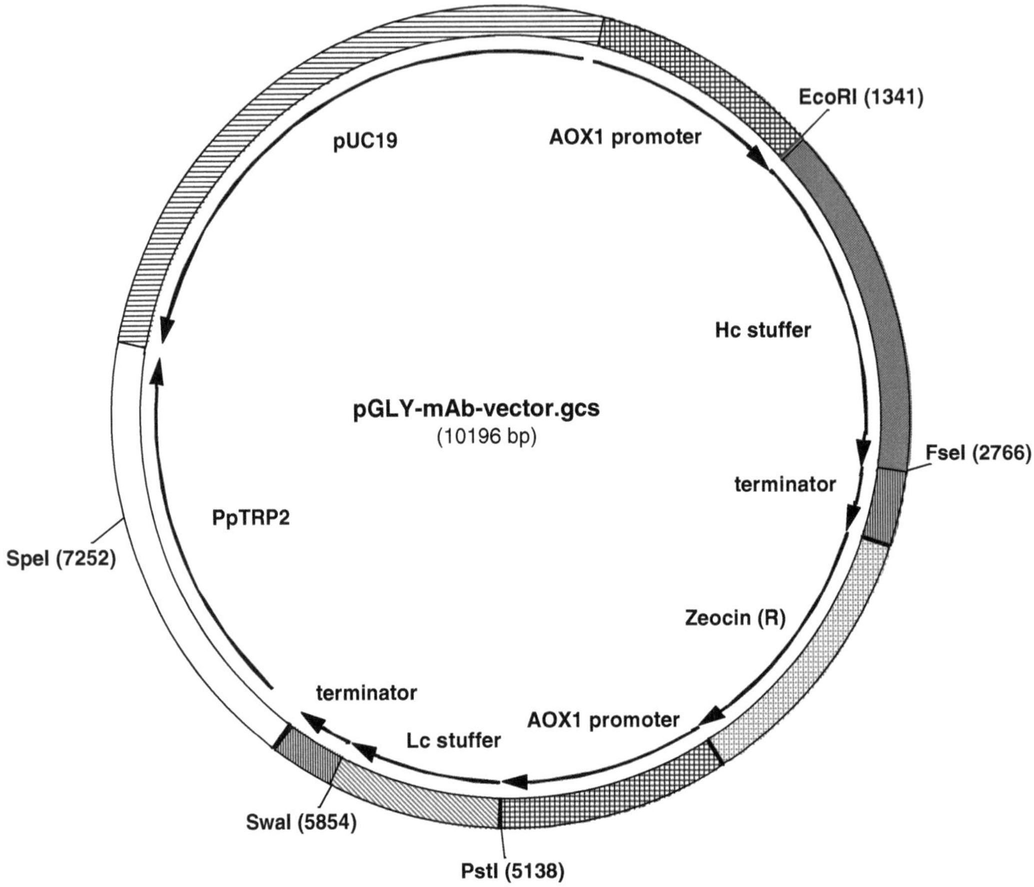

Fig. 1 Antibody expression vector, the mAb expression vector for glycoengineered *Pichia* was derived pUC19. It contains two expression cassettes under method-inducible promoter AOX1. Between the heavy- and light-chain cassette, Zeocin-resistant expression fragment was inserted in between

2.2 Antibody Expressing Strain Selection

1. BMGY is composed of 100 mM potassium phosphate, 10 g/l yeast extract (Becton Dickinson), 20 g/l peptone (Becton Dickinson), 40 g/l glycerol, 18.2 g/l sorbitol, 13.4 g/l YNB (Becton Dickinson), and 4 mg/l biotin.
2. BMMY is composed of 100 mM potassium phosphate, 10 g/l yeast extract (Becton Dickinson), 20 g/l peptone (Becton Dickinson), 20 ml/l methanol, 18.2 g/l sorbitol, 13.4 g/l YNB (Becton Dickinson), and 4 mg/l biotin per well.
3. PMT1 inhibitor is a rhodanine-3-acetic acid derivative originally identified as an inhibitor of fungal protein mannosyl transferase 1 (PMT1) (9).
4. Glycerol feed solution: 50% w/w glycerol, 5 mg/l biotin, and 12.5 ml/l PTM1 salts (65 g/l $FeSO_4{\cdot}7H_2O$, 20 g/l $ZnCl_2$, 9 g/l H_2SO_4, 6 g/l $CuSO_4{\cdot}5H_2O$, 5 g/l H_2SO_4, 3 g/l $MnSO_4{\cdot}7H_2O$, 500 mg/l $CoCl_2{\cdot}6H_2O$, 200 mg/l $NaMoO_4{\cdot}2H_2O$, 200 mg/l biotin, 80 mg/l NaI, 20 mg/l H_3BO_4).

5. Pierce ImmunoPure Polyclonal goat anti-human kappa chain (Pierce, Rockford, IL), Costar 96-well white polystyrene high-binding assay plate (Corning, Corning, NY).
6. Bio-Stack microtiter plate washer was made by BioTek Instruments (Winooski, VT).
7. ImmunoPure rabbit anti-human IgG phosphatase (Pierce).
8. Wash buffer: TBS (50 mM Tris–HCl, 150 mM NaCl, pH 7.4), 0.2% Tween 20.
9. Sixfors multifermentation system (ATR Biotech).
10. Beckman Allegra 6 centrifuge (Beckman Coulter).

2.3 Cell Labeling and Flow Cytometric Sorting

Goat anti-human IgG (H + L) Alexa 488; Goat anti-human IgG (H + L) Alexa 647 (Invitrogen):

1. Mouse anti-human IgG κ chain-APC (Invitrogen).
2. Goat F(ab′)2 anti-human IgG γchain-Alexa 488 (Invitrogen).
3. FACSAria II cell sorter with three lasers (405, 488, and 633 nm) and equipped with Diva v6.1 software (Becton Dickinson).
4. IRIS multifermenter software (ATR Biotech).
5. Sorvall Evolution RC centrifuge equipped with an SLC-6000 rotor (Thermo Scientific).

2.4 Antibody Purification from Fermentation Supernatant

1. STREAMLINE rProtein A (80–165 μm particle size), SP Sepharose High Performance (SP HP) (34 μm particle size), and Tricorn 10/200 columns (GE Healthcare).
2. Polyethersulfone membrane filters (0.2 μm pore size) (Nalgene).
3. SARTOPORE 2 (0.8 + 0.45 μm) (Sartorius, Göttingen, Germany).
4. 4–20% Tris–HCl Ready Gels and Prestained SDS–PAGE standards (broad range) (Bio-Rad Laboratories).
5. XK50/30 column (GE Healthcare).
6. SP HP (GE Healthcare).
7. AKTAexplorer 100 (GE Healthcare).

3 Methods

3.1 Expression Plasmid Construction, Yeast Transformation, and Strain Selection

1. The first step is to assemble amino acid sequences of heavy and light chain with signal sequence separately. The example shown in this chapter is anti-Her2 mAb which has identical amino acid sequence to trastuzumab (see Note 1).

2. Heavy and light chain with signal sequences at its N-terminus are reversely translated into DNA using *Pichia*-preferred codon usage by Genscript, Inc. Kozak sequence (GAAACG) is added before start codon "ATG" for both heavy and light chains. Two consecutive stop codons (TAATAG) are followed at the end of the coding regions. Additional *Eco*RI and *Fse*I sites are inserted at the 5′ and 3′ end of heavy chain, respectively, and *Pst*I and *Swa*I sites are added at the 5′ and 3′ end of the light chain to facilitate cloning (see Note 2).
3. Heavy and light chains are synthesized by Genscript, Inc. and delivered in pUC57 vector.
4. Four pieces of DNA fragments, including heavy and light chain and two vector fragments from *Pichia* mAb expression pGLY-mAb-vector, are prepared according to the method described in Note 3.
5. These four pieces of DNA fragments are ligated together by Quick Ligation Kit from New England Biolabs following the provided protocol.
6. One microliter of ligation reaction is transformed into Top10 chemical competent cells by heat shock.
7. Transformation is selected on LB agar plates which contain 50–100 μg/ml ampicillin.
8. Transformants are then screened by colony PCR with primers of 5′ AOX1/up and Hc/lp for heavy chain and using primers of 5′AOX1/up and Lc/lp for light chain (see Note 4).
9. Isolate plasmids from colony PCR-positive transformants and then perform enzymatic digestion to confirm that contains both heavy and light chains using *Eco*RI and *Fse*I for heavy chain insert and *Pst*I and *Swa*I for light chain insert.
10. Plasmid which harbors both heavy and light chain inserts and proper vector size is selected for *Pichia* transformation.
11. *Pichia* transformation with mAb expression plasmid: 10 μg of mAb expressing plasmid described above is first digested with *Spe*I for 1–2 h. The linearized plasmid DNA is then precipitated in ice-cold ethanol. DNA is pelleted by centrifugation and is subsequently washed twice with 70% ethanol. The DNA pellet is dried under vacuum and then suspended in 10 μl of sterile distilled water.
12. Preparation of glycoengineered *Pichia*-competent cells: YGLY638 (host strain expressing glycoprotein with the glycan of $Man_5GLcNAc_2$) and YGLY8316 (host strain being capable of expressing glycoproteins with bi-antennary glycan of $Gal_2GlcNAc_2Man_3GlcNAc_2$) are inoculated in 50 ml BMGY media the day before transformation (see Note 5).

13. Overnight cultures of OD_{600} in between 1 and 5 are centrifuged using 50 ml conical tubes to collect the cells. Cell pellet is washed twice with ice-cold sterile distilled water and then one more time with sterile 1 M sorbitol. Cells are suspended in 10 ml of 1 M cold sorbitol and stored on ice for transformation.
14. Electroporation of glycoengineered *Pichia* with mAb expression plasmids: Linearized mAb expression plasmid (5–10 μl) is mixed with 80–100 μl of cold competent *Pichia* cells (prepared in step 13) and incubated on ice for 5 min. The plasmid and *Pichia* cell mixture was then transferred to electroporation cuvette (see Note 6). The mAb expressing plasmid was electrotransformed into glycoengineered *Pichia* YGLY638 or YGLY8316 using preset *Pichia pastoris* program. Eight hundred microliters of recovery media YPDS was immediately added into the cuvette.
15. Two hundred microliters of the transformation mixture is plated out on YPD containing Zeocin plates to select glycoengineered *Pichia* transformants (see Note 7).

3.2 mAb Expression in 96-Deep-Well Plates and Isolation of Higher Producing Strains

1. Individual colonies transformed with the expression plasmid from selective agar plates are transferred into individual wells of a 96-well deep-well titer block containing 600 μl per well BMGY.
2. This plate, called the "seed plate," is then covered with a microporous rayon film and incubated for 48 h under standard growth conditions: 24°C, 90% humidity, in a Multitron shaking incubator from ATR Biotech at a speed of 840 rpm with a 3-mm throw.
3. Multiple "expansion plates" are made by collecting 50 μl from the seed culture and inoculating into 600 μl of fresh BMGY (see Note 8).
4. Following 48 h under standard growth conditions, expansion plates are consolidated into one or two plates and then subjected to recombinant protein induction with 600 μl fresh BMMY medium with 3 μM PMTi per well.
5. Induction proceeds under standard growth conditions for 48 h. The supernatant is collected after centrifugation in a Beckman Allegra 6 centrifuge at 2,000 × *g* for 5 min.
6. Then screening clones for mAb titer using ELISA is followed. Each well of a Costar 96-well white polystyrene high-binding assay plate is coated with 0.1 μg Pierce ImmunoPure polyclonal goat anti-human kappa chain for 1 h at room temperature.
7. The plate is washed on a Bio-Stack microtiter plate washer three times with 200 μl wash buffer.
8. Diluted sample (100 μl) is added followed by incubation for 1 h at room temperature. The plate is washed as described above.

9. One hundred microliters per well of ImmunoPure rabbit anti-human IgG phosphatase (diluted 1:10,000; Pierce, Rockford, IL) is added. The plates are again incubated for 1 h at room temperature and washed.
10. 100 μl 4-Methylumbelliferyl phosphate (4-MUP) is then added to each well and allowed to incubate for 30 min in the dark.
11. Fluorescence is read using a Tecan Genios microplate reader at 350/465 nm.
12. Expression titers are calculated by referencing experimental values against an immunoglobulin G standard.

3.3 Selection of Pichia pastoris Strains Expressing mAbs by Cell Surface Labeling

The following protocol of selecting high producing strain is based on the publication of Lin et al. (10).

1. After 36-h induction in bioreactors (for details refer to Subheading 3.4), about 4 OD_{600} of induced cells (~10^8 cells) are collected into a 1.5-ml microtube and washed twice with phosphate-buffered saline at room temperature by briefly vortexing and centrifuging.
2. The cells are then suspended in 500 μl of PBS and incubated with 5 μl (10 μg) of goat anti-human IgG (H+L) Alexa 488 or goat anti-human IgG (H+L) Alexa 647 at room temperature for 30 min.
3. The cells are washed with PBS three times and suspended in 1 ml of PBS.
4. The labeled cells are kept in the dark on ice until the sorting was completed.
5. Cell sorting is performed using an FACSAria II cell sorter with three lasers (405, 488, and 633 nm) equipped with Diva v6.1 software.
6. Doublet discrimination gates are routinely used to ensure a population of single cells for further gating (see Note 9).
7. The cells then are transferred into a 50-ml conical tube containing 5 ml of BMGY with antibiotics and incubated in a shaker at 24°C for about 5 days.

The sorted populations go through another round of fermentation, titer assay, and flow cytometric analysis with the same parameters as described above.

3.4 Fermentation of Selected mAb-Producing Strains

The following protocol for fermentation of selected mAb-producing strains is modified according to Barnard et al. and Potgieter et al. (8, 11).

1. Fed-batch fermentations of glycoengineered *Pichia* were carried out in 0.5 l bioreactors (Sixfors) under the described conditions (see Note 10).

2. Prior to inoculating the bioreactors, the cell density of the inoculums is normalized based on OD_{600} measurements in a standard spectrophotometer.
3. IRIS multi-fermenter software is used to increase the stirrer speed from 550 to 1,200 rpm linearly between 1 and 10 h of the fermentation.
4. The fermentation is executed in batch mode until the initial glycerol charge (40 g/l) is consumed (typically 18–24 h).
5. A second batch phase is initiated by the addition of 17 ml of a glycerol feed solution to the bioreactor.
6. The fermentation is again operated in batch mode until the added glycerol is consumed (typically6–8 h). The induction phase is initiated by feeding a methanol feed solution (100% w/w methanol, 5 mg/l biotin, and 12.5 ml/l PTM1 salts) at 0.6 g/h. Methanol is typically fed for 36 h prior to harvest.
7. The entire volume is removed from the reactor and centrifuged in a Sorvall Evolution RC centrifuge for 30 min at 15,800 × *g*.
8. The cell mass is discarded and the supernatant retained for purification and analysis.

3.5 Purification of Recombinant mAb from Glycoengineered Pichia Fermentation Supernatant

The following protocol is modified according to Jiang et al. (12) including the first step of protein A capture and the second step of polishing by HP SP chromatography.

1. Clarify fermentation broth from fermentation by centrifugation at 4°C, 8,500 rpm (15,810 × *g*) for 45 min, and collect the supernatant.
2. Filter the supernatant with SARTOBRAN P Sterile MidiCap, 0.65 + 0.45 μm, to further clarify the supernatant.
3. Perform protein A capture with AKTApilot at room temperature at a flow rate of 100 ml/min.
4. Column is equilibrated with 3 column volume of 20 mM Tris pH 7.0 and washed with 2 CV 20 mM Tris pH 7.0 and 3 column volume of 20 mM Tris pH 7.0.
5. Column is eluted with 6 column volume of linear gradient from 0 to 100% 50 mM sodium citrate buffer, pH 3.0.
6. Pool mAb fractions, filter-sterilize using 0.2 μm membrane, and keep at 4°C.
7. Run 4–20% SDS-PAGE using nonreducing (containing 100 mM NEM) and reducing sample buffers.
8. Measure mAb concentration by Bradford assay using commercial antibody as standard.
9. The second purification step is to polish by SP HP exchange chromatography.

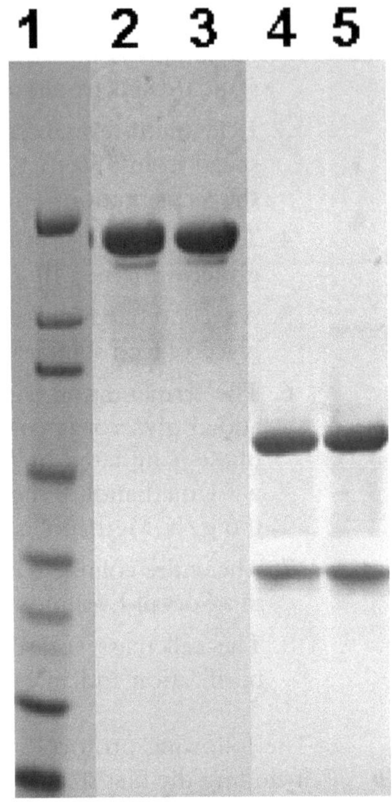

Fig. 2 Characterization of glycoengineered *Pichia* produced mAb in SDS-PAGE. Comparison of anti-HER2 produced in glycoengineered *Pichia* with CHO produced trastuzumab in reducing and nonreducing SDS-PAGE. *Lane 1*: protein marker, *lane 2*: anti-HER2 at nonreducing condition, *lane 3*: trastuzumab at nonreducing condition, *lane 4*: anti-Her2 mAb at reducing condition, *lane 5*: trastuzumab at reducing condition

10. Pack 60 ml of SP HP into XK16/30 column.
11. Take protein A capture pool and make 5× dilution with H_2O (conductivity: <~10 mS/cm) and adjust to pH 5.0.
12. Perform SP HP chromatography on AKTAexplorer 100 at room temperature at a flow rate of 5 ml/min.
13. Column is equilibrated with 5 column volume of 25 mM sodium acetate, pH 5.0 and load column with 5× diluted antibody, pH 5.0. Column is washed with 3 column of 25 mM sodium acetate, pH 5.0.
14. Column is eluted with linear gradient of 0–500 mM NaCl in 25 mM sodium acetate buffer, pH 5.0.
15. Run fractions on 4–20% SDS-PAGE gel using nonreducing (containing 100 mM NEM) and reducing sample buffers. Figure 2 has shown that glycoengineered *Pichia*-produced anti-Her2 is very similar to its CHO counterpart, trastuzumab.

4 Notes

1. Amino acid sequence of anti-Her2 is the same as trastuzumab according to Herceptin® product monograph http://www.rochecanada.com/fmfiles/re7234008/Research/ClinicalTrialsForms/Products/ConsumerInformation/MonographsandPublicAdvisories/Herceptin/herceptin-Sept14SIE.pdf).

   Anti-Her2 heavy chain with alpha mating factor predomain (underlined) as signal sequence: MRFPSIFTAVLFAASSALAEVQLVESGGGLVQPGGSLRLSCAASGFNIKDTYIHWVRQAPGKGLEWVARIYPTNGYTRYADSVKGRFTISADTSKNTAYLQMNSLRAEDTAVYYCSRWGGDGFYAMDYWGQGTLVTVSSASTKGPSVFPLAPSSKSTSGGTAALGCLVKDYFPEPVTVSWNSGALTSGVHTFPAVLQSSGLYSLSSVVTVPSSSLGTQTYICNVNHKPSNTKVDKKVEPKSCDKTHTCPPCPAPELLGGPSVFLFPPKPKDTLMISRTPEVTCVVVDVSHEDPEVKFNWYVDGVEVHNAKTKPREEQYNSTYRVVSVLTVLHQDWLNGKEYKCKVSNKALPAPIEKTISKAKGQPREPQVYTLPPSREEMTKNQVSLTCLVKGFYPSDIAVEWESNGQPENNYKTTPPVLDSDGSFFLYSKLTVDKSRWQQGNVFSCSVMHEALHNHYTQKSLSLSPGK.

   Anti-Her2 light chain with alpha mating factor predomain as signal sequence (underlined): MRFPSIFTAVLFAASSALADIQMTQSPSSLSASVGDRVTITCRASQDVNTAVAWYQQKPGKAPKLLIYSASFLYSGVPSRFSGSRSGTDFTLTISSLQPEDFATYYCQQHYTTPPTFGQGTKVEIKRTVAAPSVFIFPPSDEQLKSGTASVVCLLNNFYPREAKVQWKVDNALQSGNSQESVTEQDSKDSTYSLSSTLTLSKADYEKHKVYACEVTHQGLSSPVTKSFNRGEC.

2. Any potential restriction sites of *Eco*RI, *Swa*I, *Fse*I, and *Pst*I used for heavy-chain and light-chain cloning located in encoding region have been removed before DNA synthesis. *Spe*I used for linearizing the plasmid at encoding region is also eliminated.

3. Restriction enzymes digestion was carried out using 1 μg of each plasmid including pGLY-mAb-vector and plasmid containing heavy chain or light chain synthesized by Genscript. Four pieces of the fragments listed as the following were recovered from 1% agarose gel and purified by QIAquick Gel Extraction Kit from Qiagen after enzymatic cutting: (1) pGLY-mAb-vector cut with *Eco*RI/*Swa*I, larger piece; (2) pGLY-mAb-vector cut with *Fse*I/*Pst*I, smaller piece; (3) heavy chain piece cut from Genscript provided plasmid containing synthesized heavy chain with restriction enzymes of *Eco*RI and *Fse*I; and (4) light chain piece cut from Genscript provided plasmid containing synthesized light chain with *Pst*I and *Swa*I.

4. The same amino acid sequences could generate various DNA sequences in different codon optimization algorithm. The primers of Hc/lp and Lc/lp used in this chapter are based on the DNA sequences encoding IgG constant region the author used. Scientists should redesign these primers based on their own DNA sequence coding for heavy- and light-chain regions.
5. The growth rate of glycoengineered strains differs from wild-type *Pichia* and transformation efficiency of glycoengineered *Pichia* is 100–1,000-fold higher than wild-type *Pichia*. To achieve the best transformation efficiency, pilot study with various cell stages is needed. OD_{600} higher than 10 normally has low transformation efficiency.
6. The cuvette can be ice-cooled or kept at room temperature; no significant difference was observed in terms of *Pichia* transformation efficiency in a short period.
7. The sensitivity to Zeocin varies in different glycoengineered *Pichia* strain, for example, YGLY638, a strain which can express glycoproteins with a $Man_5GlcNAc_2$ N-glycosylation profile and is less sensitive to Zeocin comparing to YGLY8316 which is a strain capable of producing glycoproteins with bi-antennary complex glycans terminating with galactose ($Gal_2GlcNAc_2$ $Man_3GlcNAc_2$). In addition, the resistance of *Pichia* transformants to Zeocin is also dependent on the copy numbers of expressing plasmid being integrated into *Pichia* chromosome.
8. Two to eight times of expansion plates are created for every seed plate depending upon the amount of material required for subsequent analysis and expression level of antibody of interest. This two-step batch growth phase method allows the biomass to grow to saturation twice and therefore serves to normalize OD_{600} in microtiter plates.
9. When cells are labeled only with goat anti-human IgG (H+L)-Alexa 488, a gate is set on a histogram plot to collect 1% of the cells with the most intense signal. During each sort, 5,000–10,000 cells are collected into a tube containing 0.5 ml of BMGY.
10. Fermentation are carried out under the following conditions: pH 6.5, 24°C, 300 ml airflow/min, and an initial stirrer speed of 550 rpm with an initial working volume of 350 ml (330 ml BMGY medium and 20 ml inoculums at a cell density of approximately 100 OD_{600}).

References

1. Li J, Zhu Z (2010) Research and development of next generation of antibody-based therapeutics. Acta Pharmacol Sin 31:1198–1207
2. Anon (2010) Deal watch BMS acquires rights for IL-6 inhibitor. Nat Rev Drug Discov 9:10–10
3. Li HJ, Sethuraman N, Stadheim TA, Zha DX, Prinz B, Ballew N, Bobrowicz P et al (2006) Optimization of humanized IgGs in glycoengineered *Pichia pastoris*. Nat Biotechnol 24:210–215
4. Komarova TV, Kosorukov VS, Frolova OY, Petrunia IV, Skrypnik KA, Gleba YY, Dorokhov YL (2011) Plant-made trastuzumab (herceptin) inhibits HER2/Neu+ cell proliferation and retards tumor growth. PLoS One 6(3):e17541
5. Grohs BM, Niu Y, Veldhuis LJ, Trabelsi S, Garabagi F, Hassell JA, McLean MD, Hall JC (2010) Plant-produced trastuzumab inhibits the growth of HER2 positive cancer cells. J Agric Food Chem 58:10056–10063
6. Clarke SJ, Smith JT, Gebbie C, Sweeney C, Olszewski N (2009) A phase I, pharmacokinetic (PK), and preliminary efficacy assessment of ALD518, a humanized anti-IL-6 antibody, in patients with advanced cancer. J Clin Oncol 27(15 Suppl)
7. Zhang N, Liu L, Dan DC, Cummings NR, Cukan M, Jiang Y et al (2011) Glycoengineered *Pichia* produced anti-HER2 is comparable to trastuzumab in preclinical study. MAbs 3:289–298
8. Potgieter TI, Cukan M, Drummond JE, Houston-Cummings NR, Jiang YW, Li F et al (2009) Production of monoclonal antibodies by glycoengineered *Pichia pastoris*. J Biotechnol 139:318–325
9. Orchard MG, Neuss JC, Galley CM, Carr A, Porter DW, Smith P, Scopes DI et al (2004) Rhodanine-3-acetic acid derivatives as inhibitors of fungal protein mannosyl transferase 1 (PMT1). Bioorg Med Chem Lett 14:3975–3978
10. Lin S, Shen Z, Zha D, Sharkey N, Prinz B, Hamilton S et al (2010) Selection of *Pichia pastoris* strains expressing recombinant immunoglobulin G by cell surface labeling. J Immunol Methods 358:66–74
11. Barnard GC, Kull AR, Sharkey NS, Shaikh SS, Rittenhour AM, Burnina I et al (2010) High-throughput screening and selection of yeast cell lines expressing monoclonal antibodies. J Ind Microbiol Biotechnol 37:961–971
12. Jiang Y, Li F, Zha D, Potgieter TI, Mitchell T, Moore R et al (2011) Purification process development of a recombinant monoclonal antibody expressed in glycoengineered *Pichia pastoris*. Protein Expr Purif 76:7–14

Chapter 4

N-Glycosylation Humanization for Production of Therapeutic Recombinant Glycoproteins in *Saccharomyces cerevisiae*

Christelle Arico, Christine Bonnet, and Christophe Javaud

Abstract

The production of therapeutic recombinant glycoproteins deals with three main issues: cost, production capacities, and glycosylation. Nowadays, such proteins are expressed in various complex expression systems (CHO, bacteria, etc.); the processes related to those production hosts are time consuming and expensive, or the question of posttranslational modifications (as glycosylation) control is still unresolved. There is a need to find an alternative approach, while maintaining high quality level: the new system must be able to add complex *N*-glycan structures to proteins of interest. Developed in several strains of *Saccharomyces cerevisiae*, GlycodExpress™ is an innovative technology that allows production of therapeutic recombinant glycoproteins with humanized and homogeneous *N*-glycan moieties. We show how to delete mannosyltransferases involved in host N-glycosylation to obtain more than 90% of homogeneity in glycan structures. The methodology developed to select the optimal fusion between a heterologous glycosyl-enzyme and a localization sequence is also presented. Finally, the screening of the best producing strain is illustrated.

Key words *Saccharomyces cerevisiae*, Glycosylation, Glycoengineering, Homologous recombination, GlycodExpress™, YAC-express™

1 Introduction

Microorganisms have been widely used, since ancient times, for the production of fermented food or beverages. Nowadays, with advances in genetic manipulation tools, their engineering for research interest or for production of industrial bio-compound is very common.

Saccharomyces cerevisiae, a GRAS organism (1) devoid of intellectual property, is a particularly efficient and attractive system for production of therapeutic proteins. Indeed, growth in neutral and well-defined culture media is not time consuming and cheaper than in higher eukaryotic systems such as CHO cells. Moreover, yeasts are not susceptible to contaminations by viruses, endotoxins, or prions. Large-scale production allows high yields and

Alain Beck (ed.), *Glycosylation Engineering of Biopharmaceuticals: Methods and Protocols*, Methods in Molecular Biology, vol. 988, DOI 10.1007/978-1-62703-327-5_4, © Springer Science+Business Media New York 2013

proteins, with posttranslational modifications, can be more easily secreted and purified than in other production hosts. Several recombinant therapeutic proteins have been produced, and approved for commercialization, in this system: Human Serum Albumin (Recombumin and Albucult, Novozymes), insulin (Actrapid, Novo Nordisk), HBV surface antigen (Pediatrix, GlaxoSmithKline), hirudine (Revask, Avantis); and productions of other human proteins under active studies: α-amylase (2), HPV16IL (3), and immunoglobulin G (4).

However, glycan residues are generally essential for mammalian glycoproteins activity (5–8), and as proteins produced in yeast wild-type strains carry glycan structures radically different from those of the natural one, they can present less activity and changes in their immunogenicity (9, 10).

The process of N-glycosylation (11, 12) consists of a covalent linkage, in the endoplasmic reticulum, of a specific oligosaccharide ($Glc_3Man_9GlcNAc_2$) on a nascent protein, at the Asn-X-Ser/Thr consensus motif. This first step is conserved through Eukaryotes. The following maturation, occurring in the Golgi apparatus and leading to polymannosylated, complex or hybrid structures, will determine the differences between species. In *S. cerevisiae*, a set of mannosyltransferases will successively add mannose residues to the core glycan until hypermannosylation, while in Mammals, the glycan structures are complex, composed of N-acetylglucosamine, galactose, sialic acid, or fucose. Engineering yeast N-glycosylation pathway allows the production of therapeutic glycoproteins bearing homogeneous glycans of human type. This homogeneity increases protein efficiency and reproducibility between production batches (13).

Glycode has developed two innovative technologies, Glycod Express™ (Fig. 1; patent WO/2008/095797) and YAC-express (Fig. 2; patent WO/2012/013823), to enhance recombinant glycoproteins production through selective glycosylation in yeast *S. cerevisiae*.

This chapter is dedicated to GlycodExpress™ strategy: (1) *S. cerevisiae* has been modified by sequential deletion of mannosyltransferases; (2) several fusion constructions between localization sequence and glycosyl-enzyme active domain are screened; (3) expression cassette containing the selected combination is introduced into the yeast genome by homologous recombination; (4) selection of engineered strains is based on correct glycan structure and heterologous protein production capacities. This strategy offers several advantages: strong stability of the genome modifications over time, full tracking of genotypic and phenotypic changes, and possibility to reengineer the strain in an identical way, in case of spontaneous recombination.

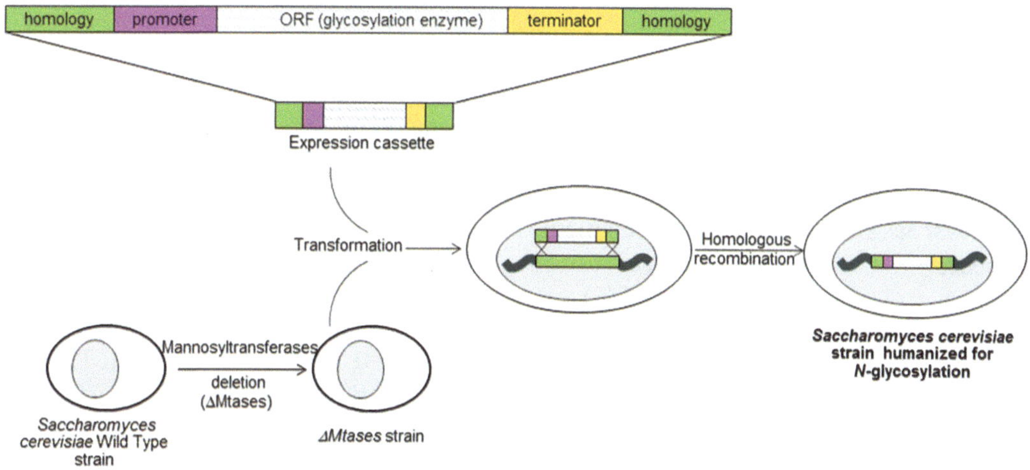

Fig. 1 GlycodExpress™ technology. Strategy, based on homologous recombination, for generating N-glycosylated-modified yeast strains of *Saccharomyces cerevisiae*. Each expression cassette was introduced in ΔMannosyltransferases (ΔMtases) strain. The integration of the expression cassette into the genome by homologous recombination allowed the expression of the heterologous ORF in the modified yeast strain

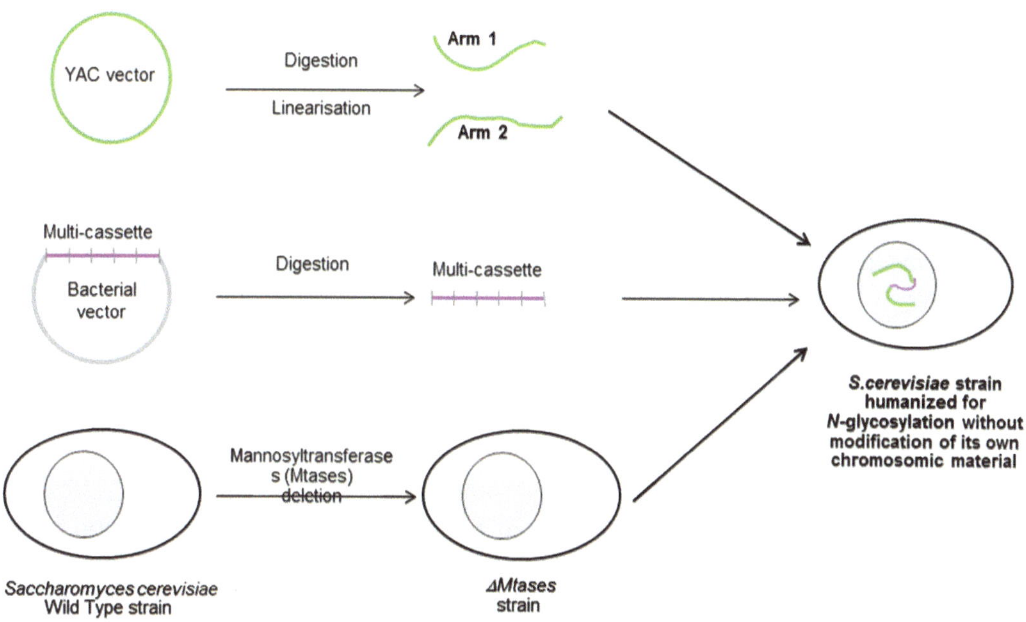

Fig. 2 YAC-Express™ technology. Strategy, based on artificial chromosome technology, for generating N-glycosylated-modified yeast strains of *Saccharomyces cerevisiae*. A multi-cassette containing all ORFs needed for a humanized *N*-glycan pathway is ligated with a digested YAC vector to obtain a linearized YAC containing the multi-cassette. The integration of the linearized YAC into the nucleus allowed the expression of the heterologous ORF in the modified yeast strain

2 Materials

2.1 Culture Medium

YPD (Yeast Peptone Dextrose; Formedium ref. CCM0110), YNB (Yeast Nitrogen Base; Formedium, ref. CYN0210), Galactose (Alpha Aesar ref. A12813), Agar (Formedium ref. AGA03).

2.2 Yeast Selection Compound

G418 (Geneticin/Kanamycin; Formedium ref. G8141S), Hygromycin (Euromedex ref. EU3250), α-Amino Adipate (Sigma ref. A0637-5G).

2.3 Yeast Transformation

TE/LiAC: Tris–HCl (pH 7.5) 10 mM; EDTA 1 mM; Lithium Acetate (pH 7.5) 100 mM sterilized.

TE/LiAc/PEG: Tris–HCl (pH 7.5) 10 mM; EDTA 1 mM; Lithium Acetate (pH 7.5) 100 mM; PEG 3350 40% sterilized.

Carrier DNA (Sonicated fish sperm DNA; Roche ref. 11467140001).

Sorbitol (Fisher ref. C0486C).

2.4 Molecular Biology

RNeasy mini Kit (RNA extraction and purification, Qiagen ref. 74104), DNAse (Macherey Nagel ref. 740963), RNAse (Sigma ref. R6513-50MG), Reverse Transcriptase (Promega ref. A3802), RNaseOUT (Invitrogen ref. E3401F), OligoDT (Invitrogen ref. 18418012), dNTP (New England Biolabs ref. N0447S), TaqPlatinium DNA polymerase (Invitrogen ref. 10966.083), Oligonucleotides (Eurofins), pESC-LEU and pESC-URA (Stratagene refs. 217452 and 217454).

2.5 EPO Production and Analysis

Acetone (Sigma ref. 179973-1L), nitrocellulose membrane (Dutscher ref. 034912), Mouse anti-human EPO monoclonal antibody (R&D systems ref. MAB2871), HRP-Goat Anti-Mouse IgG (H+L) Conjugate (Invitrogen ref. G21040).

3 Methods

3.1 Deletion of S. cerevisiae Mannosyltransferases

The first and key step to humanize yeast N-glycosylation is the deletion of sequences encoding specific mannosyltransferases, among which are *OCH1* and *MNN1* genes, responsible for hypermannosylation; the resulting strains produce glycoproteins with a core type structure $Man_{9/8}GlcNAc_2$ (14–17).

3.1.1 OCH1 Deletion

The *OCH1* α-1,6-mannosyltransferase was deleted by insertion in its locus of a kanamycin resistance cassette KANMX (18) (see Note 1). The cassette is flanked by homologous *OCH1* sequences (around 200 bp at the ATG for the 5′ end and the Stop codon for the 3′ end), essential for homologous recombination mechanism.

The transformation was processed as following: 5 mL of complete culture medium (YPD) was inoculated with the strain of interest and incubated overnight at 30°C under agitation (200 rpm). Then, the culture was diluted, in 10 mL final volume, to OD_{600} around 0.5 and re-incubated at 30°C under agitation of 200 rpm until an OD_{600} around 1 is reached; it was then centrifuged 5 min (4,000 × *g*) at 4°C. The supernatant was discarded and the cells were washed carefully in 10 mL of sterile TE/LiAC solution followed by another cycle of centrifugation. The yeast pellet was suspended in 50 μL final volume of TE/LiAC solution and incubated 15 min, without agitation, at 30°C. Those competent cells are carefully added to 300 μL of freshly prepared TE/Liac/PEG solution, mixed by vigorous shaking with 50 μg of carrier DNA and the linearized cassette (1 μg for 1 kb length), and incubated at 30°C for 30 min. Heat choc was performed at 42°C, for 20 min. After 1–2 h of relaxation in 1 mL of sorbitol (1 M), at 30°C under agitation of 200 rpm, yeasts were spread on selective agar medium, YPD containing 200 μg/mL of G418. After few days at 30°C, visible colonies were restricted on two complete medium agar plates for growth comparison between 30 and 37°C. As *OCH1* deletion has been reported to induce a thermosensitive phenotype (19), colonies that did not present any growth at the highest temperature were tested by PCR, following a protocol adapted for *S. cerevisiae* strain impaired for glycosylation (article submitted). Oligonucleotides have been designed to check the absence of *OCH1* and the presence of the resistance cassette (Table 1).

Finally, several validated strains were cultivated in microfermenter device; total secreted proteins in the culture medium were

Table 1
Sequences of primers used in this study

Target	Primer name	Primer sequence
OCH1 (Ext.)	CR25 (Fw)	5′-TTGCAGGGACAGCAACGT-3′
	EF122 (Rv)	5′-GGGCAGTTAGTACGGTATCG-3′
OCH1 (Int.)	CR03 (Fw)	5′-CCGATAGTCATCCAAGCGT-3′
	CR04 (Rv)	5′-AACTGGATCCTACGAGCATA-3′
MNN1 (Ext.)	CB306 (Fw)	5′-TGGTCTTGCTTCGTACTCTCG-3′
	CB314 (Rv)	5′-AGCGGCCTTTTTATGCTGTTGG-3′
MNN1 (Int.)	CB111 (Fw)	5′-CTTTGAGATCGTGTACCATA-3′
	CB114 (Rv)	5′-CCTCGCCACTTAGGGTGGAT-3′
LYS2 (Ext.)	CB289 (Fw)	5′-CACAACAAGAACCTTATACG-3′
	CB293 (Rv)	5′-CATCACCTCTTTTGATACCTG-3′

Fw forward, *Rv* reverse, *Ext* amplification outside of the recombination sequences, *Int* amplification inside of the recombination sequences

precipitated by acetone; and glycan analyses were performed by mass spectrometry (MALDI-TOF, Proteodynamics). Strains presenting the best glycosylation profile were screened for their production capacities as described in Subheading 3.4.

3.1.2 MNN1 Deletion

MNN1 is an α-1,3-mannosyltransferase responsible for addition of α 1–3 mannoses on the core glycan structure. Its deletion is a prerequisite to further engineering of *S. cerevisiae* N-glycosylation. The protocol described for *OCH1* deletion was applied. Homologous sequences targeted around 100 bp after the ATG for the 5′ end and before the STOP codon for the 3′ end. The deletion cassette expressed the *hph* ORF that confers resistance to hygromycin. The agar selecting medium was YPD containing 150 μg/mL of the antibiotic. As no specific phenotype is associated to *MNN1* deletion, all colonies were restricted on selective medium and tested by PCR for the absence of the targeted ORF and the presence of the cassette. The glycan analysis and the selection of the best producing strain were performed as described earlier.

Deletion of *OCH1* and *MNN1* led to a *S. cerevisiae* strain with 90% of the core glycan structure $Man_{9/8}GlcNAc_2$ common to all Eukaryotes (Fig. 3). This homogeneity was sufficient to go on with the engineering. Next step consisted in introduction of glycosyl-enzymes able to construct *N*-glycans of the mammalian type.

3.2 Screening of Glycosyl-Enzymes Activities

This step consists of expression and activity assays of heterologous glycosyl-enzymes that allow N-glycosylation humanization.

3.2.1 Design of Fusion Enzymes

N-glycosylation is a sequential process occurring all along the secretory pathway. Each glycosyl-enzyme needs to be localized in the right cellular compartment. Sequence for the enzyme of interest was deleted for its own localization region and fused to one of a *S. cerevisiae* protein. For each activity, enzymes of various origins were fused to several localization sequences as illustrated in Table 2. Then sequences were optimized for expression in budding yeast and synthesized (DNA2.0).

3.2.2 Glycosyl-Enzyme Selection

Each construction was cloned in an expression vector and transformed, following the protocol described in Subheading 3.1, in an appropriate yeast strain. The presence of the plasmid was validated by PCR on yeast colonies. A colony was grown in microfermenter device and total secreted proteins were precipitated and analyzed as described before. As a final control, the selected glycosyl-enzyme construction was validated by renewing experiment in other yeast backgrounds.

An expression cassette was designed with, in 5′ of the selected glycosyl-enzyme, a constitutive promoter functional in *S. cerevisiae* and, in 3′, an adapted terminator. This cassette was introduced in the yeast genome as shown in the next paragraph.

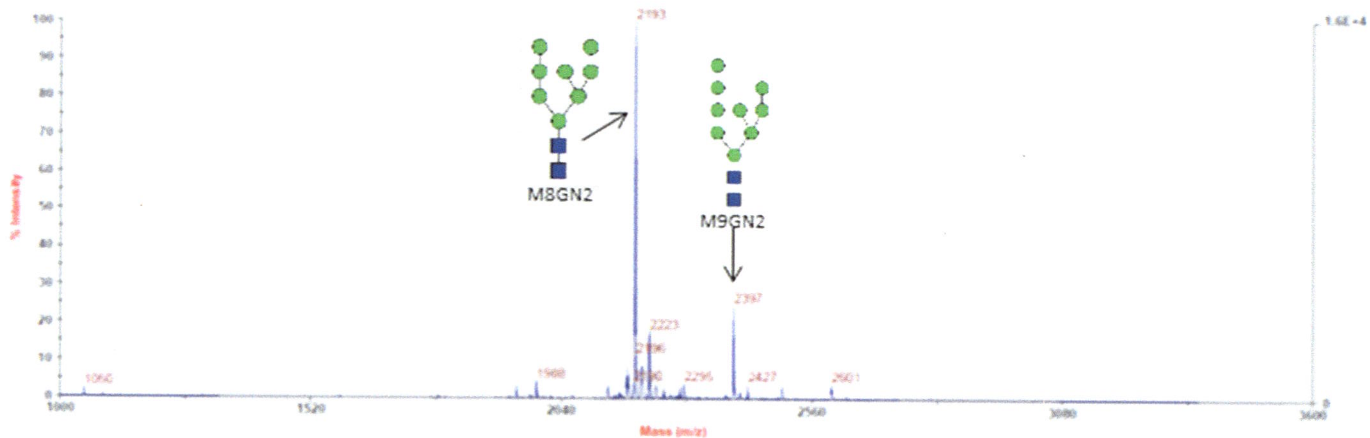

Fig. 3 MALDI-TOF MS analysis of *N*-glycans released from secreted proteins (Δoch1Δmnn1 strain). Total soluble proteins from culture media were recovered (see precipitation methods) and glycans were released by PNGase F treatment. The PSD fragmentation of the ions *m/z* 2,192 Da and *m/z* 2,396 Da defined glycan structures of the $Man_8GlcNAc_2$ (75% of the total *N*-glycans) and the $Man_9GlcNAc_2$ (17% of the total *N*-glycans) as shown in the graph (*green circle*: mannose, *blue square*: *N*-acetyl-glucosamine)

Table 2
Heterologous origin of glycosylation activities tested in fusion with different localization signals

Glyco-enzyme activity	Origine	Localization
α-1,2-mannosidase	*Tricoderma reesei* *Aspergillus saitoi* *Caenorhabditis elegans*	MNS1 (*Sc*) HDEL GLS1 (*Sc*)
UDP-GlcNAc α3-D mannoside β-1,3-acetylglucosaminyltransferase 1	*Rattus norvegicus* *Homo sapiens* *Mus musculus* *Phaeodactylum tricornutum*	MNN1 (*Sc*) VAN1 (*Sc*) KRE2 (*Sc*) Alpha-factor (*Sc*)
UDP-GlcNAc transporter	*Saccharomyces cerevisiae* *Kluyveromyces lactis* *Homo sapiens*	
Mannosyl-oligosaccharide 1,3-1,6 α-mannosidase	*Caenorhabditis elegans* *Xenopus tropicalis* *Homo sapiens* *Drosophila melanogaster* *Mus musculus*	MNN10 (*Sc*) KTR2 (*Sc*) MNN1 (*Sc*) MNN11 (*Sc*) MNN9 (*Sc*)
Mannosyl (α-1,6)-glycoprotein β-1,2-N-acetylglucosaminyltransferase	*Homo sapiens* *Caenorhabditis elegans* *Xenopus tropicalis* *Ixodes scapularis*	MNN2 (*Sc*) KRE2 (*Sc*) MNN9 (*Sc*) MNN10 (*Sc*)
UDP-Gal β1GlcNAc β1,4-galactosyltransferase	*Homo sapiens* *Mus Musculus* *Rattus norvegicus*	KRE2 (*Sc*) MNN11 (*Sc*) MNN1 (*Sc*)
UDP-galactose transporter	*Homo sapiens* *Schizosaccharomyces pombe* *Rattus norvegicus*	
UDP-glucose 4-epimerase	*Homo sapiens* *Drosophila melanogaster* *Mus musculus* *Dictyostelium discoideum*	
N-acetylneuraminic acid synthase	*Neisseria meningitidis*	
CMP-acetylneuraminic acid synthase	*Neisseria meningitidis*	
ST3 β-galactosidase α-2,3-sialyltransferase 4	*Homo sapiens* *Xenopue laevis* *Mus musculus*	MNN10 (*Sc*) MNN11 (*Sc*) MNN1 (*Sc*)
CMP-sialic acid transporter	*Homo sapiens* *Mus musculus* *Rattus norvegicus*	

(*Sc*) sequences from *Saccharomyces cerevisiae*

Table 3
Selection marker usable for targeted integration of expression cassette in *Saccharomyces cerevisiae*

Targeted gene	Selection
ADE2	Red colonies Auxotrophy for adenine
URA3	Resistance to 5-fluoro-orotic acid Auxotrophy for uracil
TRP1	Resistance to 5-fluoro-anthranilic acid Auxotrophy for tryptophan
LYS2	Resistance to α-Amino Adipate Auxotrophy for lysine
CAN1	Resistance to canavanin Auxotrophy for arginine
MET17	Growth on methylmercure containing medium Auxotrophy for methionine
MET28	Growth on chromate or selenate containing medium Auxotrophy for methionine
FPR1	Resistance to rapamycin
FCY1	Resistance to 5-fluoro cytosine
OSH3	Resistance to myriocin

3.3 Genomic Integration

3.3.1 Choice of the Target

Each expression cassette containing a glycosyl-enzyme was integrated into a specific locus in the yeast genome. Effects of the deletion or disruption of various markers are described for *S. cerevisiae* (Table 3). The targeted integration of the expression cassette induced resistance and/or auxotrophy that allowed the selection of the modified strain.

On both sides of an expression cassette, assigned to a specific locus, homologous sequences to the target locus were added to allow homologous recombination (see Note 2).

3.3.2 Yeast Transformation and Selection

LYS2 gene was inactivated by inserting an expression cassette containing one of the selected glycosyl-enzymes. Integration of this sequence into the genome of the yeast was accomplished following the same protocol than for the *OCH1* deletion.

The flanking regions consisted of about one hundred bases and allowed integration of the glycosyl-cassette within the *LYS2* gene between bases 121 and 700 after the ATG. Strains with integrated expression cassette were selected on a medium containing

1.5 g/L of α-Amino Adipate in YNB with lysine, without ammonium sulfate (see Note 3). A second selection step was performed to compare the growth on complete YNB against LYS DO YNB media. A lack of growth on selective medium identified positive-modified clones for cassette integration in the specific locus.

These clones were validated using the specific PCR protocol mentioned earlier. Oligonucleotides (Table 1) were selected so as to check, on one hand, the presence of the cassette and, on the other hand, the integration into *LYS2* gene.

Positive clones were recovered and their RNA extracted and purified with a commercial Kit. Each RNA sample was split into two parts: one being treated with RNase for 30 min at room temperature (control for no DNA contamination during extraction) and the other left untreated. Reverse transcription was performed on all of the RNA samples, including the RNase-treated negative control, to confirm the expression of the integrated glycosyl-enzyme.

As explained for Δ*och1* validation, extracellular secreted proteins were recovered by acetone precipitation and purified glycans were analyzed.

3.4 Best Producing Strain Screening

3.4.1 Expression of a Model Protein

The nucleotide sequence of the human EPO optimized for the expression in *S. cerevisiae* was cloned into commercial expression vectors, pESC-LEU or pESC-URA (see Note 4). The EPO expression was under the inducible GAL1-10 promoter. This promoter is one of the strongest known in *S. cerevisiae* and is commonly used for expression of recombinant proteins. The presence of galactose in the culture media, without any trace of glucose, raised the expression of recombinant protein about 1,000 times (20).

All modified strains were transformed by the EPO vector by heat-shock method to test their production capacities. Positive clones, validated by PCR screening, were cultivated in microplates and induced for production.

3.4.2 Analysis

Secreted proteins were recovered and transferred onto a nitrocellulose membrane in order to proceed with detection by the anti-hEPO antibody. The membrane was saturated with a blocking solution (PBS, 5% fat milk) for 1 h and then put into contact with the anti-hEPO antibody solution (dilution 1:1,000) for 1 h. After three rinses with 0.05% Tween 20-PBS the membrane was put into contact with the secondary anti-mouse-HRP antibody in order to proceed with colorimetric detection (Fig. 4). Comparison of spot intensities revealed production capacity of each clone. The production analysis of clones derived from the same strain allowed the detection of the best producer.

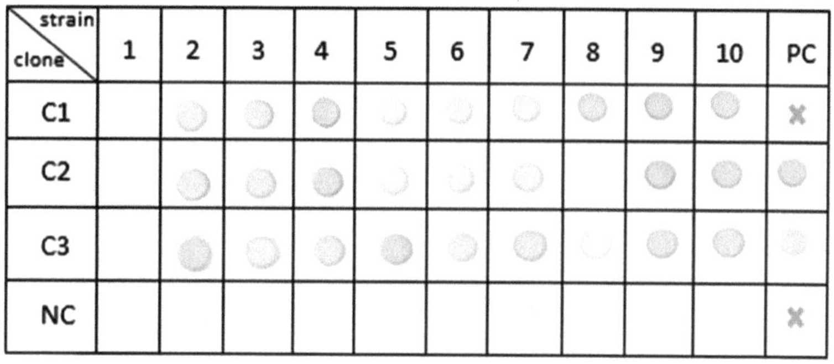

Fig. 4 Dot blot analysis of secreted rEPO to determine which strain has the ability to produce a heterologous protein. *NC* negative control (empty vector, column 1–10); *PC* positive control (known producing clones)

4 Notes

1. Recycling cassettes. To avoid presence of antibiotic resistance in yeast producing host, recycling cassettes based on Cre-Lox or Flp-FRT recombination mechanisms have been developed by Glycode. These systems require two steps: an integration of a cassette to a specific locus then induction of cleavage process.
2. Auxotrophic markers. The integration method is limited by the number of available markers in *S. cerevisiae* (Table 3). Glycode has developed YAC-express™ technology to go further on the engineering of posttranslational modifications.
3. α-Amino Adipate selection. The α-Amino Adipate is converted to α-semialdehyde Amino Adipate by the α-Amino Adipate reductase activity of Lys2p/Lys5p. Yeast cells use α-Amino Adipate only on nitrogen-poor environment. Therefore, selective media must not contain ammonium sulfate. Media free of other nitrogen sources may include uracil, adenine, methionine, lysine, tryptophan, and leucine without altering α-Amino Adipate toxicity.
4. Secretion signal screening for each new heterologous protein. For each new heterologous protein to be produced in Glycode's modified strains, various secretion sequences were tested to define the optimal combination between a secretion factor and the protein of interest: pre-pro sequence of the alpha-factor, K1 killer preprotoxin sequence, signal sequences of invertase, alpha-amylase, *INU1*, *SUC2*, *PHO5*, *MEL1*, etc.

Acknowledgments

C.A., C.B., and C.J. want to thank the entire Glycode team for its high technical and scientific assistance. Special thanks to Ludivine Perrocheau and Jean-Noël Chaize, respectively, head of Biochemistry and Downstream Process Development and head of Fermentation Process for the helping discussions and the critical reading of this manuscript.

References

1. http://www.fda.gov/Food/FoodIngredientsPackaging/ucm078956.htm
2. Liu Z, Tyo K, Martinez JL, Petranovic D, Nielsen J (2012) Different expression systems for production of recombinant proteins in *Saccharomyces cerevisiae*. Biotechnol Bioeng 109(5):1259–1268, http:/dx.doi.org/10.1002/bit.244009
3. Kim HJ, Lee SJ, Kim HJ (2010) Optimizing the secondary structure of human papillomavirus type 16L1 mRNA enhances L1 protein expression in *Saccharomyces cerevisiae*. J Biotechnol 150:31–36
4. Rakestraw JA, Sazinsky SL, Piatesi A, Antipov E, Wittrup KD (2009) Directed evolution of a secretory leader for the improved expression of heterologous proteins and full length antibodies in *Saccharomyces cerevisiae*. Biotechnol Bioeng 103:1192–1201
5. Higuchi M, Oh-eda M, Kuboniwa H, Tomonoh K, Shimonaka Y, Ochi N (1992) Role of sugar chains in the expression of the biological activity of human erythropoietin. J Biol Chem 267:7703–7709
6. Fujihara J, Yasuda T, Kunito T, Fujii Y, Takatsuka H, Moritani T, Takeshita H (2008) Two *N*-linked glycosylation sites (Asn18 and Asn106) are both required for full enzymatic activity, thermal stability, and resistance to proteolysis in mammalian deoxyribonuclease I. Biosci Biotechnol Biochem 72:3197–3205
7. Jez J, Antes B, Castilho A, Krainer M, Wiederkum S, Grass J, Rüker F, Woisetschläger M, Steinkellner H (2012) Significant impact of single N-glycan residues on the biological activity of Fc-based antibody-like fragments. J Biol Chem 287:24313–24319
8. Böhm S, Schwab I, Lux A, Nimmerjahn F (2012) The role of sialic acid as a modulator of the anti-inflammatory activity of IgG. Semin Immunopathol 34:443–453
9. Kniskern PJ, Hagopian A, Burke P, Schultz LD, Montgomery DL, Hurni WM, Ip YC, Schulman CA, Maigetter RZ, Wampler DE, Kubek D, Sitrin RD, West DJ, Ellis RW, Miller WJ (1994) Characterization and evaluation of a recombinant hepatitis B vaccine expressed in yeast defective for *N*-linked hyperglycosylation. Vaccine 12:1021–1025
10. Lam JS, Mansour MK, Specht CA, Levitz SM (2005) A model vaccine exploiting fungal mannosylation to increase antigen immunogenicity. J Immunol 175(11):7496–7503
11. Yan A, Lennarz WJ (2005) Unraveling the mechanism of protein *N*-glycosylation. J Biol Chem 280(5):3121–3124
12. Lehle L, Strahl S, Tanner W (2006) Protein glycosylation, conserved from yeast to man: a model organism helps elucidate congenital human diseases. Angew Chem Int Ed Engl 45:6802–6818
13. Schiestl M, Stangler T, Torella C, Cepeljnik T, Toll H, Grau R (2011) Acceptable changes in quality attributes of glycosylated biopharmaceuticals. Nat Biotechnol 29:310–312
14. Nakayama K, Nagasu T, Shimma Y, Kuromitsu J, Jigami Y (1992) OCH1 encodes a novel membrane bound mannosyltransferase: outer chain elongation of asparagine-linked oligosaccharides. EMBO J 11:2511–2519
15. Nakanishi-Shindo Y, Nakayama K, Tanaka A, Toda Y, Jigami Y (1993) Structure of the *N*-linked oligosaccharides that show the complete loss of alpha-1,6-polymannose outer chain from *och1*, *och1 mnn1*, and *och1 mnn1 alg3* mutants of *Saccharomyces cerevisiae*. J Biol Chem 268:26338–26345
16. Lehle L, Eiden A, Lehnert K, Haselbeck A, Kopetzki E (1995) Glycoprotein biosynthesis in *Saccharomyces cerevisiae*: ngd29, an *N*-glycosylation mutant allelic to *och1* having a defect in the initiation of outer chain formation. FEBS Lett 370:41–45
17. Nakayama K, Nakanishi-Shindo Y, Tanaka A, Haga-Toda Y, Jigami Y (1997) Substrate specificity of alpha-1,6-mannosyltransferase that initiates *N*-linked mannose outer chain elongation in *Saccharomyces cerevisiae*. FEBS Lett 412:547–550

18. Wach A, Brachat A, Pöhlmann R, Philippsen P (1994) New heterologous modules for classical or PCR-based gene disruptions in *Saccharomyces cerevisiae*. Yeast 10:1793–1808
19. Nagasu T, Shimma Y, Nakanishi Y, Kuromitsu J, Iwama K, Nakayama K, Suzuki K, Jigami Y (1992) Isolation of new temperature-sensitive mutants of *Saccharomyces cerevisiae* deficient in mannose outer chain elongation. Yeast 8:535–547
20. Johnston M (1987) A model fungal gene regulatory mechanism: the GAL genes of *Saccharomyces cerevisiae*. Microbiol Rev 51:458–476

Chapter 5

Engineering the Baculovirus Genome to Produce Galactosylated Antibodies in Lepidopteran Cells

Sylvie Juliant, Marylêne Lévêque, Pierre Cérutti, Annick Ozil, Sylvie Choblet, Marie-Luce Violet, Marie-Christine Slomianny, Anne Harduin-Lepers, and Martine Cérutti

Abstract

Nowadays, recombinant proteins are used with great success for the treatment of a variety of medical conditions, such as cancer, autoimmune, and infectious diseases. Several expression systems have been developed to produce human proteins, but one of their most critical limitations is the addition of truncated or nonhuman glycans to the recombinant molecules. The presence of such glycans can be deleterious as they may alter the protein physicochemical properties (e.g., solubility, aggregation), its half-life, and its immunogenicity due to the unmasking of epitopes.

The baculovirus expression system has long been used to produce recombinant proteins for research. Thanks to recent methodological advances, this cost-effective technology is now considered a very promising alternative for the production of recombinant therapeutics, especially vaccines. Studies on the lepidopteran cell metabolism have shown that these cells can perform most of the posttranslational modifications, including *N*- and *O*-glycosylation. However, these glycan structures are shorter compared to those present in mammalian proteins. Lepidopteran *N*-glycans are essentially of the oligomannose and paucimannose type with no complex glycan identified in both infected and uninfected cells. The presence of short *N*-glycan structures is explained by the low level of *N*-acetylglucosaminyltransferase I (GNT-I) activity and the absence of several other glycosyltransferases, such as GNT-II and β1,4-galactosyltransferase I (β1,4GalTI), and of sialyltransferases.

In this chapter, we show that the glycosylation pathway of a lepidopteran cell line can be modified via infection with an engineered baculovirus to "humanize" the glycosylation pattern of a recombinant protein. This engineering has been performed by introducing in the baculovirus genome the cDNAs that encode three mammalian glycosyltransferases (GNT-I, GNT-II, and β1,4GalTI). The efficiency of this approach is illustrated with the construction of a recombinant virus that can produce a galactosylated antibody.

Key words Lepidopteran cells, Baculovirus, Recombinant proteins, Glycosylation, Vaccines, Therapeutic proteins

Alain Beck (ed.), *Glycosylation Engineering of Biopharmaceuticals: Methods and Protocols*, Methods in Molecular Biology, vol. 988, DOI 10.1007/978-1-62703-327-5_5, © Springer Science+Business Media New York 2013

1 Introduction

The baculovirus/insect cell expression system has long been limited to the production of recombinant proteins for basic research with some applications in diagnosis and in drug discovery to investigate the pharmacological properties of specific recombinant proteins, such as membrane (1) or nuclear receptors (2). Currently, this cost-effective system is employed as an interesting alternative for the production of human therapeutics. Recently, two human vaccines produced in insect cells (Cervarix, GlaxoSmithKline for the prevention of HPV-associated cervical cancer; and Provenge, Dendreon for prostate cancer) have been approved by EMA and FDA and are commercially available. Many others, such as FluBlok (Protein Sciences Corporation, seasonal influenza vaccine) and Diamyd (Diamyd inc., type 1 diabetes), are in advanced stages of clinical trials (*for review*, see ref. 3). The advantages of the baculovirus/insect cell expression system are (1) its safety: baculoviruses are arthropod-specific viruses with a restricted host range limited to some invertebrates; (2) the rapidity of generation of stable recombinant viruses; and (3) the low-cost and efficient production process using serum-free culture media to obtain high yields of recombinant proteins at the industrial scale (4).

During their replication, baculoviruses produce large amounts of two proteins that are not essential for the virus replication in tissue culture: polyhedrin (PH) and P10 (P10). PH forms occlusion-bodies, named polyhedra, in which viral particles are embedded. These structures are essential for the protection of the virus in the natural environment, but they are not required in vitro. The function of P10 is less clear, but it is generally assumed to play a role in cell lysis. These original properties have been exploited for using these viruses as expression vectors (5). Specifically, the foreign gene of interest is replaced by homologous recombination of one of these nonessential genes and thus its transcription is driven by the strong *PH* or *P10* promoter. Two baculoviruses are commonly used for the production of recombinant proteins: *Ac*MNPV, isolated from the Alfalfa Looper *Autographa californica*, and *Bm*MNPV, isolated from the silkworm *Bombyx mori*. The general principle of this technology is described in Fig. 1. Recombinant baculoviruses are generated after homologous recombination between a noninfectious defective viral DNA (linearized DNA (6) or Bacmid-derived systems (7)) and a transfer vector. The specific integration of the foreign gene into a nonessential gene is easily promoted using viral flanking sequences that are present on both side of the expression cassette. During this recombination event, the inactivated gene present on the defective viral DNA is also repaired, thus allowing the formation of infectious particles.

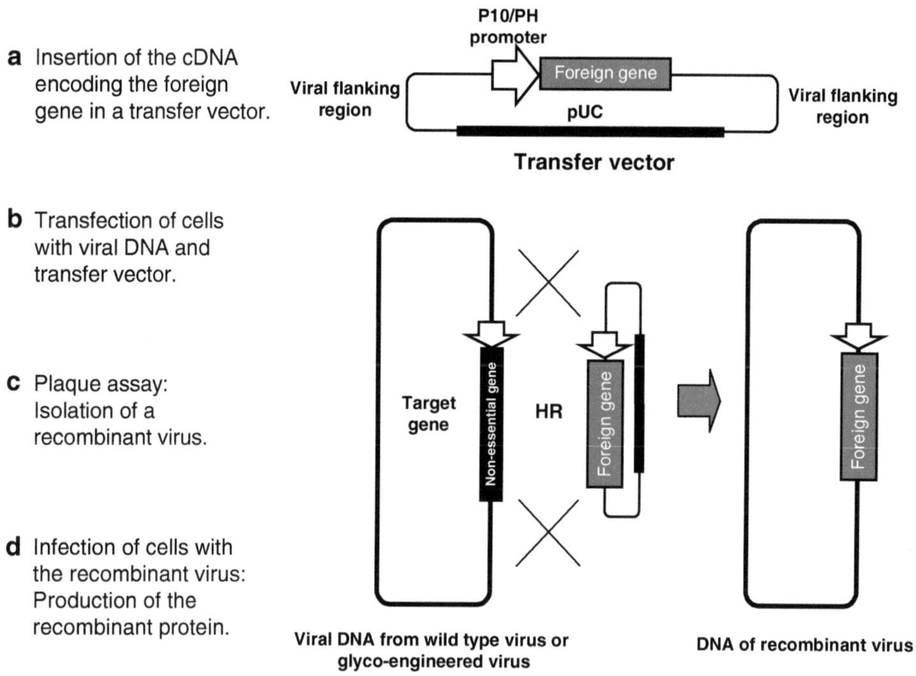

Fig. 1 Schematic representation of the different steps involved in the construction of a recombinant baculovirus. (**a**) The foreign gene is inserted downstream of a viral promoter (e.g., the promoter of the *P10* gene or of *polyhedrin*, PH, gene). (**b**) *Sf9* cells are transfected with the transfer vector and viral DNA. Recombinant viruses are then generated via homologous recombination between the viral DNA and the viral flanking regions present on both sides of the transfer vector. (**c**) Recombinant viruses are purified by plaque assay. (**d**) Infection of insect cells with this virus results in the production of the foreign protein

Specific expression cassettes can be constructed. In Fig. 2, a generic "universal" baculovirus expression cassette designed to express the heavy or light chain of an antibody is presented (8).

Furthermore, the identification of many other nonessential genes in addition to the *PH* and *P10* genes in the viral genome (9) combined with the use of different viral or cellular promoters that present specific transcriptional patterns and expression rates allows the insertion and the controlled expression of several genes into a single virus (10). This highly efficient and flexible system has been employed successfully to produce complex proteins (e.g., immunoglobulins) (8) or multiprotein complexes (e.g., virus-like-particles (VLPs)) (9). These VLPs, which can be naked or enveloped particles, are highly effective in vaccine preparations (3).

Lepidopteran cells can perform most of the posttranslational modifications. Particularly, glycosylation, which is critical for the protein structure, biological activity, and immunogenicity, has been extensively studied in these cells. Marchal et al. (11) showed that *N*-glycan processing in lepidopteran and mammal cells is similar during the early steps of the glycosylation pathway with the synthesis of GlcAcβ1,2Manα1,3(Manα1,6)Manβ1,4GlcNAcβ1,4(Fucα1,6)GlcNAc. However, *N*-glycans are then not processed to sialylated

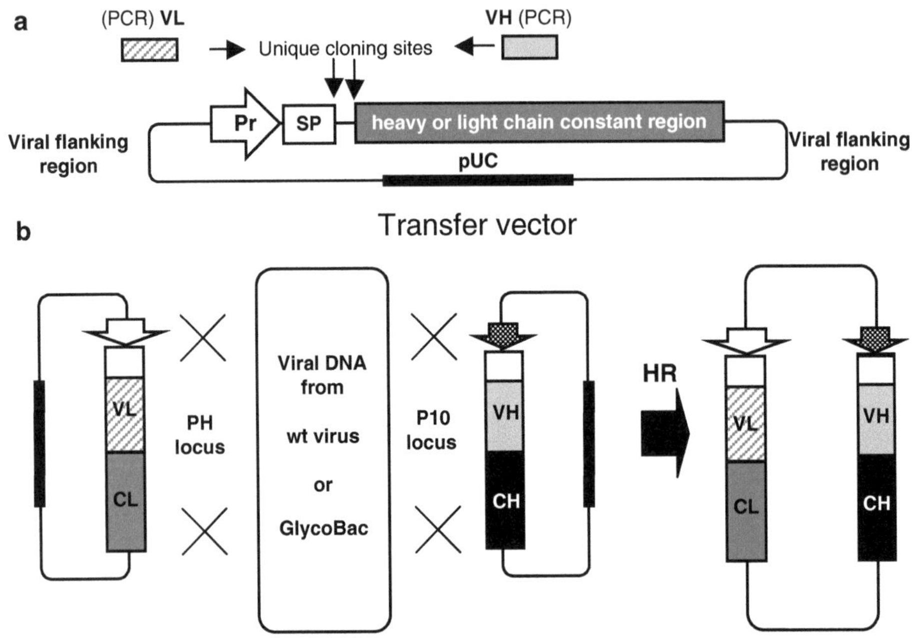

Fig. 2 Schematic representation of the different steps involved in the construction of a baculovirus expressing a recombinant antibody. (**a**) General organization of specific transfer vectors used to express the heavy and light chains of an antibody. The cDNAs encoding the heavy and light chain variable regions (VH and VL) are inserted in frame with an upstream sequence encoding an immunoglobulin signal peptide (SP) and downstream with the cDNA encoding the human heavy or light constant region (CH or CL). (**b**) *Sf*9 cells are transfected with (1) the two transfer vectors and (2) a defective noninfectious viral DNA (Bacmid-derived or linearized viral DNA) that co-expresses (GlycoBac) or not (wt virus) the three additional glycosyltransferases: GNT-I, GNT-II, and β1,4 GalTI. Infectious baculoviruses generated after homologous recombination are cloned by plaque assay. Infection of insect cells with this virus results in the secretion of the recombinant antibody in the culture medium. *Pr* viral promoter. *PCR* polymerase chain reaction, amplification of VH and VL regions

complex-type structures and the most frequently identified structures in glycoproteins synthesized by both infected and noninfected insect cells are oligomannose-type $Man_{9-5}GlcNAc_2$ and more specifically paucimannosic glycans $Man_{3-2}(Fuc)_{0-1}GlcNAc_2$ (Fig. 3). Analysis of the enzymatic potential of several insect cell lines showed that some glycosyltransferase activities are deficient or missing. Thus, the GNT-I activity is very low compared to mammalian cells (11, 12), whereas the GNT-II (12), β1,4GalTI, and sialyltransferase activities are missing (13). Moreover, the identification of *N*-acetylglucosaminidase activity (14) associated with the Golgi apparatus of some lepidopteran cells (e.g., *Sf*9 cells) explains the absence of complex glycans.

Furthermore, in humans, core α1,6-fucosylation plays an essential role in various biological processes, such as cell adhesion, modulation of growth factor receptors, microbial and viral infections, cancer, and atherosclerosis (15). For instance, α1,6-fucosylation of *N*-glycans in human IgG_1 modulates antibody-dependent

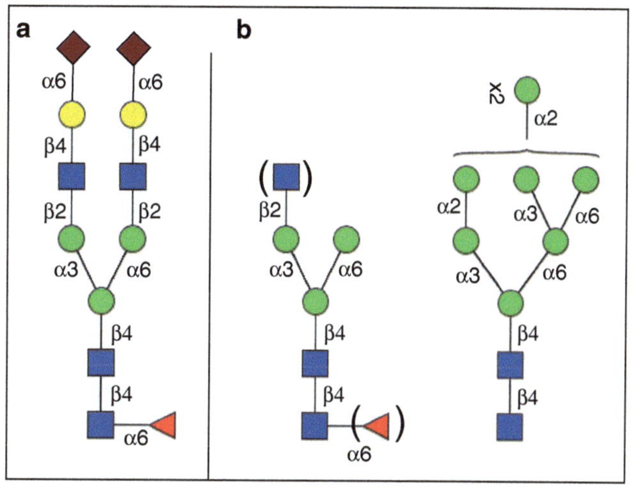

Fig. 3 Representation of typical *N*-glycan structures synthesized (**a**) in mammalian cells and (**b**) in *Sf*9 cells. No complex glycan can be identified in both infected and noninfected lepidopteran cells. Sialic acid (*brown diamond*); Galactose (*yellow circle*); Mannose (*green circle*); *N*-acetylglucosamine (*blue square*); Fucose (*red triangle*)

cell-mediated cytotoxicity (ADCC) in vivo and in vitro (16). Analysis of the fucosyltransferase activity in several lepidopteran cell lines showed distinct types of fucosylation in each cell line. Specifically, core α1,3- and α1,6-fucosyltransferases were found in the *Mamestra brassicae* (13, 17, 18) and *Trichoplusia ni* (High Five™) cell lines (19), whereas only α1,6-fucosyltransferase was identified in *B. mori* and *Sf*9 (*Spodoptera frugiperda*) cells (13, 17, 20). *Sf* 9 and High Five™ cells are commonly used as host cell lines for the production of recombinant proteins. High Five™ cells are generally more efficient at expressing and secreting recombinant proteins than *Sf*9 cells. However, the presence in recombinant proteins of the α1,3-linked fucose residue, a potential allergenic epitope, may be a limitation to the use of High Five™ cells for the production of therapeutics (21).

Insect cell glycan structures can be "humanized" by integrating the missing glycosyltransferases into the cell genome (22) or the viral genome (23, 24). Using the second approach, we have constructed a new baculovirus vector that allows the co-expression of mammalian GNT-I, GNT-II, and β1-4GalTI (Patent FR2891552). In order to obtain a stable genetic construct without any duplicated sequences, the expression of the genes encoding these three glycosyltransferases is under the control of heterologous RNA polymerase II promoters, such as the actin 3 promoter from *B. mori* cells (25), the promoter of the gene encoding the gp64/67 protein of the *Orgya pseudostugata* baculovirus (*Op*MNPV) (26), and the P9 promoter of the densovirus *Junonia coenia* (27) (Fig. 4). Three new

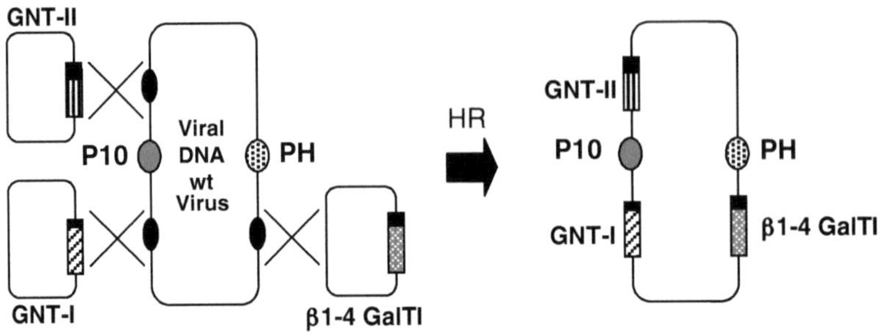

Glycosyltransferase	Promoter	Target gene
Human N-acetylglucosaminyltransferase I (GNT-I)	Actin 3 from *B. mori*	Chitinase/Cathepsin
Human N-acetylglucosaminyltransferase II (GNT-II)	P9 densovirus *Jc*	pp34
Bovine β1,4-galactosyltransferase (β1,4-GalTI)	gp64/67 *Op*MNPV	EGT

Fig. 4 Construction of a recombinant baculovirus that co-expresses three additional glycosyltransferases. Three nonessential genes were identified and three new transfer vectors targeting these genes were designed. Heterologous RNA polymerase II promoters from (1) cells (*Bombyx mori* cells), such as the actin 3, promoter or from (2) insect viruses, such as the promoter of gene encoding the gp64/67 protein of the baculovirus *Op*MNPV, or the *P9* promoter of the densovirus *Junonia coenia*, were used to drive the expression of each glycosyltransferase gene. These expression cassettes were introduced into the viral genome by homologous recombination (HR) as described in Fig. 1. Recombinant viruses were purified by plaque assay and the specific glycosyltransferase activities determined (data not shown)

specific transfer vectors that allow homologous recombination of these three genes into three nonessential baculovirus genes were constructed. Cells infected with this glyco-engineered baculovirus express high level of glycosyltransferases activities allowing the production of galactosylated proteins, for instance, mono- and di-galactosylated antibodies (Fig. 5).

In the following sections, we will describe the procedure to generate a highly stable recombinant virus to express a galactosylated antibody. The same approach could be used to produce sialylated proteins. However, insect cells cannot synthesize the nucleotide-sugar CMP-NeuAc (22). Therefore, at least two additional enzymatic activities (NeuAc synthase and CMP-NeuAc synthase) should be co-expressed with the sialyltransferase of interest.

Fig. 5 Analysis of the *N*-glycosylation pattern of a recombinant antibody produced after infection of *Sf*9 cells with a recombinant virus generated from a wild-type baculovirus (*experiment A*, *recMab*/*Man*) or from the glyco-engineered baculovirus (*experiment B*, *recMab*/*Gal*). (**a**, **b**) Analysis of *N*-glycans by MALDI-TOF-MS. Purified antibodies were digested with PNGase F to release *N*-glycans. After purification and permethylation, glycans were analyzed by MALDI-TOF-MS. The proposed structures of the major *N*-glycans of antibodies are presented with their relative proportions. These structures are based on the composition and biosynthetic pathway knowledge. Galactose (*yellow circle*); Mannose (*green circle*); *s*-acetylglucosamine (*blue square*); Fucose (*red triangle*). (**c**) The presence of β-galactosyl residues on the heavy chain (H) can also be revealed by western blotting using the specific biotinylated lectin (in this case biotinylated RCA_{120}). 1 μg of each purified antibodies was loaded in each lane

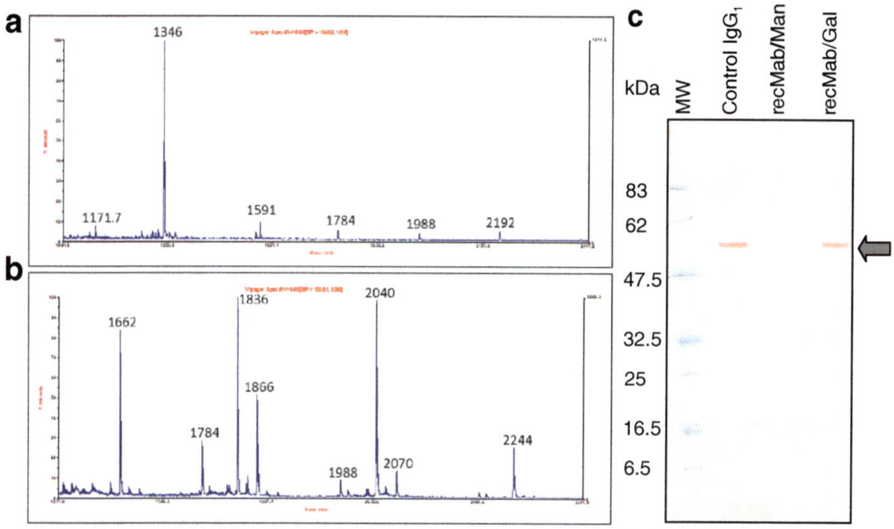

Observed m/z Experiment A	Assignment	Proposed structures	Relative proportions
1171.7	$Hex_3HexNAc_2+Na^+$	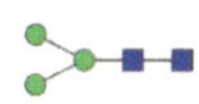	1.28 %
1346	$Hex_3(dHex)HexNAc_2+Na^+$	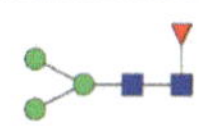	89.34 %
1591	$Hex_3(dHex)HexNAc_3+Na^+$	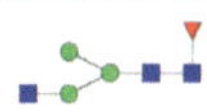	4.84 %
1784	$Hex_6HexNAc_2+Na^+$	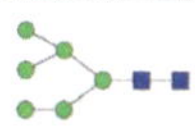	2.31 %
1988	$Hex_7HexNAc_2+Na^+$	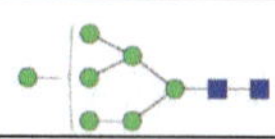	1.27 %
2192	$Hex_8HexNAc_2+Na^+$	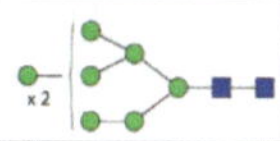	0.96%

Observed *m/z* Experiment B	Assignment		Proposed structures	Relative proportions
1662	$Hex_3HexNAc_4+Na^+$	G0	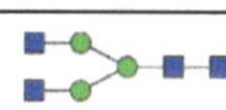	19.38 %
1784	$Hex_6HexNAc_2+Na^+$		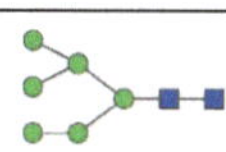	6.48 %
1836	$Hex_3(dHex)HexNAc_4+Na^+$	G0F	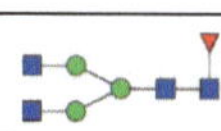	24.72 %
1866	$Hex_4HexNAc_4+Na^+$	G1	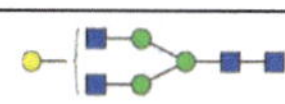	11.70 %
1988	$Hex_7HexNAc_2+Na^+$		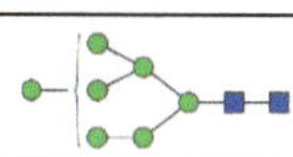	2.03 %
2040	$Hex_4(dHex)HexNAc_4+Na^+$	G1F	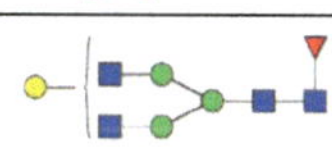	26.05 %
2070	$Hex_5HexNAc_4+Na^+$	G2	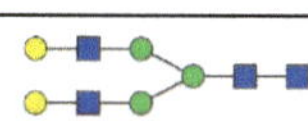	3.23 %
2244	$Hex_5(dHex)HexNAc_4+Na^+$	G2F	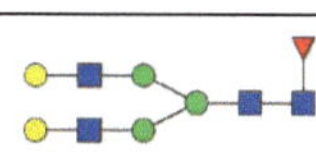	6.41 %

2 Materials

2.1 Cells

1. *Sf*9 cells (ATCC) (28), an insect cell line that was established from *S. frugiperda* ovaries (fall armyworm).
2. Protein production in serum-free medium is performed using *Sf*9 cells adapted to serum-free conditions.

2.2 Media and Supplements

2.2.1 Complete Medium

TC100 medium (GIBCO®, Life Technologies Corp., Carlsbad, CA, USA) supplemented with 5% heat-inactivated (56°C for 30 min) fetal calf serum (FCS) and 3.33 g/l lactalbumin hydrolysate (Becton Dickinson and Company, Franklin Lakes, NJ, USA).

2.2.2 Serum-Free Media

1. In-house developed SB5/6 medium, a serum-free culture medium derived from the TC100 medium.
2. Other serum-free culture media, such as Insect-XPRESS™ (Biowhittaker® Lonza Rockland, ME, USA), Sf900II™SFM and Sf900III™SFM (protein- and animal origin-free medium) (GIBCO®), or Ex-Cell®405, 420, and ExCell®Titer High™ (Sigma-Aldrich, St. Louis, USA) may be used (4).

2.2.3 Supplements

1. Pluronic F68 (GIBCO®).
2. 250× Cholesterol Lipid Concentrate (GIBCO®).

2.3 Virus

2.3.1 Viral Clone

A. californica multiple nuclear polyhedrosis virus (*Ac*MNPV), clone 1, 2 (29).

2.3.2 Viral Amplification

1. Plastic T-flasks. Sizes: 25 and 75 cm^2 (BD-Falcon, Bedford, MA, USA).
2. Sterile plastic serological pipettes; sizes: 1, 2, 5, 10 ml.
3. Sterile plastic cell scrapers.
4. Disposable plastic centrifuge tubes; sizes: 15 or 50 ml.

2.3.3 Plaque Assay

1. Complete medium (see Subheading 2.2.1).
2. TC100 medium (GIBCO®) without any supplement (M0).
3. Sterile plastic cell scrapers.
4. 6-well plates (BD-Falcon).
5. Sea-Plaque® agarose (Lonza).
6. Neutral red (Sigma): 0.05% solution in M0 medium.
7. Sterile plastic serological pipettes; sizes: 1, 2, 5, 10, 25 ml.
8. Sterile Pasteur pipettes.
9. Pipet-Aid® (BD-Falcon).

2.4 Molecular Biology

2.4.1 Enzymes and Cloning Vectors

1. Restriction enzymes, T4 DNA ligase, alkaline phosphatase (New England Biolabs, Ipswich, MA, USA or Roche Applied Science, Mannheim, Germany).
2. $VENT_R$® DNA polymerase (New England Biolabs) or Advantage™ PCR Kit (Clontech Laboratories Inc., Mountain View, CA, USA).
3. pGem®Teasy vector system (Promega BioSciences, San Luis Obispo, CA, USA).
4. Low melting temperature Nusieve® GTG agarose (Lonza).
5. QIAquick® Gel Extraction Kit (Qiagen, GmbH Hilden, Germany).
6. NucleoBond®AX100 Kit (Macherey-Nagel, GmbH & Co. KG, Düren, Germany).

2.4.2 Transfer Vectors

A generic transfer vector is described in Fig. 1. A specific transfer vector designed for the expression of the heavy or light chain of an antibody is represented in Fig. 2. These plasmids are amplified in *E. coli* XL1-Blue competent cells (Stratagene, Agilent, Santa Clara, CA, USA), purified through Nucleobond®AX100 columns as described by the manufacturer and stored at −20°C in 70% ethanol.

2.4.3 Materials

1. Centrifuge: Heraeus Pico 17 (Thermo Fisher Scientific, Rockford, IL, USA).
2. Automatic pipettes P20, P100, P200, and P1000 (Gilson S.A.S., Villiers le Bel, France).

2.5 Transfection of Insect Cells

1. Complete medium (see Subheading 2.2.1).
2. TC100 medium (GIBCO®) without any supplement (M0).
3. DOTAP (*N*-(1-(2,3-dioleoyloxy)propyl)-*N*,*N*,*N*-trimethylammonium methylsulfate), a liposomal transfection reagent (Roche Applied Science).
4. Sterile plastic cell scrapers.
5. Cryogenic vials, 2 ml.
6. Sterile plastic serological pipettes; sizes: 1 and 2 ml.
7. Polystyrene sterile tubes, 8 ml (BD-Falcon).
8. Centrifuge: Heraeus Multifuge 3SR (Thermo Fisher Scientific).

2.6 Production

2.6.1 Monolayer Culture

1. Plastic T-flasks; sizes: 25 and 75 cm^2 (BD-Falcon).
2. Sterile plastic serological pipettes; sizes: 1, 2, 5, 10 ml.

2.6.2 Roller Bottle Culture

1. Disposable 850 cm^2 plastic roller bottles.
2. Sterile plastic serological pipettes; sizes: 1, 2, 5, 10, 25, 50 ml.

3. Roller bottle incubator (Bellco Glass, Vineland, NJ, USA).
4. Centrifuge: Beckman Coulter Allegra X-15R centrifuge (Beckman Coulter, Brea, CA, USA).

2.7 Purification of the Recombinant Antibody

1. Chemicals (Sigma): Tris, sodium hydroxide, sodium chloride, sodium phosphate monobasic, sodium phosphate dibasic, citric acid, sodium acetate.
2. Protein A Sepharose HiTrap rProtein A FF 5 ml (GE Healthcare, Waukesha, WI, USA).
3. Protein A binding and washing buffer: 20 mM sodium phosphate, pH 7.0.
4. Protein A elution buffer: 0.1 M sodium citrate, pH 3.5.
5. SP Sepharose (GE Healthcare, HiTrap SP FF 5 ml).
6. SP Sepharose binding and washing buffer: 20 mM sodium acetate, pH 5.4.
7. SP Sepharose elution buffer: 0–0.3 M NaCl in 20 mM sodium acetate, pH 5.4.
8. Filter paper (PRAT DUMAS Velin Quality 07, Couze Saint Front, France).
9. Liquid filtration device: Stericup and Steritop 0.45 μm with Durapore filter membrane (Merck Millipore Corp., Billerica, MA, USA).
10. Diafiltration device: Pall Filtron Centramate (Pall Corp. Port Washington, NY, USA) equipped with a 30 kDa Pall Omega membrane in suspended screen channel format.
11. Akta Purifier 10 (GE Healthcare).
12. Concentration device: Amicon Ultra-15, PLTK Ultracel-PL 30 K (Millipore).
13. Sterilization device: Millex HV filter units, 0.45 μm, with Durapore membrane (Millipore).

2.8 Analysis of Glycans

2.8.1 Lectin Labelling

1. 2× Bis-acrylamide solution (2%, w/v) (QBiogen, MP Biomedicals, Solon, OH, USA).
2. 4× Acrylamide solution (40%, w/v) (QBiogen).
3. Pre-stained Protein Markers, Broad Range (7–175 kDa) (New England Biolabs).
4. Control human IgG_1 (Sigma).
5. Transfer apparatus EBU-4000 (C.B.S. Scientific Company, San Diego, CA, USA).
6. Transfer buffer: 25 mM Tris (Sigma), 192 mM Glycine (Sigma), 20% Methanol (Sigma).

7. Nitrocellulose membranes: Optitran BA-S 83, pore size 0.2 μm (Schleicher and Schüell, Dassel, Germany).
8. Ponceau Red concentrate (Sigma).
9. Polyvinylpyrrolidone K30 (Fluka, Chemika/Biochemika, Buchs, Switzerland).
10. Lectin RCA_{120} (*Ricinus communis* agglutinin), biotin conjugate (Vector Laboratories, Burlingame, CA, USA) (see Note 1).
11. Blocking Reagent (Roche Applied Science).
12. Tris-buffered Saline (TBS), 20× (Amresco, LLC, Solon, OH, USA).
13. Tween 20 (Sigma).
14. Buffer I: TBS/0.05% Tween 20, 1 mM $MgCl_2$ (Sigma), 1 mM $MnCl_2$ (Sigma), 1 mM $CaCl_2$ (Sigma), pH 7.5.
15. Horseradish peroxidase (HRP)-conjugated NeutraAvidin™ (Pierce, Thermo Fisher Scientific).
16. Diaminobenzidine (DAB) Substrate Kit (Pierce).

2.8.2 Mass Spectra

1. Chemicals: Sodium dodecyl sulfate (SDS) (BioRad, Hercules, CA, USA), 2-mercaptoethanol (BioRad), Nonidet NP40 (Pierce), *N*-glycosidase F (PNGase F) (Roche Diagnostics GmbH, Manheim, Germany), trifluoroacetic acid (TFA) (Thermo Fisher Scientific), dimethyl sulfoxide (DMSO) (Fluka), NaOH powder (Sigma-Aldrich), iodomethane (Sigma-Aldrich), chloroform (Thermo Fisher Scientific), methanol (Thermo Fisher Scientific), 2,5-dihydroxybenzoic acid (DHB) (Bruker Daltonics, Bellerica, MA, USA), acetonitrile "HiPerSolv Chromanorm®" (AcN) (BDH-Prolabo, VWR Radnor, PA, USA), and acetic acid "GPR Rectapur" (BDH-Prolabo).
2. Nonporous graphitized "Extract-Clean Carbon 150 mg/4 ml" columns (Grace Davison Discovery Sciences, Deerfield, IL, USA).
3. Extract-Clean C18 columns, 500 mg/4 ml (Grace Davison Discovery Sciences).
4. MALDI-TOF spectrometer (Voyager) from Applied Biosystems® Life technologies.

3 Methods

3.1 Cell Culture

*Sf*9 cells are grown at 28°C in complete medium. Cultures are split every 4 days, with a subcultivation ratio of 1:5. *Important*: cells must be infected during the exponential phase of growth (see Note 2).

3.2 Cloning the cDNAs Encoding the Heavy and Light Chains of an Antibody in Specific Transfer Vectors

Figure 2 shows a generic "universal" baculovirus expression cassette designed to express the heavy and light chains of an antibody. This cassette consists of (1) a viral (*P10* or *PH*) promoter, (2) a sequence encoding an immunoglobulin signal peptide, (3) two unique restriction sites to allow the insertion of the VH or VL sequences of any antibody in frame with the upstream signal peptide sequence, and (4) a downstream sequence that encodes the mouse or human heavy or light chain constant region. These cassettes are flanked by viral sequences that direct the integration of the foreign genes into a specific locus (here, the *P10* or *PH* locus). The VH and VL sequences of the antibody of choice may be amplified using DNA isolated from B cells, such as mouse hybridoma cells or immortalized human B cells. For example, the cDNA encoding the VH and VL domains of a mouse antibody produced by mouse hybridoma cells can be amplified by PCR, using the primers described in ref. 30. The PCR products are then cloned in the pGemTeasy plasmid and controlled by sequencing (Eurofins MWG Operon, Ebersberg, Germany). A second run of PCR amplification is then performed to introduce restriction sites on both 5′ and 3′ sides of these cDNAs to allow their insertion in frame with the signal peptide sequence and the constant regions present in the transfer vectors as described above.

3.3 Generation of Recombinant Viruses

*Sf*9 cells are co-transfected with the heavy and light chain transfer vectors and the purified, defective noninfectious viral DNA isolated from wild-type or glyco-engineered baculoviruses (see Fig. 2).

1. Inoculate 2×10^6 *Sf*9 cells in a T-flask (25 cm^2) with M0 (see Note 3) and allow cells to adhere to the bottom of the flask for 1 h.
2. Dispense 3 ml of M0 medium in two cryotubes (see Note 4) and add 40 μl of DOTAP to the first tube. In the second one, add 500 ng of viral DNA and 5 μg of each transfer vectors. Mix gently.
3. Combine the content of the two cryotubes into an 8 ml polystyrene tube (see Note 5) and gently mix (see Note 6) by inverting the tube. Incubate the mixture at room temperature for 30 min with gentle mixing every 10 min.
4. Wash cells with 3 ml of M0 medium once and then lay the transfection mixture on the cell monolayer and incubate cells at 28°C for 4 h.
5. Discard the supernatant and replace it with 4 ml of complete medium. Put cells back in the incubator at 28°C.
6. Four days post-transfection, spin down infected cells by low speed centrifugation ($1{,}900 \times g$ for 10 min) and harvest the supernatant containing the infectious recombinant viruses. Store samples at +4°C until use.

3.4 Isolation of Recombinant Viral Clones by Plaque Assay (See Note 7)

1. Resuspend exponentially growing *Sf9* cells in M0 medium and plate them in 6-well plates at a density of 0.8×10^6 cells/well; allow cells to adhere to the plate surface for at least 1 h.
2. Prepare tenfold dilutions from 10^{-1} to 10^{-7} of the cell supernatant containing the infectious recombinant viruses in complete medium (see Note 8).
3. Remove the culture medium, add 0.5 ml of each viral dilution (two wells for each dilution), and incubate cells at room temperature for 1 h.
4. Meanwhile prepare an agarose overlay (these quantities are for two 6-well plates) by mixing 22.5 ml of pre-warmed complete medium (see Note 8) (at 37°C in a water bath) with 7.5 ml of melted Sea-Plaque® agarose cooled at 50–55°C. This preparation is kept at 37°C in a water bath.
5. After 1 h, remove the viral supernatant and deposit slowly 2 ml of agarose overlay in each well; make sure to cover completely the cell layer. Allow the agarose overlay to solidify at room temperature and then put the plates back in the incubator at 28°C.
6. After 4 days, add 1 ml of neutral red solution per well and incubate at room temperature for 1 h.
7. Remove the neutral red solution with a pipette and incubate at 28°C overnight. "White" plaques of dead cells indicate cells that have been infected by a single virus, whereas uninfected (living) cells will be stained red.
8. Pick isolated plaques with a Pasteur pipette and deposit them in cryotubes containing 1 ml of complete medium (see Note 9). Store tubes at +4°C.
9. At this stage, it is also possible to count the plaques and to calculate the titre of the viral suspension (i.e., the number of infectious viral particles/ml) as follows: number of plaques/well × 2 × dilution factor = titre. The titre is expressed in plaque forming unit (PFU)/ml.

3.5 Virus Amplification

1. Plate 2×10^6 *Sf9* cells in 25 cm^2 T-flasks in M0 medium (serum free).
2. After 1 h at room temperature, discard the culture medium and add the viral suspension at a multiplicity of infection (m.o.i, number of infectious particles per cells) of 2 PFU/cell.
3. After 1-h adsorption at room temperature, replace the viral suspension with fresh complete medium.
4. Incubate cells at 28°C for 5 days without changing medium.
5. Pellet cells by centrifugation at $1,900 \times g$ for 15 min and collect the supernatants; store them at 4°C.

3.6 Production of Recombinant Antibody

1. Inoculate *Sf9* cells adapted to grow in serum-free medium (SB5/6 or other, see Subheading 2.2.2) in roller bottles at a density of 0.6×10^6 cells/ml and infect them immediately at a m.o.i of 2 PFU/cells.
2. After 3–4 days incubation at 28°C, pellet cells by centrifugation at $4{,}700 \times g$ for 15 min and collect the supernatants.
3. Supernatants are stored at −80°C until use.

3.7 Purification of the Recombinant Antibody

3.7.1 Pretreatment of Cell Culture Supernatant

1. Adjust the pH of the crude supernatants at 8.2 with 2 M NaOH; a precipitate will form.
2. Filter through filter paper.
3. Concentrate (20 times) the supernatant against 20 mM sodium phosphate pH 7.0 using a concentration/diafiltration device.
4. Filter through 0.45 μm Durapore membranes to remove the insoluble material that can appear during the concentration step.

3.7.2 Protein A Sepharose

1. Load the concentrates on the Protein A column/5 ml (2 ml/min).
2. Wash with a 40-bed volume (i.e., 40 times the total volume of the material in the column) of 20 mM sodium phosphate pH 7.0.
3. Elute with 25 ml 0.1 M sodium citrate pH 3.5.

3.7.3 SP Sepharose (See Note 10)

1. Load the eluate from Protein A Sepharose on the SP Sepharose column/5 ml (2 ml/min).
2. Wash with a 40-bed volume of 20 mM sodium acetate pH 5.4.
3. Elute with a NaCl gradient (0–0.3 M NaCl in 20 mM sodium acetate pH 5.4).
4. Concentrate with Amicon Ultra 30 K devices.
5. Filter/sterilize the concentrates through Millex HV 0.45 μm filters.

3.8 Analysis of Glycans

Glycans can be analyzed by using many different techniques. We chose to present here two of them: (1) a simple and fast technique (lectin labelling) that exploits the fact that lectins bind very specifically to glycan structures (here we used the lectin of *R. communis* (RCA), which binds to β-galactosyl residues) and (2) a more complex and time-consuming method that is based on the analysis of mass spectra and allows the identification of glycan structures.

3.8.1 Lectin Labelling

1. Resuspend 1 μg of purified recombinant antibodies in denaturing sample buffer that contains 5% SDS and 2% β-mercaptoethanol, boil them for 5 min, load on a 12% polyacrylamide gel (31), and perform electrophoresis at 150 V for 2.30–3 h (till the bromophenol blue dye is at about 1 cm from the end of the gel).

2. After electrophoresis, transfer proteins onto a nitrocellulose membrane using a transfer apparatus operated at 0.8 mA/cm^2 for 2 h (32). The efficiency of the transfer can be controlled by staining the membrane with Ponceau Red solution for 5 min. Destain the membrane with distilled water.

   All the subsequent steps are performed under gentle stirring, at room temperature.

3. Incubate the membrane in 2% polyvinylpyrrolidone K30 in TBS Buffer for 2 h.
4. Wash twice in TBS/0.05% Tween 20 for 10 min.
5. Rinse in Buffer I.
6. Incubate with the specific biotinylated lectin diluted at 0.2 μg/ml in Buffer I overnight (see Note 1).
7. Wash with TBS/0.05% Tween 20 for 10 min/three times.
8. Incubate in a solution of 0.5% Blocking Reagent in TBS/0.05% Tween 20 for 30 min to saturate the nonspecific sites.
9. Wash with TBS/0.05% Tween 20 for 10 min/three times.
10. Incubate with 0.02 μg/ml HRP-NeutrAvidin™ in TBS/0.05% Tween 20 for 1 h.
11. Wash in TBS/0.05% Tween 20 for 10 min/three times.
12. Reveal with the DAB Substrate Kit according to the manufacturer's instructions (see Fig. 5c).

3.8.2 Mass Spectra (33)

N-Glycan Release

1. Add 5 μl of 20% SDS to 200 μl of 1 mg/ml sample, mix.
2. Add 2 μl β-mercaptoethanol, mix, and incubate at 100°C for 20 min.
3. Leave the mixture at room temperature for 20 min.
4. Add 21 μl of 10% Nonidet P40 (v/v), mix.
5. Add 220 μl of 50 mM sodium phosphate buffer, pH 7.5, mix.
6. Add 3 μl of 1 U μl^{-1} PNGase F, mix, and incubate at 37°C overnight.

N-Glycan Purification

1. Add 4.5 μl of 10% TFA solution (v/v).
2. Condition a column of 150 mg of nonporous graphitized carbon with 5 ml methanol and wash with 10 ml of 0.1% TFA solution (v/v).
3. Load the sample onto the column and remove salts with 10 ml of 0.1% TFA.
4. Recover the glycans in a glass tube (4 ml vial) by elution with 3 ml of 25% acetonitrile (AcN) in 0.1% TFA.
5. Evaporate the eluted AcN under a stream of nitrogen and freeze-dry.

Permethylation

1. Place the glass tube containing the freeze-dried sample in a vacuum vessel saturated with argon atmosphere (see Note 11).
2. Add 200 μl DMSO using an oven-dried (100°C overnight) glass pipette.
3. Add approximately 25 mg NaOH powder with an oven-dried (100°C overnight) spatula.
4. Add 300 μl iodomethane using an oven-dried (100°C overnight) glass pipette and flush the tube with a stream of argon.
5. Mix vigorously on a vortex at room temperature for 35 min.
6. Quench the reaction by adding 1 ml of 5% acetic acid under a chemical fume hood, mix vigorously (see Note 12).
7. Add 600 μl chloroform. Mix vigorously and allow the mixture to settle in two layers.
8. Transfer the lower chloroform phase into a new glass tube.
9. Repeat steps 7 and 8 twice.
10. Wash the chloroform phase three times with one volume of water and discard the aqueous phases.
11. Dry the chloroform phase under a stream of nitrogen in a hood.

Purification of Permethylated Glycans

1. Dissolve the sample in 200 μl methanol (see Note 13).
2. Condition the C_{18} column (500 mg) with 5 ml methanol and then with 10 ml milliQ water.
3. Load the sample onto the column and elute sequentially with 10 ml milliQ water, 3 ml 10% AcN, and 3 ml 80% AcN.
4. Collect the 80% AcN fraction and evaporate it under a stream of nitrogen in a hood.
5. Freeze-dry the sample.

Preparation of Samples for MALDI-TOF-MS Analysis

1. Dissolve samples (purified and permethylated glycans) in 30 μl of methanol/water (1:1; v/v) solution.
2. Mix 1 μl of sample solution with 1 μl of freshly made 10 mg/ml DHB matrix solution directly onto the MALDI target.
3. Allow the mixture to dry at room temperature.

MALDI-TOF-MS Analysis

1. Acquire the spectra by submitting each spot to multiple laser shots (600 minimum) over a range of $m/z = 1{,}000\text{–}5{,}000$. The laser spot should be moved to different areas of the target during the acquisition and spectra should be averaged.
2. The procedure to assign molecular ions can be carried out with a computer using tools that are available online, such as Glycan Mass (expasy.org/glycanmass) or the free software GlycoWorkBench (glycoworkbench.org http://www.glycoworkbench.org).

The objective is to match the experimental masses ($M + Na^+$) with the calculated masses. For example, the monoisotopic *m/z* of the biantennary *N*-glycan (G2) which contains four GlcNAc residues, three mannose residues, and two galactose residues is 2,047.04; with the Na^+ adduct it becomes 2,070.03 (Fig. 5).

4 Notes

1. Be careful, RCA_{120} is toxic, use gloves, and work under a chemical hood. DO NOT discard RCA solutions down the laboratory sink.
2. It is very important to use cells in exponential growth, baculovirus is unable to replicate in cells in stationary phase.
3. It is recommended to resuspend the cells in serum-free medium, in these conditions cells adhere firmly to the plastic dish.
4. We prefer cryotubes to microcentrifuge tubes as it is easier to work with them and they can be closed more tightly to avoid contamination.
5. To obtain optimal transfection efficiency, use *polystyrene tubes.* DOTAP and DOTAP–DNA complexes have a propensity to stick to polypropylene surfaces, leading to loss of material.
6. Do not vortex or centrifuge.
7. During homologous recombination, some illegitimate recombination may occur, leading to the generation of unstable viral particles. Therefore, it is very important to isolate viral clones and to characterize them by using different analytical techniques, such as PCR and Southern blot assays. Here, we present what we consider as the most efficient technique for isolating viral clones, the plaque assay.
8. Formation of plaques is faster when the agarose overlay is prepared with complete medium (medium with serum).
9. Baculovirus suspensions are stored at +4°C for at least 2 years *in complete medium* (medium with serum) with no change in titre. Storage of the virus in serum-free medium results in a substantial decrease in virus titre. It is also possible to extract the DNA from viral particles secreted in the cell culture supernatant and to store this DNA at −20°C as a precipitate in 70% ethanol. Infectious viruses are easily regenerated after transfection of *Sf9* cells with the viral DNA as described in Subheading 3.3.
10. Depending on the purity required, it is possible to complete the purification of recombinant antibodies by Q Sepharose chromatography and/or a size exclusion chromatography.
11. Do not wait after stopping freeze-drying more than 1 h to limit hydratation.

12. Be careful because reaction is exothermic, so add acetic acid slowly.
13. No more than 200 μl to optimize interaction with C_{18} phase.

Acknowledgments

This work was initiated with Pr. Gérard Devauchelle and we wish to dedicate it to his memory. Gérard Devauchelle died on 17 February 2010. We miss his friendship and enthusiasm. Marie-Christine Slomianny, Anne Harduin-Lepers, Martine Cérutti, Marylêne Lévêque, Sylvie Choblet, and Marie-Luce Violet were supported by Centre National de la Recherche Scientifique. Sylvie Juliant, Pierre Cérutti, and Annick Ozil were supported by Institut National de la Recherche Agronomique. This publication has been funded with support from the French National Research Agency under the program "Investissements d'avenir" Grant Agreement LabEx MAbImprove: ANR-10-LABX-53. Additionally, "Recombinant baculovirus expressing several heterologous genes" has been patented under patent number FR2891552 (Cérutti, M., Juliant, S., Mutuel, D., Devauchelle, G.).

References

1. Schneider EH, Seifer R (2010) Sf9 cells: a versatile model system to investigate the pharmaceutical properties of G protein-couples receptors. Pharmacol Ther 128:387–418
2. Condreay JP, Ames RS, Hassan NJ et al (2006) Baculoviruses and mammalian cell-based assays for drug screening. Adv Virus Res 68: 255–286
3. Mena JA, Kamen AA (2011) Insect cell technology is a versatile and robust vaccine manufacturing platform. Expert Rev Vaccines 10: 1063–1081
4. Agathos N (2007) Development of serum-free media for lepidopteran insect cell lines. Methods Mol Biol 388:155–186
5. Miller LK (1988) Baculovirus as gene expression vectors. Annu Rev Microbiol 42: 177–199
6. Kitts PA, Possee RD (1993) A method for producing recombinant baculovirus expression vectors at high frequency. Biotechniques 14:810–817
7. Luckow VA, Lee SC, Barry GF et al (1993) Efficient generation of infectious recombinant baculoviruses by site-specific transposon-mediated insertion of foreign genes into a baculovirus genome propagated in *Escherichia coli*. J Virol 67:4566–4579
8. Cérutti M, Golay J (2012) Lepidopteran cells, an alternative for the production of recombinant antibodies? MAbs 4(3):294–309
9. Noad RJ, Stewart M, Boyce M et al (2009) Multigene expression of protein complexes by iterative modification of genomic Bacmid DNA. BMC Mol Biol 10:87–99
10. Sokolenko S, George S, Wagner A (2012) Co-expression vs. co-infection using baculovirus expression vectors in insect cell culture: benefits and drawbacks. Biotechnol Adv 30: 766–781
11. Marchal I, Mir AM, Kmiecik D et al (1999) Use of inhibitors to characterize intermediates in the processing of *N*-glycans synthesized by insect cells: a metabolic study with Sf9 cell line. Glycobiology 9:645–654
12. Altmann F, Kornfeld G, Dalik T et al (1993) Processing of the asparagine-linked oligosaccharides in insect cells. N-acetylglucosaminyltransferase I and II activities in cultured lepidopteran cells. Glycobiology 3:619–625
13. Lopez M, Coddeville B, Langridge J et al (1997) Microheterogeneity of the oligosaccharides carried by the N-glycosylation sites of the bovine lactoferrin expressed in Mamestra brassicae using a baculovirus vector. Glycobiology 7:635–651

14. Altmann F, Schwihla H, Staudacher E et al (1995) Insect cells contain an unusual, membrane bound (beta)-N-acetylglucosaminidase probably involved in the processing of proteins *N*-glycans. J Biol Chem 270:17344–17349
15. Becker DJ, Lowe JB (2003) Fucose: biosynthesis and biological function in mammals. Glycobiology 13:41R–53R
16. Shields RL, Lai J, Keck R et al (2002) Lack of fucose on human IgG1 N-linked oligosaccharide improves binding to human Fc gamma RIII and antibody-dependent cellular toxicity. J Biol Chem 277:26733–26740
17. Staudacher E, Kubelka V, März L (1992) Distinct *N*-glycan fucosylation potentials of three lepidopteran cell lines. Eur J Biochem 207:987–993
18. Kubelka V, Altmann F, Kornfeld G et al (1994) Structures of the N-linked oligosaccharides of the membrane glycoproteins from three lepidopteran cell lines (Sf-21, IZD-Mb-0503, Bm-N). Arch Biochem Biophys 308:148–157
19. Hancock K, Narang S, Pattabhi S et al (2008) False positive reactivity of recombinant, diagnostic, glycoproteins produced in High FiveTM insect cells: effect of glycosylation. J Immunol Methods 330:130–136
20. Voss T, Ergülen E, Ahorn H et al (1993) Expression of interferon w1 in Sf9 cells. No evidence for complex-type N-linked glycosylation or sialylation. Eur J Biochem 217:913–919
21. Prenner C, Mach L, Glössl J et al (1992) The antigenicity of the carbohydrate moiety of an insect glycoprotein, honey-bee (*Apis mellifera*) venom phospholipase A2. The role of alpha 1,3-fucosylation of the asparagines-bound N-acetylglucosamine. Biochem J 284:377–380
22. Aumiller JJ, Mabashi-Asazuma H, Hillar A et al (2012) A new glycoengineered insect cell line with an inducibly-mammalianized protein N-glycosylation pathway. Glycobiology 22: 417–428
23. Jarvis DL, Finn EE (1996) Modifying the insect cell *N*-glycosylation pathway with immediate early baculovirus expression vectors. Nat Biotechnol 14:1288–1292
24. Wagner R, Liedtke S, Kretzschmar E et al (1996) Elongation of the N-glycans of fowl plaque virus hemagglutinin expressed in Spodoptera frugiperda (Sf9) cells by coexpression of human beta 1,2-N-acetylglucosaminyltransferase I. Glycobiology 6:165–175
25. Mounier N, Prudhomme JC (1986) Isolation of actin genes in Bombyx mori: the coding sequence of a cytoplasmic actin gene expressed in the silk gland is interrupted by a single intron in an unusual position. Biochimie 68:1053–1061
26. Ahrens CH, Russell RL, Funk CJ et al (1997) The sequence of the Orgyia pseudotsugata multinucleocapsid nuclear polyhedrosis virus genome. Virology 229:381–399
27. Shirk PD, Bossin H, Furlong RB (2007) Regulation of Junonia coenia densovirus P9 promoter expression. Insect Mol Biol 16: 623–633
28. Vaughn JL, Goodwin RH, Tompkins GJ et al (1977) The establishment of two cell lines from the insect Spodoptera frugiperda (Lepidoptera; Noctuidae). In Vitro 13:213–217
29. Croizier G, Croizier L, Quiot J-M (1988) Recombination of *Autographa californica* and *Rachiplusia ou* nuclear polyhedrosis viruses in *Galleria mellonella* L. J Gen Virol 69:177–185
30. Chardès T, Villard S, Ferrières G (1999) Efficient amplification and direct sequencing of mouse variable regions from any immunoglobulin gene family. FEBS Lett 452:386–394
31. Laemmli UK (1970) Cleavage of structural proteins during assembly of the head of the bacteriophage T4. Nature 227:680–685
32. Towbin H, Staehelin T, Gordon J (1979) Electrophoretic transfer of proteins from polyacrylamide gels to nitrocellulose sheets: procedure and some applications. Proc Natl Acad Sci U S A 76:4350–4354
33. Morelle W, Michalski J-C (2007) Analysis of protein glycosylation by mass spectrometry. Nat Protoc 7:1585–1602

Part II

Glycoanalytics

Chapter 6

NanoLC Chips MS/MS for the Characterization of *N*-Glycopeptides Generated from Trypsin Digestion of a Monoclonal Antibody

Elsa Wagner-Rousset, Christine Schaeffer-Reiss, Audrey Bednarczyk, Nathalie Corvaïa, Alain Van Dorsselaer, and Alain Beck

Abstract

In the field of therapeutic recombinant proteins, monoclonal antibodies (mAbs) have achieved a rising success with more than 30 mAbs that have reached the market in the past 20 years. From a structural standpoint, one of the most important posttranslational modifications affecting antibodies is by far glycosylation. Furthermore, glycosylation of mAbs directly impacts on their biological activity and safety and therefore needs to be well characterized. Glycoprotein analysis requires high-resolution separation techniques that can provide detailed structural analysis able to discriminate between glycoforms of various abundances. This chapter describes a protocol for nanoLC-Chip-MS/MS analysis of a proteolytic digest of the heavy chain of a recombinant mAb. The use of graphitized carbon column instead of classical C18 reversed-phase material is shown to be well suited to detect low abundant glycoforms and to provide in one shot information regarding both the oligosaccharide structure and the amino acid sequence of its peptide moiety.

Key words Glycosylation analysis, N-linked glycopeptide, NanoLC-Chip-MS/MS, Ion trap

1 Introduction

Glycosylation is one of the most important posttranslational protein modifications and has a crucial impact on the pharmacological and pharmacokinetic properties of therapeutic monoclonal antibodies (mAbs) like effector functions, immunogenicity, plasmatic clearance (1, 2), and resistance toward proteases (3).

Immunoglobulin G (IgG)-type mAbs have at least one N-glycosylation site located at the Asn-Xxx-Ser/Thr (Xxx is not Pro) consensus sequence of each heavy-chain CH_2 domain of the Fc portion (4). Glycans of recombinant IgG subtype antibodies depend on the expression systems and manufacturing processes. They belong to the bi-antennary complex type. A conserved

Alain Beck (ed.), *Glycosylation Engineering of Biopharmaceuticals: Methods and Protocols*, Methods in Molecular Biology, vol. 988, DOI 10.1007/978-1-62703-327-5_6, © Springer Science+Business Media New York 2013

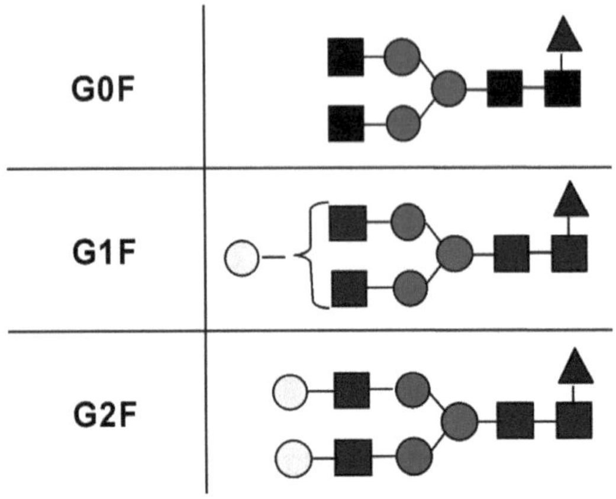

Fig. 1 Structures of the three most abundant recombinant antibody *N*-glycans produced in CHO or NSO host systems. The structures are represented using the symbol nomenclature outlined by the Consortium for Functional Glycomics (http://www.functionalglycomics.org/). Theoretical monoisotopic masses used for data interpretation of MS/MS spectra are GlcNac = 203.08 Da; Man, Gal = 162.05 Da; Fuc = 146.06 Da

heptasaccharide core is composed of two *N*-acetylglucosamine (GlcNAc), three mannose (Man), and two GlcNAc residues that are β-1,2 linked to α-6 Man and α-3 Man, forming two arms. Additional fucose (Fuc), galactose (Gal), and *N*-acetylneuraminic acid (NANA) residues may be present. Among their considerable heterogeneity (1, 5), three N-linked glycans are the most abundant structures described in CHO- or NS0-produced mAbs and are usually called G0F, G1F, and G2F (see Fig. 1) depending on the number of additional galactose residues on their branches.

Mass spectrometry-based (MS) approaches have been extensively used to characterize N-glycosylation of recombinant glycoproteins (6–8). These techniques can be classified in different levels of complexity delivering also different degree of information. From the top-down to the bottom-up approaches the level of fine details increases. As depicted in Fig. 2, a first strategy includes the analysis of intact or reduced mAbs by mass measurement with electrospray ionization (ESI) time-of-flight (TOF) instrument. The method provides in one shot masses of ≈150 kDa weight glycoforms in heterogeneous mixtures with accuracy lower than 100 ppm permitting a preliminary global characterization of the antibody. The second approach—ESI-MS/MS direct analysis of released N-linked glycans—provides a fine characterization by tandem mass spectrometry (MS/MS) of the most abundant carbohydrates reported for mAbs. The third strategy deals with glycopeptide characterization by LC-Chip-MS/MS and is further detailed in this chapter.

Successful use of nanoLC-Chip-MS-based approaches have recently been reported in glycomics studies. Indeed, miniaturization

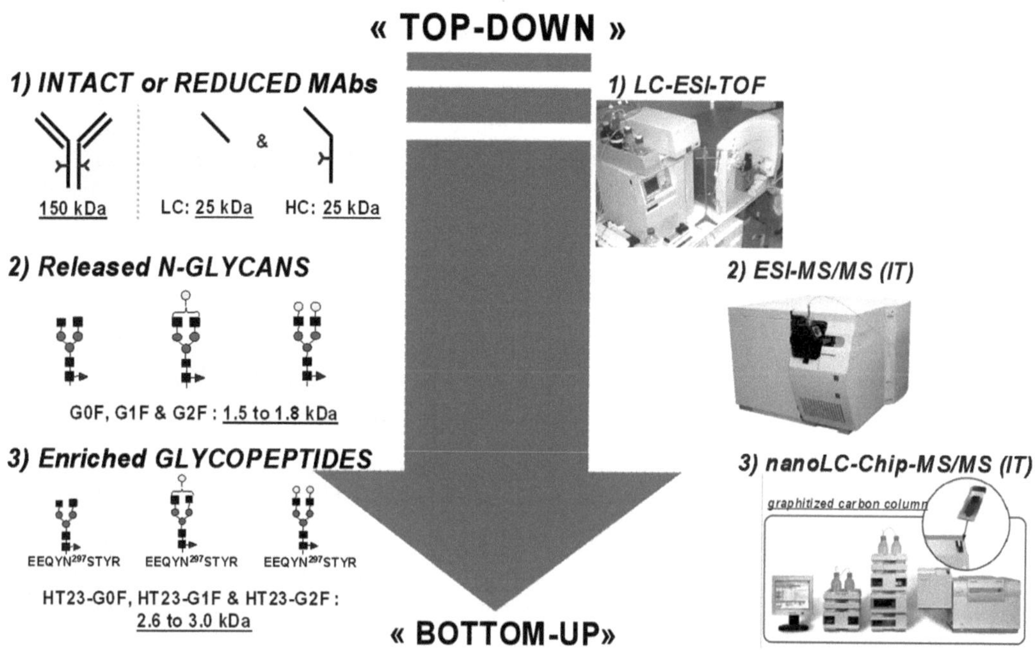

Fig. 2 From top-down to bottom-up strategies for mAb glycosylation characterization. To each level of glycosylated structures (presented on the *left* side with their respective molecular weight) is associated an MS method (*right*)

became a general trend in LC-MS of glycoconjugates and glycans (9, 10). When compared to classical ones, the lower size systems have smaller dead volumes allowing nanoLC-Chip to show better chromatographic performance (reproducibility, peak resolution, peak shape, sensitivity, spray stability). Moreover the embedding microsprayers are much easier to use than glass nanospray tips and they provide superior robustness and stability with improved ionization efficiency and sensitivity. NanoLC-Chip-MS packed with graphite material has been used for separation of oligosaccharides (11, 12). In the present chapter, we describe how we have performed nanoLC-MS/MS analyses of the proteolytic digest from the heavy chain of a recombinant antibody, A2CHM, with graphitized carbon LC-chips (43 mm length). Benefits from such a technique will be highlighted showing the comparison of results obtained with a C18 LC-Chip (43 mm length).

2 Materials

2.1 Buffers

Ultrapure water produced from a Milli-Q Water System™ (Millipore).

All chemicals and reagents were of analytical grade and purchased from Sigma-Aldrich or VWR International.

2.2 mAb

The recombinant humanized antibody A2CHM analyzed in this study was expressed in a CHO cell line and purified using standard manufacturing procedures as previously described (13).

2.3 Reduction and Alkylation of mAbs

1. *Buffer*. Tris–HCl 100 mM, 2 mM EDTA, guanidine HCl 6 M, pH 8.
2. *Reducing reagent*. 0.5 M dithiothreitol.
3. *Alkylation reagent*. 0.5 M iodoacetamide.

2.4 Separation of Light and Heavy Chains by Size Exclusion Chromatography

1. *LC equipment*. HPLC system consisting in two 510 pumps, a 717 auto-injector, and a 490 UV detector (Waters).
2. TSK gel filtration column G3000SW, particle size 10 μm, i.d. 7.5 mm × 30 cm.
3. *Mobile phase*. 3 M guanidine hydrochloride in 50 mM sodium phosphate pH 6.2.
4. Vivaspin™ centrifugal concentrators (Sartorius) of 10 kDa nominal molecular weight cutoff for light chains and 30 kDa for heavy chains.
5. Acetic acid 10% (v/v) aqueous solution.

2.5 Tryptic Digestion

1. *Proteolytic digestion buffer*. Tris–HCl 50 mM, 1 mM $CaCl_2$, pH 7.5.
2. Rapigest™ (Waters) 1% (v/v) aqueous solution.
3. *Protease*. Trypsin Sequencing Grade™ solution (Roche), 0.5 μg/μL in water.

2.6 NanoLC-Chip-MS/MS Experiments

NanoLC-Chip-MS/MS system consisting of an Agilent 1100 capillary HPLC-system™ (Agilent Technologies) coupled to an HCTplus™ ion trap (Bruker Daltonics) equipped with a microfluidic (chip-cube) interface including a microfluidic chip device for chromatographic separations.

1. *Microfluidic chip device* (Agilent Technologies). 40 nL trap column plus (1) a 43 mm × 75 μm, 5 μm reverse-phase C18 analytical column; or (2) a 150 mm × 75 μm, 5 μm reverse-phase C18 column; or (3) a 43 mm × 75 μm, 5 μm porous graphitized carbon column. Nanoelectrospray tip of 2 mm length with conical shape (100 μm o.d. × 8 μm i.d.).
2. The HPLC system is equipped with an autosampler (Agilent) with an injection loop of 8 μL and a temperature cooler maintaining the samples at 10°C.
3. *LC mobile phase*. Solvent A: 2% (v/v) acetonitrile, 0.1% (v/v) formic acid in water. Solvent B: 2% (v/v) water, 0.1% (v/v) formic acid in acetonitrile.

4. *Software.* Complete system controlled by ChemStation™ (Agilent Technologies) and Esquire Control™ (Bruker Daltonics).

3 Methods

3.1 mAb Sample Preparation for NanoLC-Chip-MS/MS Analysis

3.1.1 Reduction and Alkylation of mAbs

1. Lyophilize 1 mg of mAb and dissolve in 1 mL reduction buffer.
2. Flush the sample with N_2 (see Note 1).
3. Incubate the mAb solution with 10 μL DTT 0.5 M for 2 h at 37°C.
4. Add 50 μL iodoacetamide 0.5 M for alkylation which occurs 1 h in the dark at room temperature. Quench the reaction by freezing the sample on ice (see Note 2).

3.1.2 Separation of Light and Heavy Chains by SEC

1. Load 100 μL of mixture of mAb heavy and light chains on the size exclusion chromatography (SEC) column.
2. Set the flow rate at 1 mL/min.
3. Perform isocratic elution.
4. Monitor the elution at 280 nm and collect separately the light and heavy chains. A typical chromatogram of both separated chains is illustrated in Fig. 3.
5. Repeat steps 1–4 and pool the fractions of interest.
6. Concentrate and desalt the collected fractions with the Vivaspin™ centrifugal filter devices. Wash four times with 2 mL acetic acid 10% (see Note 3).

3.1.3 Tryptic Digestion

1. Lyophilize a volume of alkylated light or heavy chain corresponding to an amount of 80 μg of protein.
2. Dissolve in 100 μL digestion buffer.
3. Add 10 μL Rapigest™ 1% aqueous solution to speed up and enhance the quality of the enzymatic reaction.
4. Incubate the light or heavy chain with 10 μL trypsin solution (1:16; enzyme: substrate) at 37°C for 2 h.
5. Quench the reaction by adding 0.4 μL TFA.

3.1.4 NanoLC-Chip MS/MS Analysis of Tryptic N-Glycopeptides

1. Equilibrate the column by running through solvent A at a flow rate of 300 nL/min during 10 min.
2. Set up the mass spectrometer and check the stable spray with elution buffer. For selective MS detection of *N*-glycopeptides, the voltage applied to the capillary cap was set to 1,750 V. The mass spectrometer transmission parameters were optimized for high *m/z* set optimal around 1,300.

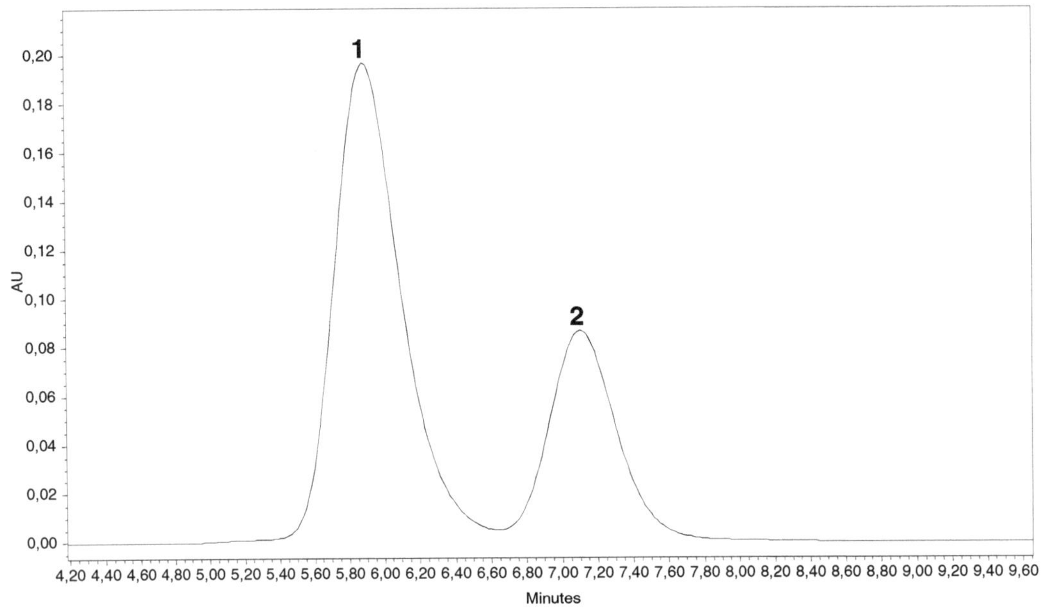

Fig. 3 Typical UV chromatogram at 280 nm of both heavy (1) and light (2) chains separated by size exclusion chromatography

3. Inject samples (2–3 μL) onto the column and simultaneously start both the chromatography gradient and the mass spectrometer data collection. Load the sample into the enrichment column at a flow rate of 3.75 μL/min with solvent A during 3 min. The analytical column is eluted typically at a flow rate of 300 nL/min by a two-step linear gradient going from 3 to 20% of solvent B in 10 min and from 20 to 35% of solvent B in 30 min.
4. A full MS scan (*m/z* 1,000–2,000) followed by MS/MS scans (*m/z* 50–3,000) are acquired in a data-dependent manner (see Note 4). Fragmentation amplitude is set at 1 V. The MS/MS scanning is performed in the ultrascan resolution mode at a scan rate of 26,000 *m/z* per second. A total of three scans are averaged to obtain an MS/MS spectrum.

As a typical separation of peptides, the total ion chromatograms (TIC) obtained after nanoLC-MS experiments on two types of nanoLC-Chip are illustrated in Fig. 4A1, B1. Chromatographic performances of graphitized carbon column versus C18 column are compared concerning glycopeptide analysis. Proteolytic peptides are numbered from N-terminal side of the antibody and called "HT" for tryptic peptides from the heavy chain. Peptide HT23 from A2CHM is carrying the N-glycosylation site of the Fc portion. The ion currents corresponding to HT23-G0F, HT23-G1F, and HT23-G2F (2+) are extracted showing that glycopeptides are not fully separated whatever is the column (Fig. 4A2, B2).

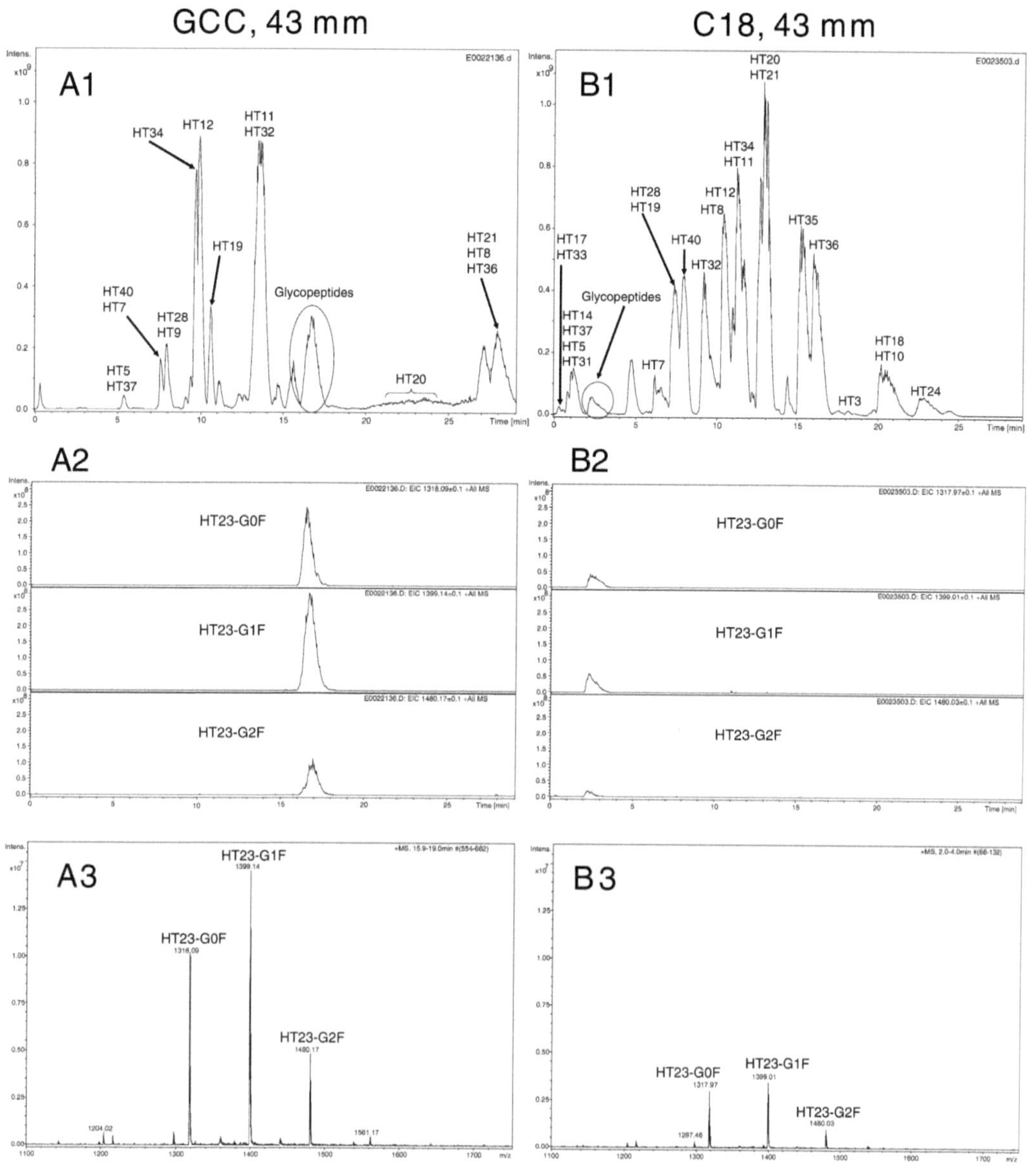

Fig. 4 Comparison of the nanoLC-Chip-MS analysis performed on a graphitized carbon column (A) and a C18 column (B). From the top downwards: the annotated base peak chromatograms (A1, B1); the extracted ion chromatograms of HT23-G0F, HT23-G1F, and HT23-G2F (2+) (A2, B2); and the average MS spectra corresponding to the area of glycopeptide elution (A3, B3)

Glycopeptides are better retained on the graphitized carbon column with retention time ranging from 16 to 19 min compared to the RP C18 column with much shorter retention times (2–8 min). Spectra recorded in the elution time range of glycopeptides are combined resulting in an average mass spectrum, which displayed nine glycoforms for each nanoLC-Chip type (Fig. 4A3, B3). Even if those spectra look qualitatively identical, the glycopeptide signal is more intense with the graphitized carbon nanoLC-Chip than with reversed-phase column, which therefore improves significantly the quality of further MS/MS experiments, especially on minor species. Indeed hydrophilic species are known to be better retained on graphitized carbon phase. The potential glycopeptide losses during the enrichment step on the trapping column are thus less critical for the graphitized carbon LC-Chip than for C18 LC-Chips. Moreover, glycopeptides retained on the graphitized carbon phase are eluted with a higher acetonitrile percentage leading to a better ionization. In conclusion, the use of graphitized carbon LC-Chip versus C18 column clearly improves the detection of glycopeptides allowing MS/MS experiments to be undertaken in optimized conditions.

3.2 Data Interpretation

1. For tryptic peptide assignment, mass data collected during analysis are processed using DataAnalysis™ (Bruker Daltonics), converted into a Mascot Generic File (.mgf) and searched against an *in-house* database using a local Mascot™ server (Matrix Science).
2. Typical glycopeptide diagnostic ions corresponding to glycan fragmentation ($[HexNAc+H]^+$ and $[HexHexNAc+H]^+$) are extracted from MS/MS ion current and are used to determine the retention time of glycopeptides. Corresponding MS/MS spectra are manually interpreted taking into account the three types of fragment ions classically generated for glycopeptides (neutral sugar loss, non-glycosylated peptide fragments, and glycan backbone internal fragments). Looking at the tandem mass spectra of glycopeptides, we observe predominantly fragment ions resulting from the fragmentation of glycosidic bonds as already described (14). Ions corresponding to the fragmentation of the peptide backbone are much less intense and, for some cases, it is possible to observe fragment ions corresponding to y- and b-type ions of the peptide backbone (15). As an example, the MS/MS spectrum of HT23-G2F is shown in Fig. 5. The doubly charged molecular ion at *m/z* 1,480.43 displays a series of intense, single protonated ion fragments. Most of the fragments correspond to successive neutral sugar losses. Two fragmentation pathways are observed. One consists of ions, which conserved the core-fucosyl region intact and lost successively neutral sugars. The second one consists in ions, which lost the core-fucose plus additional monosaccharides. The combination of both types of information allows us to ascertain the structure of the N-linked glycan. In addition, we observe in the low *m/z* range ions corresponding to Y and B glycan backbone fragments and to internal glycan fragments according to the nomenclature of

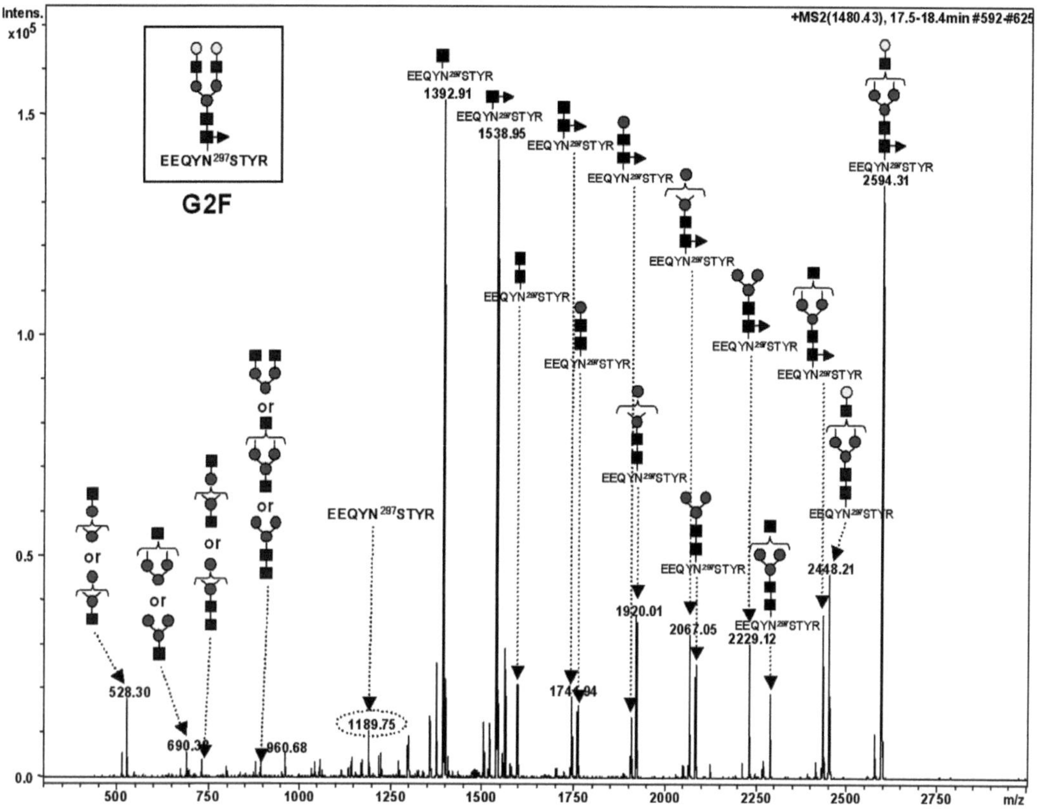

Fig. 5 MS/MS spectrum of the glycopeptide HT23-G0F after nanoLC-Chip ion trap analysis of the tryptic mixture of A2CHM heavy chain

Domon and Costello (16). For example, ion at *m/z* 528.26 identified as [2Hex+HexNAc+H)+is a fragment typically observed in MS/MS spectra of *N*-glycans and associated glycopeptides. This ion does not give information about the glycan structure itself. Nevertheless it refers to diagnostic ions useful to assign glycopeptides in complex mixtures (15). Moreover, the MS/MS spectrum presents a single protonated ion at *m/z* 1,189.73 corresponding to the mass of peptide HT23 without glycosylation. In addition weak intense y-type ions of this peptide are detected in the low mass range of the MS/MS spectrum providing additional information regarding the peptide amino acid sequence.

Data interpretation of the MS/MS spectra corresponding to the nine detected glycopeptides of the heavy-chain tryptic digest allows identifying 12 potential glycoforms which are presented in Fig. 6.

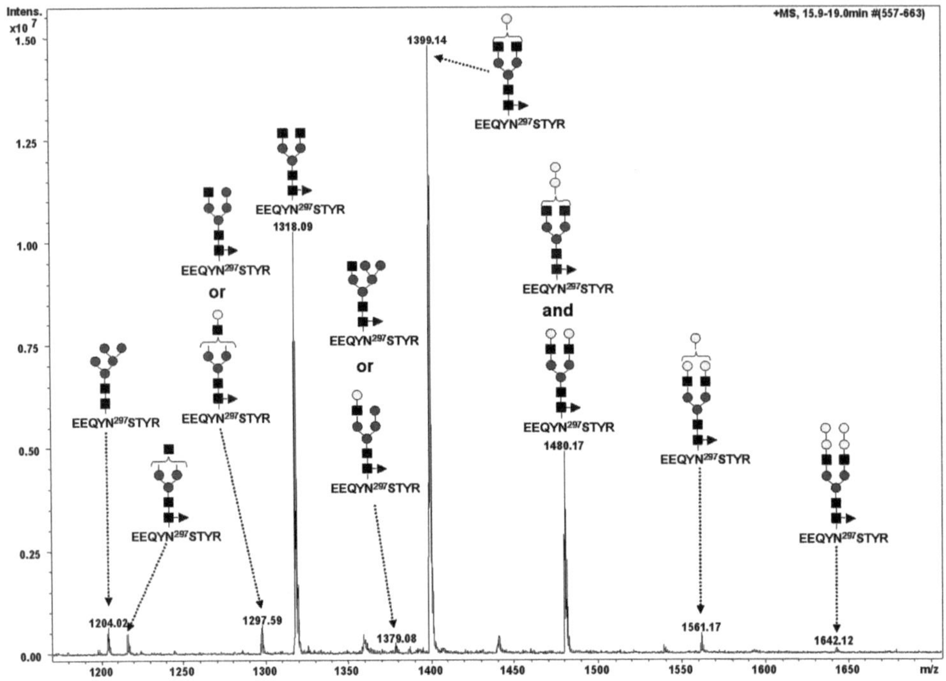

Fig. 6 Average mass spectrum in the area of elution of glycopeptides after nanoLC-chip ion trap analysis with a 43 mm length graphitized column. Nine glycopeptides corresponding to 12 potential glycoforms of A2CHM were detected and further fragmented by MS/MS

4 Notes

1. It is important to flush the sample with N_2 in order to inert the atmosphere and avoid undesirable reactions to occur like unspecific oxidation of certain amino acids.
2. The alkylation step may be difficult to control since over-alkylations are commonly observed. Cooling sample on ice permits to quench the reaction and to limit such a modification.
3. Concentrations of separated products are determined by UV spectroscopy at 280 nm following the Beer–Lambert law. The molar absorptivity is calculated from the amino-acid sequence.
4. MS and MS/MS data are acquired in a data-dependent manner:
   - The system is operated with automatic switching between MS and MS/MS modes. The three most abundant species of each *m/z* range are selected to be further isolated and fragmented.
   - For selective tandem MS experiments on *N*-glycopeptides, inclusion lists are created. A list of masses is set in order to automatically isolate and fragment the doubly protonated ion of each detected glycopeptide.

References

1. Jefferis R (2005) Glycosylation of recombinant antibody therapeutics. Biotechnol Prog 21: 11–16
2. Beck A, Wagner-Rousset E, Bussat MC, Lokteff M, Klinguer-Hamour C, Haeuw JF et al (2008) Trends in glycosylation, glycoanalysis and glycoengineering of therapeutic antibodies and Fc-fusion proteins. Curr Pharm Biotechnol 9: 482–501
3. Raju TS, Scallon BJ (2006) Glycosylation in the Fc domain of IgG increases resistance to proteolytic cleavage by papain. Biochem Biophys Res Commun 341:797–803
4. Takahashi N, Ishii I, Ishihara H, Mori M, Tejima S, Jefferis R et al (1987) Comparative structural study of the N-linked oligosaccharides of human normal and pathological immunoglobulin G. Biochemistry 26:1137–1144
5. Jefferis R (2006) Criteria for selection of IgG isotype and glycoform of antibody therapeutics. BioProcess Int 4:40–43
6. Srebalus Barnes CA, Lim A (2007) Applications of mass spectrometry for the structural characterization of recombinant protein pharmaceuticals. Mass Spectrom Rev 26:370–388
7. Dell A, Morris HR (2001) Glycoprotein structure determination by mass spectrometry. Science 291:2351–2356
8. Mechref Y, Novotny MV (2002) Structural investigations of glycoconjugates at high sensitivity. Chem Rev 102:321–369
9. Mechref Y, Novotny MV (2006) Miniaturized separation techniques in glycomic investigations. J Chromatogr B Analyt Technol Biomed Life Sci 841:65–78
10. Zamfir A, Vakhrushev S, Sterling A, Niebel HJ, Allen M, Peter-Katalinic J (2004) Fully automated chip-based mass spectrometry for complex carbohydrate system analysis. Anal Chem 76:2046–2054
11. Ninonuevo M, An H, Yin H, Killeen K, Grimm R, Ward R et al (2005) Nanoliquid chromatography-mass spectrometry of oligosaccharides employing graphitized carbon chromatography on microchip with a high-accuracy mass analyzer. Electrophoresis 26:3641–3649
12. Ninonuevo MR, Park Y, Yin H, Zhang J, Ward RE, Clowers BH et al (2006) A strategy for annotating the human milk glycome. J Agric Food Chem 54:7471–7480
13. Beck A, Bussat MC, Zorn N, Robillard V, Klinguer-Hamour C, Chenu S et al (2005) Characterization by liquid chromatography combined with mass spectrometry of monoclonal anti-IGF-1 receptor antibodies produced in CHO and NS0 cells. J Chromatogr B Analyt Technol Biomed Life Sci 819:203–218
14. Demelbauer UM, Zehl M, Plematl A, Allmaier G, Rizzi A (2004) Determination of glycopeptide structures by multistage mass spectrometry with low-energy collision-induced dissociation: comparison of electrospray ionization quadrupole ion trap and matrix-assisted laser desorption/ionization quadrupole ion trap reflectron time-of-flight approaches. Rapid Commun Mass Spectrom 18:1575–1582
15. Wuhrer M, Catalina MI, Deelder AM, Hokke CH (2007) Glycoproteomics based on tandem mass spectrometry of glycopeptides. J Chromatogr B Analyt Technol Biomed Life Sci 849:115–128
16. Domon B, Costello CE (1988) A systematic nomenclature for carbohydrate fragmentations in FAB-MS/MS spectra of glycoconjugates. Glycoconjugate J 5:397–409

Chapter 7

Cetuximab Fab and Fc *N*-Glycan Fast Characterization Using IdeS Digestion and Liquid Chromatography Coupled to Electrospray Ionization Mass Spectrometry

Marie-Claire Janin-Bussat, Laure Tonini, Céline Huillet, Olivier Colas, Christine Klinguer-Hamour, Nathalie Corvaïa, and Alain Beck

Abstract

Antibodies and related products represent one of the fastest growing areas of new drug development within the pharmaceutical industry. Monoclonal antibodies (mAbs) undergo many posttranslational modifications (PTMs) that must be extensively characterized. Here we described a rapid mass spectrometry (MS) method for the characterization of cetuximab glycosylation. The reported analytical technique is based on the use of a cystein protease, immunoglobulin-degrading enzyme of *Streptococcus pyogenes* that allows a fast limited proteolysis of the mAb with low material consumption. The resulting large fragments are analyzed by ultrahigh-performance liquid chromatography combined to an electrospray ionization mass spectrometer and a time-of-flight analyzer (ESI–TOF). Cetuximab is a potent chimeric mouse/human antibody worldwide approved for the treatment of colon and head and neck cancers. This antibody, produced by SP2/0 murine myeloma cells, is N-glycosylated both in the Fc and Fab moieties, which have been shown to impact on safety and PK/PD and considered as a critical quality attribute. The method can also be applied for biosimilars, biobetters, and next-generation antibodies and Fc-fusion proteins.

Key words Monoclonal antibody, Biosimilar, Biobetter, Mass spectrometry, High-performance liquid chromatography, Middle-up and -down, Immunoglobulin-degrading enzyme of *Streptococcus pyogenes*, Hydrophobicity, Fab and Fc domains, Oligosaccharides, Charge variants, Sialic acids, Gal-α-1,3-Gal

1 Introduction

In the last decade, monoclonal antibodies (mAbs) have emerged as one of the most promising drug classes in the biopharmaceutical industry. More than 40 mAbs and related products (Fab fragments, Fc-fusion proteins and peptides, radio-immunoconjugates, antibody–drug conjugates, bispecific antibodies) have been approved and several hundred are in clinical trials in various therapeutic indications including oncology and inflammatory diseases. These G-type immunoglobulins (IgGs) are large tetrameric

Alain Beck (ed.), *Glycosylation Engineering of Biopharmaceuticals: Methods and Protocols*, Methods in Molecular Biology, vol. 988,
DOI 10.1007/978-1-62703-327-5_7, © Springer Science+Business Media New York 2013

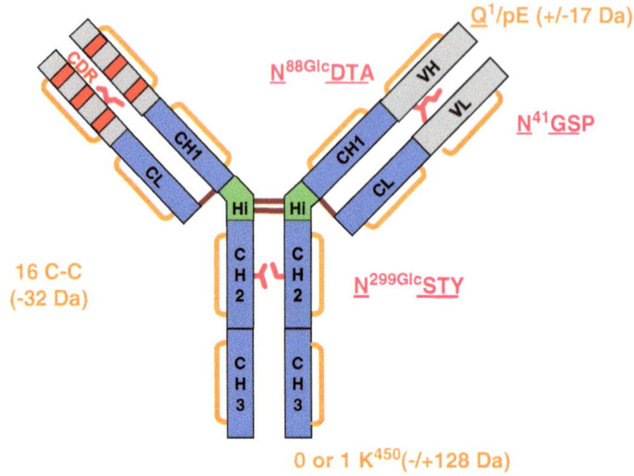

Fig. 1 Cetuximab schematic structure (IgG1k). Three N-glycosylation consensus sites with the common definition (-Asn-X-Ser or Thr-, X is not a Pro) are present in cetuximab (Asn88 and 299 of the heavy chain; Asn41 of the light chain). Nevertheless, experimental mass spectrometry data show that only both the Asn of the heavy chain are glycosylated, which is in agreement with the following more restrictive definition of consensus sites: (-Asn-X-Ser or Thr-Y-, X and Y are not Pro). The scheme indicates also the other main posttranslational modifications (PTMs) that are observed by mass spectrometry (heavy chain N-terminal pyroGlu cyclization and C-terminal Lys cleavage; 16 disulfide bridges as expected for a chimeric IgG1 mAb)

glycoproteins (150 kDa range) composed of two light and heavy chains. Each light chain contains one variable domain (VL) and one constant domain (CL), whereas each heavy chain consists of one variable domain (VH) and three constant domains (CH1, CH2, and CH3) (Fig. 1). Each light chain is linked to a heavy chain through a disulfide bond, and the two heavy chains are connected in the hinge region through at least two disulfide bonds, depending on the IgG isotype (1).

All IgGs show common posttranslational modifications (PTMs) (2) light- and heavy-chain pyroglutamic acid cyclization for mAbs containing N-terminal glutamine (3) or glutamic acid residues (4), heavy-chain N-glycosylation and C-terminal lysine processing (5). Heavy-chain N-glycosylation depends on the nature of the eukariotic cell line used for manufacture of the mAb, and particular glycoforms can be involved in important immune cytotoxic functions. Glycosylation can also play a role in the plasmatic half-life via binding to neonatal Fcgamma receptor (FcγRn) (6, 7).

A conserved N-glycosylation site is typically present at Asn297, in the CH2 domain of the Fc portion (consisting of CH2 and CH3 domains of both heavy chains), but 20% of human IgGs have an additional site of glycosylation in the CH1 domain of the Fd portion (consisting in VH and CH1 domains) (8, 9). It is the case for cetuximab, a chimeric human/mouse antibody produced in the SP2/0 murine myeloma cell line, glycosylated on both Asn88 and Asn 299 (Fig. 1). These oligosaccharides located at the two glycosylation sites are a complex mixture of different glycans, making them challenging for structural characterization. However, this characterization is necessary since glycoforms that are not commonly biosynthesized in humans may potentially be immunogenic and must be identified, quantified, and limited for therapeutic use. In particular, glycoforms containing *N*-glycolylneuramic acid (NGNA) or galactose-α-1,3-galactose (Gal-α-1,3-Gal) structures, observed on proteins produced in the mouse-derived cell lines SP2/0 and NS0, can adversely affect safety and half-life (10, 11). In 2007, Qian et al. (12) published a study about *N*-glycans present on both cetuximab Fc and Fab fragments. They identified 21 different glycoforms with around 30% capped by at least one Gal-α-1,3-Gal residue and 12% capped by an NGNA residue and traces of oligomannose. Importantly both Gal-α-1,3-Gal and NGNA were found only in the Fab moieties in contrast to the Fc fragment, for which only typical IgGs G0F, G1F, and G2F glycoforms were identified. In a report on cetuximab-induced anaphylaxis, preexisting IgEs specific for this galactose-α-1,3-galactose epitope were detected in patients treated with the mAb. Using a solid-phase immunoassay (ImmunoCAP), these IgEs were found to bind to SP2/0-produced cetuximab and F(ab')2 fragments and not the Fc fragment (13).

Over past years, mass spectrometry has become an essential analytical tool for PTM analysis due to its superior structural characterization resolution over other analytical techniques that are presently available. The development of ESI–TOF-MS technology with high mass resolution, high sensitivity, and broad mass range (up to hundreds of kDa) has transformed the analysis of large heterogeneous biomolecules into a routine task (14–16). While mass measurement of the intact molecule gives a general representation of the protein as a whole, it does not provide locations of the modifications. Cleaving the protein into smaller fragments before mass analysis can be more straightforward to obtain structural resolution (17). In particular, limited proteolysis of mAbs consisting on the cleavage of each heavy chain at a single site can lead to large fragments and is usually done using endoproteinase Lys-C or papain. However, unspecific cleavage of heavy chains can be obtained, complicating the interpretation of data. Immunoglobulin-degrading enzyme of *Streptococcus pyogenes* (IdeS), a bacterial cysteine protease, seems to be an enzyme of choice for limited proteolysis. Indeed, this enzyme cleaves IgG heavy chains between the two glycines in close proximity to the hinge domain of the immunoglobulin type 1 constant sequence (CPPCPAPELLG/G) (Fig. 2), with a very high specificity (18–23).

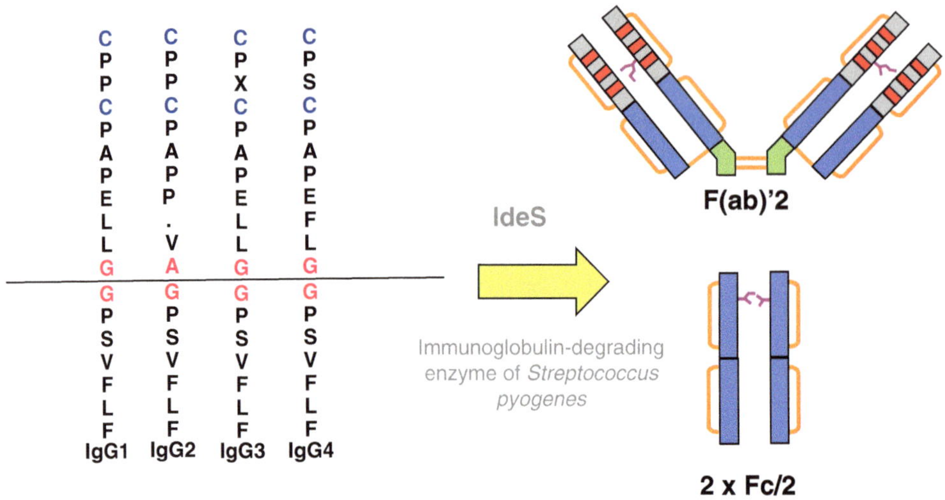

Fig. 2 Heavy-chain cleavage site of IdeS according to the human IgG isotype. IdeS cleaves IgG heavy chains between two glycines (IgG1, 3, 4) or between a glycine and an alanine (IgG2)

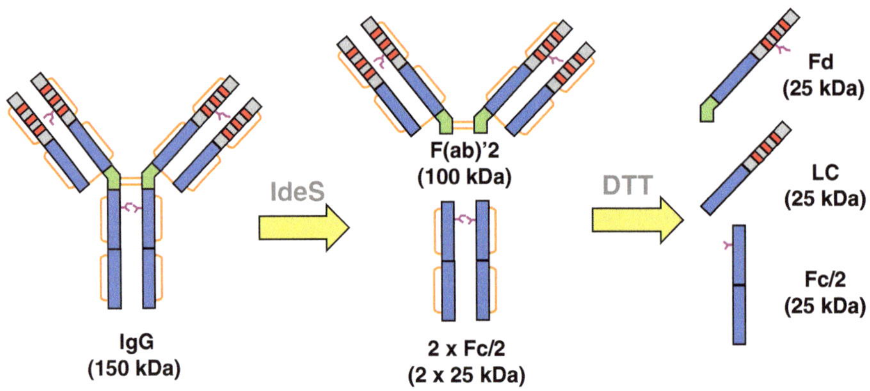

Fig. 3 Generation of three fragments of near 25 kDa after IdeS digestion and DTT reduction. The IdeS digestion of the 150 kDa IgG leads first to F(ab')2 (100 kDa) and Fc (50 kDa). After reduction of disulfide bridges with DTT, the three Fc/2, LC, and Fd of 25 kDa are obtained

Here we use IdeS digestion on cetuximab in order to obtain, after reduction of disulfide bonds with DTT, three fragments of around 25 kDa consisting on the light chain (LC), the Fd domain, and the Fc/2 domain (composed by CH2 and CH3 domains of one single heavy chain) (Fig. 3). Reversed-phase liquid chromatography with a diphenyl column (24, 25) is used for the separation of the fragments prior to mass analysis with electrospray ionization (ESI) and time-of-flight (TOF) detection, according to a "middle-up" method (26). An extensive characterization of oligosaccharides located on both Fd and Fc/2 domains of cetuximab is performed, and a particular attention is given to those containing NGNA or Gal-α-1,3-Gal structures considered as critical quality attributes (CQA).

2 Materials

2.1 Chemical

1. IdeS (FabRICATOR® Kit, Genovis, Lund, Sweden).
2. IgGZERO™ (Genovis, Lund, Sweden).
3. *Rapi*Gest™ SF (Waters, Saint-Quentin-en-Yvelines, France).
4. Peptide *N*-glycosidase F, PNGase F (Biolabs, Ipswich, England).
5. α2-3,6,8,9-Neuraminidase (Calbiochem, San Diego, USA).
6. Sodium dibasic phosphate dodecahydrate, $Na_2HPO_4 \cdot 12H_2O$ (VWR, Leuven, Belgium).
7. Sodium monobasic phosphate dihydrate, $NaH_2PO_4 \cdot 2H_2O$ (VWR, Leuven, Belgium).
8. Sodium chloride, NaCl (VWR, Leuven, Belgium).
9. Dithiothréitol, DTT (Sigma Aldrich, Saint-Quentin-Fallavier, France).
10. Cesium Iodide, CsI (Sigma Aldrich, Saint-Quentin-Fallavier, France).
11. Acetonitrile, CH_3CN (Carlo Erba, Val de Reuil, France).
12. Trifluoroacetic acid, TFA (Fluka, Buchs, Swiss).
13. Tris(hydroxymethyl)aminomethane hydrochloride, Tris–HCl (Sigma Aldrich, Saint-Quentin-Fallavier, France).
14. Ethylenediaminetetraacetic acid, EDTA (Prolabo, Paris, France).
15. MilliQ water.

2.2 Equipment

1. Mass spectrometer: LCT Premier™ (Waters, Saint-Quentin-en-Yvelines, France), with ESI and TOF analyzer.
2. Ultrahigh-performance liquid chromatography (UHPLC) system: Ultra Performance Liquid Chromatography Acquity™ (Waters, Saint-Quentin-en-Yvelines, France).
3. Diphenyl column: 2.1 mm ID × 100 mm length, 1.9 μm particle size, 140 Å pore size (Restek, Bellefonte, USA) (see Note 1).
4. Mobile phase A (MPA): 0.05/100 (v/v) TFA/H_2O.
5. Mobile phase B (MPB): 0.05/100 (v/v) TFA/ACN.

3 Methods

3.1 Solution Preparation

3.1.1 Digestion Buffer 1: 10 mM Sodium Phosphate, 150 mm NaCl pH 7.4 (100 mL)

1. Weight 8.78 g NaCl.
2. Add 50 mL 20 mM sodium phosphate, pH 7.4.
3. Add milli-Q water to have a final volume of 100 mL.

3.1.2 Digestion Buffer 2: 50 mM Sodium Phosphate, 150 mm NaCl pH 6.6 (100 mL)

1. Weight 4.39 g NaCl.
2. Add 50 mL 100 mM sodium phosphate, pH 6.6.
3. Add milli-Q water to have a final volume of 100 mL.

3.1.3 0.01 M Sodium Acetate Buffer pH 5 (100 mL)

1. Measure 25 mL of a 0.02 M solution of acetic acid.
2. Adjust pH to 5 with 0.02 M solution of sodium acetate.
3. Half dilution.

3.1.4 3 M Tris–HCl Buffer pH 8 Containing 1 mM EDTA (100 mL)

1. Weight 38 mg of EDTA.
2. Weight 36.33 g of Tris–HCl.
3. Add 50 mL of milli-Q water.
4. Adjust pH to 8 with HCl.
5. Add milli-Q water necessary to obtain a final volume of 100 mL.

3.2 Sample Preparation

Protocols were described for 20 mg/mL mAb concentration approximately; otherwise the pH must be adjusted for each digestion.

3.2.1 IdeS Hydrolysis (See Note 2)

1. Mix the 100 μg solution of mAb with 100 U of IdeS.
2. Add 45 μL of digestion buffer 2 to obtain a sample pH of 6.6.
3. Incubate the sample at 37°C for 30 min.

3.2.2 IgGZERO™ Deglycosylation (See Note 3)

1. Mix the 100 μg solution of mAb with 100 U of IgGZERO™.
2. Add 45 μL of digestion buffer 1 to obtain a sample pH of 7.4.
3. Incubate the sample at 37°C for 30 min.

3.2.3 PNGase F Deglycosylation

1. Mix the 100 μg solution of mAb with 0.1% of *Rapi*Gest™ SF.
2. Add 100 mM ammonium bicarbonate buffer pH 8 to obtain a sample pH of 8.
3. Add 1 M DTT solution (1.5 μl) to have a final concentration of 20 mM.
4. Incubate the sample at 80°C for 15 min.
5. Cool the sample.

6. Add 1 μL of PNGase F.
7. Incubate the sample at 37°C for 6 h.
8. Add 1 μL of PNGase F.
9. Incubate the sample at 37°C overnight.

3.2.4 α2-3,6,8,9-Neuraminidase Digestion

1. Mix the 100 μg solution of mAb with 20 μL of 0.1 M acetate buffer pH 5 to obtain a sample pH of 5.
2. Add 1 μL of α2-3,6,8,9-neuraminidase corresponding to at least 5 mU (milliunit).
3. Incubate the sample at 37°C for 15 h.

3.2.5 DTT Reduction with Tris–HCl Buffer pH 8 (See Note 4, (27))

1. Freeze-dry the 100 μg solution of mAb.
2. Add 70 μL of 3 M Tris–HCl buffer pH 8 containing 1 mM EDTA.
3. Add 2 μL of a 1 M DTT solution to have a final concentration of 25 mM approximately.
4. Incubate the sample at 56°C during 45 min.
5. Quench the reduction by adding acetic acid.

3.3 Analysis Conditions

3.3.1 UHPLC Conditions

- Column temperature: 60°C.
- Flow: 0.4 mL/min.

The initial mobile phase is 5% MPB for 4 min, and then three-stage gradients are applied: 5% MPB per minute from 25 to 40% MPB, 1% MPB per minute from 25 to 40% MPB, and 5% MPB per minute from 40 to 80% MPB, and finally the % MPB is performed at 80% for 2 min.

3.3.2 ESI–TOF Experiments

1. MS analysis is performed in a W-positive ion mode on a Waters LCT Premier™ equipped with an ESI source.
2. Source and desolvatation temperatures are set to 120°C and 300°C, respectively.
3. Desolvatation and source voltages are set to 2,800 V and 120 V, respectively.
4. Nitrogen gas flow rates are set at 50 L/h for the cone and 600 L/h for desolvatation.
5. Calibration is carried out using a 2 g/L solution of CsI in 1-propanol/water (50/50: v/v) for the protein mass measurements. CsI forms 12 charged clusters from 900 to 4,000 *m/z*.
6. MS data acquisition and processing are performed using MassLynx™ 4.1 (Waters, Saint-Quentin-en-Yvelines, France). Ions are scanned over a *m/z* range of 300–3,500.
7. Online recalibration is based upon CsI clusters.
8. Deconvolution of the protein mass spectra is performed with Transform™.

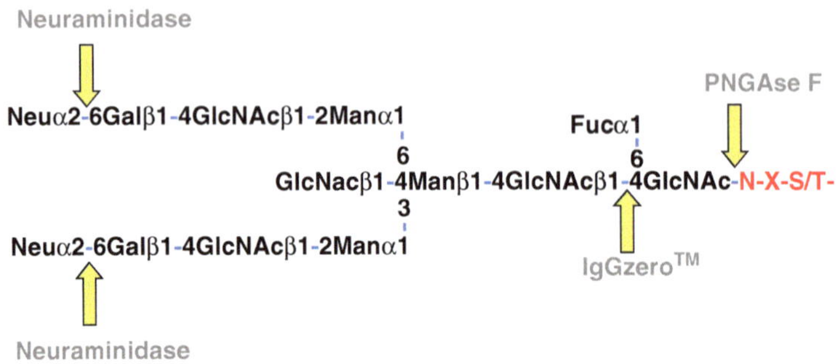

Fig. 4 Hydrolysis of the conserved glycans attached to asparagine residue on the IgG heavy-chain consensus N-glycosylation site by the enzyme IgGZERO™. Cleavage sites by PNGase F and neuraminidase are also indicated

3.3.3 Data Interpretation

1. *LC-ESI–TOF of cetuximab after IdeS and IgGZERO digestions and DTT reduction.* The solution of mAb is directly injected into the HPLC column, with a volume of injection of 10 μL (approximately 10 μg of mAb). The *N*-glycan cleavage sites of PNGAse F and IgGZERO™ and neuraminidase are reported in Fig. 4. Two distinct masses are observed for the Fc/2 fragment populations separated by 128.25 Da (Fig. 5a). By comparison of the two experimental masses with the theoretical mass of the deglycosylated Fc/2 fragment, the mass of 24,137.13 Da is assigned to Fc/2 fragment with C-terminal lysine cleavage whereas the one of 24,265.38 Da is assigned to Fc/2 fragment without this cleavage. In Fig. 5b, on the deconvoluted mass spectrum corresponding to the eluted Fd fragment, after treatment with IgGZERO™, several distinct masses are observed. This mass heterogeneity is due to the remaining presence of glycans on the fragment despite the step of deglycosylation. Under the conditions used for deglycosylation with IgGZERO™, the removal of Fd fragment glycans is not efficient.
2. *LC-ESI–TOF of cetuximab after IdeS and PNGase F digestions and DTT reduction (Fig. 6).* Both *N*-glycan profiles of deglycosylated Fc/2 and Fd fragments are observed. On the deconvoluted mass spectrum obtained for the deglycosylated Fc/2 (Fig. 6a), two peaks are observed. These two peaks, corresponding to two populations separated by 127.87 Da, are attributed to both Fc/2 fragments with and without C-terminal lysine cleavage. In Fig. 6b, peaks corresponding to both glycosylated and non-glycosylated Fd fragments are observed (see Note 5). A mass of 25,449.13 Da is obtained for the non-deglycosylated Fd fragment.
3. *LC-ESI–TOF of cetuximab after IdeS digestion and DTT reduction (Fig. 7).* After IdeS digestion and reduction of disulfide bonds with DTT in the presence of Tris 3 M buffer, 10 μg of

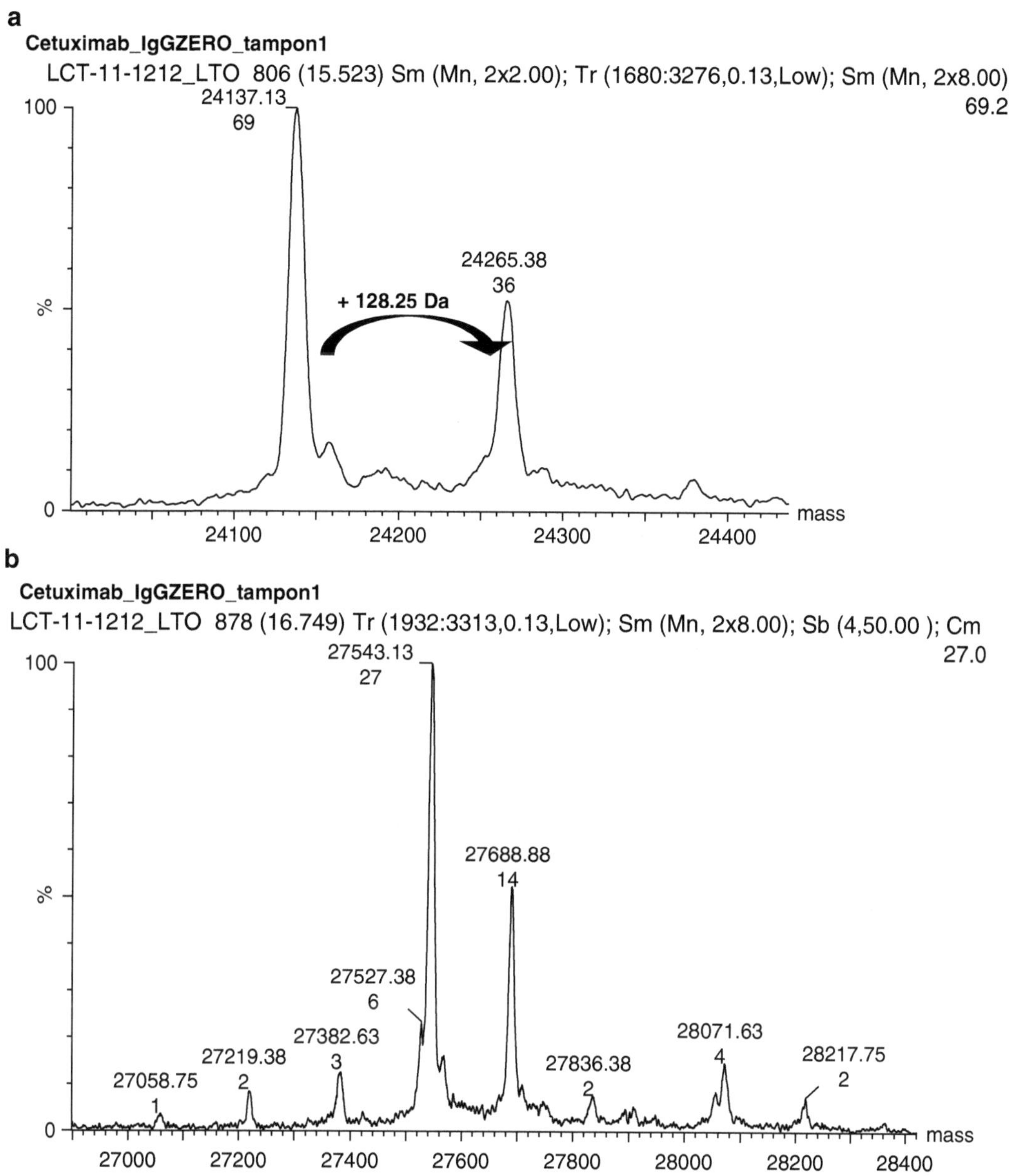

Fig. 5 Deconvoluted mass spectrum of both cetuximab Fc/2 and Fd fragments deglycosylated with IgGZERO™. 10 μg of cetuximab digested by IgGZERO™ and IdeS, reduced by DTT, were injected into the diphenyl UHPLC column. Panels (**a**) and (**b**) correspond, respectively, to deconvoluted mass spectra of both deglycosylated Fc/2 and Fd fragments (transform algorithm from MassLynx™). For the Fc/2 fragments, on trace (**a**), two peaks are obtained, corresponding to Fc/2 fragment populations with and without C-terminal lysine cleavage, separated by 128.25 Da. On trace (**b**), a complex deconvoluted mass spectrum is obtained for the Fd fragment due to the remaining presence of glycans on the fragment despite the step of deglycosylation

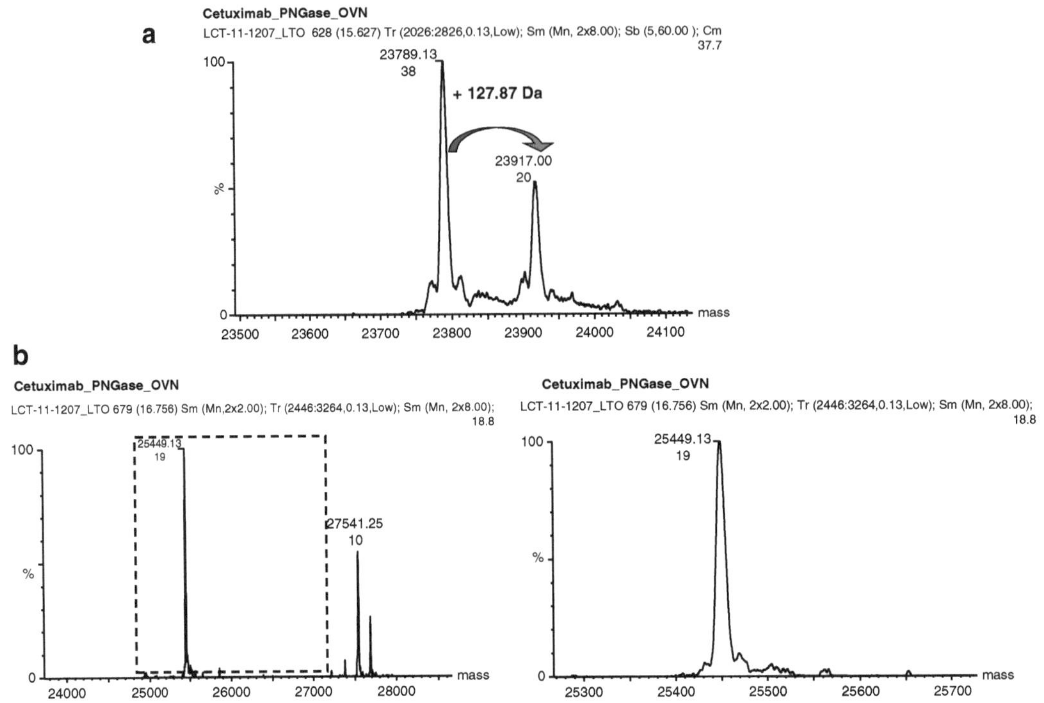

Fig. 6 Deconvoluted mass spectrum of both cetuximab Fc/2 and Fd fragments deglycosylated with PNGase F. 10 μg of cetuximab digested by IdeS and PNGase F, reduced by DTT, were injected into the diphenyl column. Panels (**a**) and (**b**) correspond, respectively, to deconvoluted mass spectra of both deglycosylated Fc/2 and Fd fragments (transform algorithm from MassLynx™ 4.1). On trace (**a**), two peaks are obtained, corresponding to Fc/2 fragment populations with and without C-terminal lysine cleavage, separated by 127.87 Da. On trace (**b**), both glycosylated and non-glycosylated Fd fragments are observed. A mass of 25,449.13 Da is obtained for the non-glycosylated Fd fragment

mAbs are injected into the column. In Fig. 7a six major Fc/2 fragment populations are observed. For comparison with the mass of the deglycosylated Fc/2 fragments obtained in the mass of the glycoform present on each Fc/2 fragment population is calculated. The three glycan masses of 1,441.87, 1,607.00, and 1,766.00 Da are assigned for the Fc/2 fragment populations with C-terminal lysine cleavage and three glycans of 1,443.38, 1,607.00, and 1,765.13 Da for the Fc/2 fragment populations without C-terminal lysine cleavage (Table 1). These experimental masses correspond to G0F, G1F, and G2F glycans, the core fucosylated biantennary oligosaccharides with zero, one, or two Gal residues, that are typical for IgGs.

Figure 7b shows a complex glycan profile of the Fab domain, with eleven peaks corresponding to populations of glycosylated Fd fragments. The theoretical and the experimental masses of the N-linked glycans are listed in Table 2,

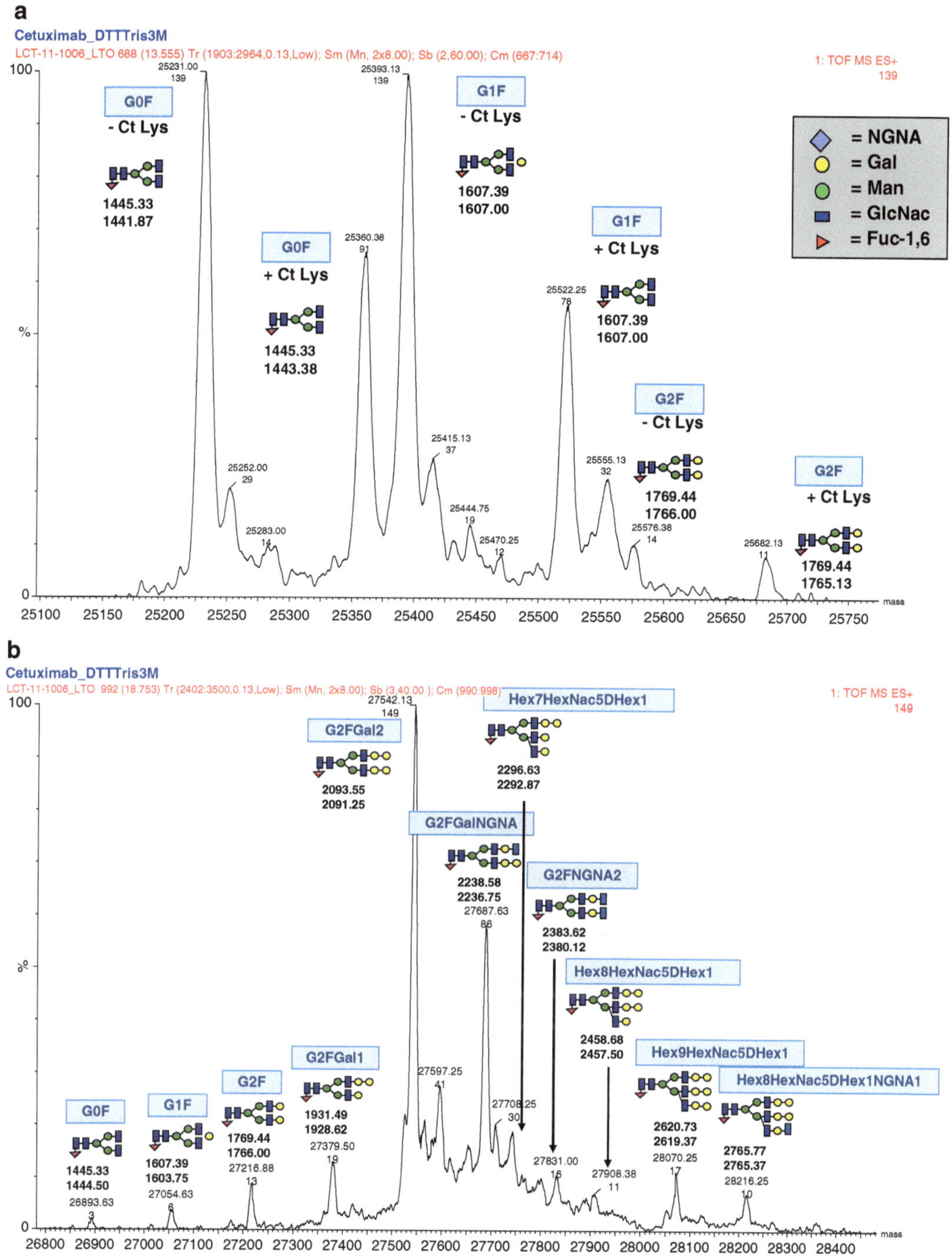

Fig. 7 Deconvoluted mass spectrum of both cetuximab Fc/2 and Fd fragments. 10 μg of cetuximab digested by IdeS and reduced by DTT were injected into the diphenyl UHPLC column. Panels (**a**) and (**b**) correspond, respectively, to deconvoluted mass spectra of both Fc/2 and Fd fragments (transform algorithm from MassLynx™ 4.1). On trace (**a**), six peaks are observed for the glycosylated Fc/2 fragment, corresponding to both populations with and without C-terminal lysine cleavage, with the core fucosylated biantennary oligosaccharides G0F, G1F, and G2F (calculated mass/experimental mass), typical of IgGs. On trace (**b**), 11 glycans (calculated mass/experimental mass) on Fd fragment are attributed. Seven glycans contain a Gal-α-1,3-Gal structure and three contain at least one NGNA-type sialic acid A, representing 89.38% and 31.28% of the whole glycans, respectively

Table 1
Theoretical masses of Fc/2 fragment oligosaccharides and experimental masses calculated from the Fc/2 fragment ESI–TOF MS spectrum

Fragment	Oligosaccharide	Symbol	Oligo structure	Theoretical MS (Da)	Measured MS (Da)	MS difference (Da)
Fc/2 with C-terminal lysine cleavage	Hex3HexNac4DHex1	G0F	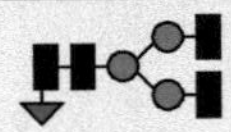	1,445.33	1,441.87	−3.46
Fc/2 without C-terminal lysine cleavage	Hex3HexNac4DHex1	G0F	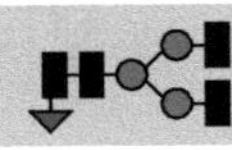	1,445.33	1,443.38	−1.45
Fc/2 with C-terminal lysine cleavage	Hex4HexNac4DHex1	G1F	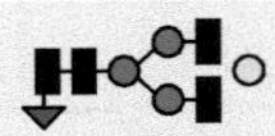	1,607.39	1,604.00	−3.39
Fc/2 without C-terminal lysine cleavage	Hex4HexNac4DHex1	G1F	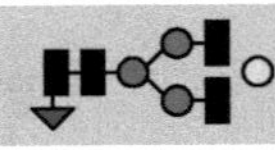	1,607.39	1,605.25	−2.14
Fc/2 with C-terminal lysine cleavage	Hex5HexNac4DHex1	G2F	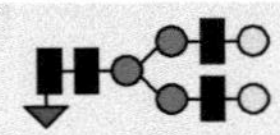	1,769.44	1,766.00	−3.44
Fc/2 without C-terminal lysine cleavage	Hex5HexNac4DHex1	G2F	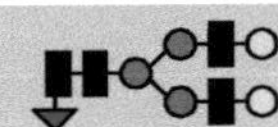	1,769.44	1,765.13	−4.31

Table 2
Theoretical masses of Fd fragment oligosaccharides and their experimental masses calculated from the Fd fragment ESI–TOF MS spectrum

Oligosaccharide	Symbol	Oligo structure	Theoretical MS (Da)	Measured MS (Da)	MS difference (Da)
Hex3HexNac4DHex1	G0F	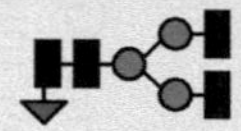	1,445.33	1,444.50	−0.83
Hex4HexNac4DHex1	G1F	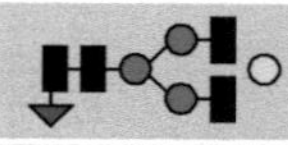	1,607.39	1,603.75	−3.64
Hex5HexNac4DHex1	G2F	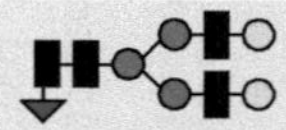	1,769.44	1,766.00	−3.44
Hex6HexNac4DHex1	G2FGal1	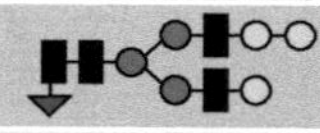	1,931.49	1,928.62	−2.87
Hex7HexNac4DHex1	G2FGal2	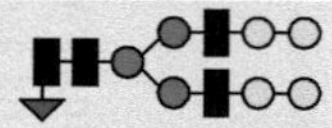	2,093.55	2,091.25	−2.30
Hex6HexNac4DHex1NGNA1	G2FGalNGNA	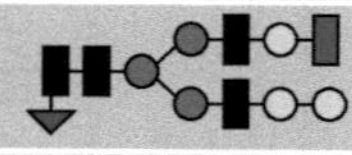	2,238.58	2,236.75	−1.83
Hex7HexNac5DHex1		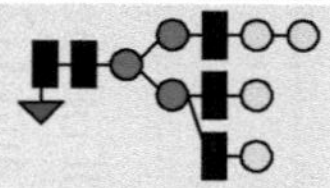	2,296.63	2,292.87	−3.76
Hex5HexNac4DHex1NGNA2	G2FNGNA2	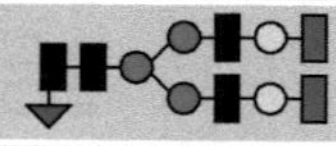	2,383.62	2,380.12	−3.50
Hex8HexNac5DHex1		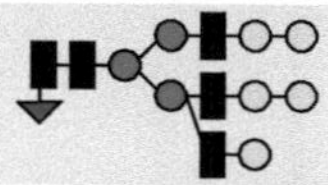	2,458.68	2,457.50	−1.18
Hex9HexNac5DHex1		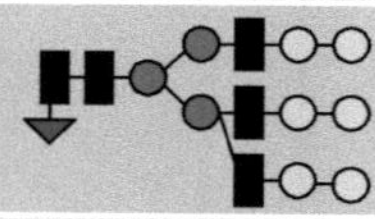	2,620.73	2,619.37	−1.36
Hex8HexNac5DHex1NGNA1		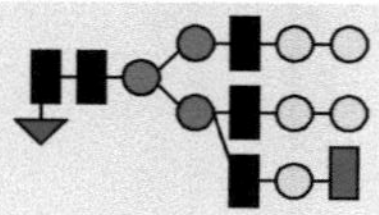	2,765.77	2,765.37	−0.40

Table 3
Proportion of each Fd fragment oligosaccharide calculated from chromatographic peak intensities (see Note 6)

Oligosaccharide	Intensity	Percentage	Structure
G0F	3	0.84	
G1F	6	1.68	
G2F	13	3.63	
G2F Gal1	19	5.31	Gal-α-1,3-Gal
G2F Gal2	149	41.62	Gal-α-1,3-Gal
G2F GalNGNA	86	24.02	Gal-α-1,3-Gal and NGNA
Hex7HexNac5DHex1	28	7.82	Gal-α-1,3-Gal
G2FNGNA2	16	4.47	NGNA
Hex8HexNac5DHex1	11	3.07	Gal-α-1,3-Gal
Hex9HexNac5DHex1	17	4.75	Gal-α-1,3-Gal
Hex8HexNac5DHex1NGNA1	10	2.79	Gal-α-1,3-Gal and NGNA

Structure	Total percentage
Gal-α-1,3-Gal	89.38
NGNA	31.28

and all of them are fucosylated at the proximal GlcNac. G0F, G1F, and G2F glycans are observed, with low intensities. The most intense peak corresponds to an oligosaccharide G2F-Gal2 with a terminal Gal-α-1,3-Gal structure, showed as responsible for hypersensitivity reaction after administration of cetuximab. Six other glycans containing this Gal-α-1,3-Gal structure are found and represent 89.38% of total glycans (Table 2, Note 6). One of those is found with high intensity and contains an NGNA sialic acid, and four are found with low intensities, containing triantennary oligosaccharides. Totally, three glycans containing an NGNA sialic acids are observed, representing 31.28% of total glycans (Tables 2 and 3).

4. *LC-ESI–TOF of cetuximab after neuraminidase treatment (Fig. 8a)*. After neuraminidase treatment, and at the end of reduction with DTT and Tris 3 M buffer, 10 μg of mAbs are approximately injected into the column. In Fig. 8a, on the deconvoluted mass spectrum obtained for the Fc fragment treated with neuraminidase, six major peaks are observed.

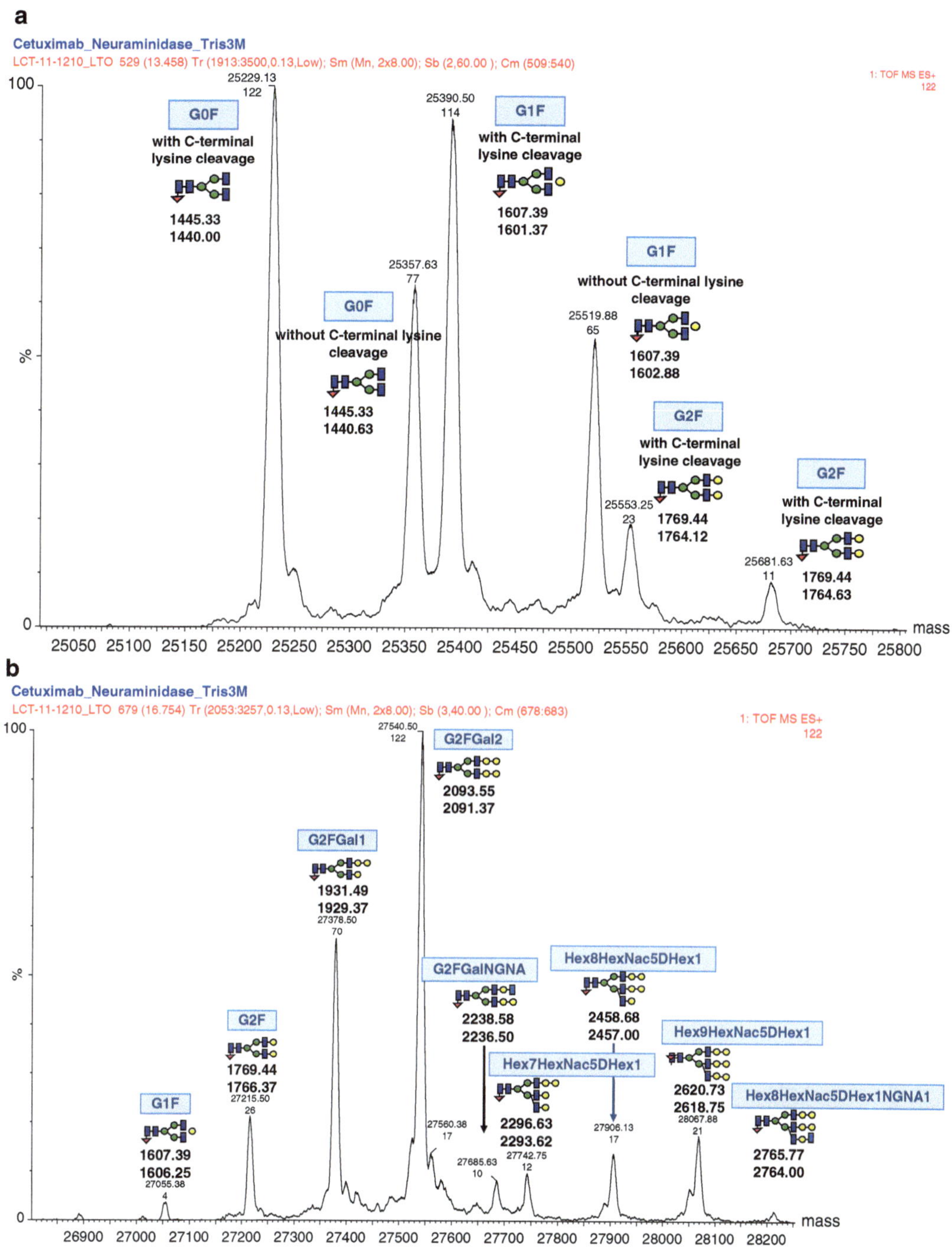

Fig. 8 Deconvoluted mass spectrum of both cetuximab Fc/2 and Fd fragments digested by neuraminidase. 10 μg of cetuximab digested by IdeS and neuraminidase, and reduced by DTT, were injected into the diphenyl UHPLC column. Panels (**a**) and (**b**) correspond to deconvoluted mass spectra of both Fc/2 and Fd fragments, respectively (transform algorithm from MassLynx™ 4.1)

Table 4
Theoretical masses of Fd fragment oligosaccharides, after neuraminidase treatment, and their experimental masses calculated from the Fd fragment ESI–TOF MS spectrum

Oligosaccharide	Symbol	Oligo structure	Theoretical MS (Da)	Measured MS (Da)	Ms difference (Da)
Hex4HexNac4DHex1	G1F	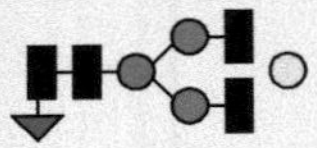	1,607.39	1,606.25	−1.14
Hex5HexNac4DHex1	G2F	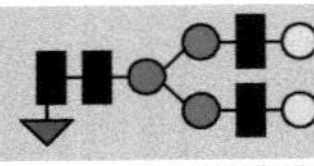	1,769.44	1,766.37	−3.07
Hex6HexNac4DHex1	G2FGal1	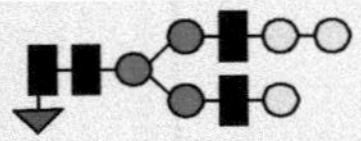	1,931.49	1,929.37	−2.12
Hex7HexNac4DHex1	G2FGal2	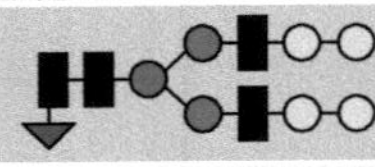	2,093.55	2,091.37	−2.18
Hex6HexNac4DHex1NGNA1	G2FGalNGNA	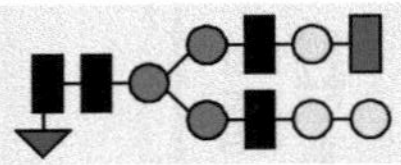	2,238.58	2,236.50	−2.08
Hex7HexNac5DHex1		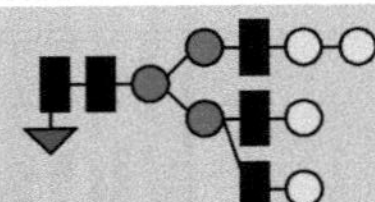	2,296.63	2,293.62	−3.01
Hex8HexNac5DHex1		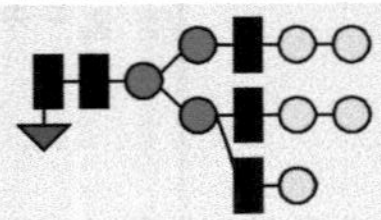	2,458.68	2,457.00	−1.68
Hex9HexNac5DHex1		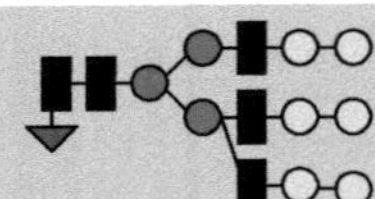	2,620.73	2,618.75	−1.98
Hex8HexNac5DHex1NGNA1		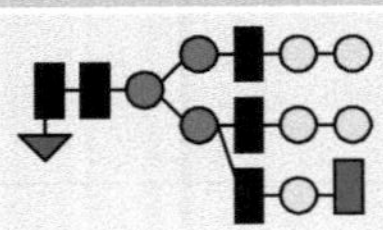	2,765.77	2,764.00	−1.67

Table 5
Proportion of each Fd fragment oligosaccharide, after neuraminidase treatment, calculated from chromatographic peak intensities (see Note 6)

Oligosaccharide	Intensity	Percentage	Structure
G1F	4	1.40	
G2F	26	9.12	
G2F Gal1	70	24.56	Gal-α-1,3-Gal
G2F Gal2	122	42.81	Gal-α-1,3-Gal
G2F GalNGNA	10	3.51	Gal-α-1,3-Gal and NGNA
Hex7HexNac5DHex1	12	4.21	Gal-α-1,3-Gal
Hex8HexNac5DHex1	17	5.97	Gal-α-1,3-Gal
Hex9HexNac5DHex1	21	7.37	Gal-α-1,3-Gal
Hex8HexNac5DHex1NGNA1	3	1.05	Gal-α-1,3-Gal and NGNA
Structure			**Total percentage**
Gal-α-1,3-Gal			89.48
NGNA			4.56

These peaks can be attributed to both Fc fragments with and without C-terminal lysine cleavage, with the three core fucosylated biantennary oligosaccharides G0F, G1F, and G2F not containing sialic acids, and not altered by neuraminidase treatment. Figure 8b shows the deconvoluted mass spectrum obtained for the Fd fragment treated with neuraminidase. The major peak corresponds to Fd fragment with G2FGal2 glycan. By comparison with the deconvoluted mass spectrum obtained for the Fd fragment untreated with neuraminidase (see Fig. 7b), we observe a strong diminution of the peak attributed to Fd fragment with G2FGalNGNA glycan, while the peak corresponding to Fd fragment with G2FGal1 has a higher intensity. After neuraminidase treatment, the sialic acid NGNA containing on G2FNGNA glycans is cleaved, leading to the obtention of G2FGal1 glycans. By the same way, we observed a diminution of the peak corresponding to Fd fragment with Hex8HexNac5DHex1NGNA1 glycan. The peak attributed to Fd fragments with G2FNGNA2 glycans totally disappears. The glycoform containing a Gal-α-1,3-Gal structure represents 89.48% of the whole glycans, while the remaining glycans containing an NGNA structure represent 4.56% (Tables 4 and 5).

4 Notes

1. Separation of mAb fragments by reversed-phase liquid chromatography can be difficult due to the high hydrophobicity of these molecules and the multiple protein conformations present during the separation. Often used as stationary phase, silica-based stationary phase with alkyl ligates (C4 and C8) provides good chromatographic separation but the selectivity obtained with this type of phase cannot be sufficient. Due to the high interaction surface with proteins offered by phenyl compounds, the biphenyl stationary phase allows better selectivity during separation of proteins and higher hydrophobic retention. Ren et al., in 2007 (24), illustrated the interest of a diphenyl column by showing that a C8 column failed to separate two IgGs of similar hydrophobicity while a diphenyl column was able to resolve these to IgGs, under the same chromatographic conditions.
2. Complete cleavage of mAb heavy chains with IdeS is possible with digestion duration of 30 min. This allows a fast analysis, in comparison to the use of other proteases extensively used for limited proteolysis like pepsin (28) or papain (29) which require several hours. Endopeptidase Lys-C is one of the proteases used for limited proteolysis that requires a lower time of digestion but cleaves heavy chains on several lysines, leading to multiple fragments. The action of the protease was successfully tested on the four human IgG isotypes. Depending on the isotype, IdeS cleaves the hinge region of the heavy chain between two glycines or between glycine and alanine residues (Fig. 2). Starting from a 150 kDa glycoprotein, after IdeS digestion and reduction of the disulfide bridges with DTT, three fragments of around 25 kDa are obtained (see Fig. 3).
3. IgGZERO™ enzyme cleaves the chitobiose core on the glycans (Fig. 4) in only 30 min, a short time of digestion compared to other glycosidase enzymes widely used, as PNGase. After IgGZERO™ treatment, an *N*-acetylglucosamine and a fucose remain linked to Asn 297 or Asn 88. A mass of 349.14 Da (GlcNAc-Fuc) has to be added to theoretical masses calculated for deglycosylated fragments.
4. DTT reduction is widely performed in the presence of guanidine hydrochloride 6 M or urea 4 M as chaotropic agents that help to solubilize hydrophobic proteins during reduction. Unfortunately, in most of the cases, guanidine forms adducts with proteins leading to the observation of additional populations of proteins with 59 Da guanidine adducts, after MS data interpretation. In solution, urea and ammonium cyanate are in equilibrium. As a result of the reaction of proteins with cyanate, carbamylation at the NH2-terminus of lysine and cysteine residues can occur with the

use of urea as denaturing agent, leading to 43 Da adducts (30). DTT reduction in the presence of Tris–HCl 3 M buffer pH 8 (27) allows a reduction of disulfide bonds without the presence of any additional populations of proteins containing adducts. However, despite a time of reduction of 45 min at 56°C, depending on the mAb studied, the reduction of disulfide bonds can be incomplete with the use of Tris–HCl buffer, and an error of mass of several daltons between theoretical and experimental masses can be obtained.

5. Under native conditions, PNGase F only releases glycans from the Fc domains on asparagine 297. The exact numbering is linked to the lengths of the complementary domain region (CDR) and may vary. It is thought that higher levels of galactosylation and sialylation on Fd fragment N-linked oligosaccharides could increase accessibility to enzyme of deglycosylation but Jefferis and colleagues (8) in 2005 showed different results. Incomplete deglycosylation of cetuximab Fd fragment is obtained but is sufficient for the study since it allows the calculation of experimental masses of the oligosaccharides present on Fd glycosylated fragment. To obtain a complete deglycosylation, the enzyme treatment has to be performed with a detergent to release glycans on both Fc/2 and Fd domains. Hexadecyltrimethylammonium bromide (CTAB) (11), or Nonidet P-40, can be used. However, detergents have to be removed from samples before mass spectrometry analysis.
6. The three IgG fragments Fc/2, LC, and Fd are all IgG subdomains. Therefore, we make the hypothesis that an equivalent ionization for the three compounds is obtained for semiquantitative purposes. As a consequence, the intensities of the peaks present on the deconvoluted mass spectrum can be used to calculate the percentage of each population representing the peaks. To confirm the mass spectrometry peak intensity of each fragment with an N-linked oligosaccharide, the techniques of capillary electrophoresis (31) or normal-phase liquid chromatography can be used.

References

1. Janin-Bussat MC, Wagner-Rousset E, Klinguer-Hamour C, Corvaïa N, Van Dorsselaer A, Beck A (2010) Antibody glycans characterization. In: Kontermann RE, Dübel S (eds) Antibody engineering. Springer, Berlin, pp 635–656
2. Yan B, Valliere-Douglass J, Brady L, Steen S, Han M, Pace D et al (2007) Analysis of post-translational modifications in recombinant monoclonal antibody IgG1 by reversed-phase liquid chromatography/mass spectrometry. J Chromatogr A 1164:153–161
3. Yu L, Vizel A, Huff MB, Young M, Remmele RL Jr, He B (2006) Investigation of N-terminal glutamate cyclization of recombinant monoclonal antibody in formulation development. J Pharm Biomed Anal 42:455–463
4. Chelius D, Jing K, Lueras A, Rehder DS, Dillon TM, Vizel A et al (2006) Formation of pyroglutamic acid from N-terminal glutamic acid in

immunoglobulin gamma antibodies. Anal Chem 78:2370–2376

5. Sundaram S, Matathia A, Qian J, Zhang J, Hsieh MC, Liu T et al (2011) An innovative approach for the characterization of the isoforms of a monoclonal antibody product. MAbs 3:505–5012
6. Beck A, Wagner-Rousset E, Bussat MC, Lokteff M, Klinguer-Hamour C, Haeuw JF et al (2008) Trends in glycosylation, glycoanalysis and glycoengineering of therapeutic antibodies and Fc-fusion proteins. Curr Pharm Biotechnol 9:482–501
7. Janin-Bussat MC, Strub JM, Wagner-Rousset E, Colas O, Klinguer-Hamour C, Corvaïa N et al (2010) Structural characterization of antibodies by mass spectrometry. In: Kontermann RE, Dübel S (eds) Antibody engineering. Springer, Berlin, pp 613–634
8. Jefferis R (2005) Glycosylation of recombinant antibody therapeutics. Biotechnol Prog 21: 11–16
9. Huang LH, Biolsi S, Bales KR, Kuchibhotla U (2006) Impact of variable domain glycosylation on antibody clearance: an LC/MS characterization. Anal Biochem 349:197–207
10. Beck A, Cochet O, Wurch T (2010) GlycoFi's technology to control the glycosylation of recombinant therapeutic proteins. Expert Opin Drug Discov 5:95–111
11. Lammerts van Bueren JJ, Rispens T, Verploegen S, Palen-Merkus T, Stapel S, Workman LJ et al (2011) Anti-galactose-alpha-1,3-galactose IgE from allergic patients does not bind alpha-galactosylated glycans on intact therapeutic antibody Fc domains. Nat Biotechnol 29:574–576
12. Qian J, Liu T, Yang L, Daus A, Crowley R, Zhou Q (2007) Structural characterization of N-linked oligosaccharides on monoclonal antibody cetuximab by the combination of orthogonal matrix-assisted laser desorption/ionization hybrid quadrupole-quadrupole time-of-flight tandem mass spectrometry and sequential enzymatic digestion. Anal Biochem 364:8–18
13. Chung CH, Mirakhur B, Chan E, Le QT, Berlin J, Morse M et al (2008) Cetuximab-induced anaphylaxis and IgE specific for galactose-alpha-1,3-galactose. N Engl J Med 358: 1109–1117
14. Dillon TM, Bondarenko PV, Speed RM (2004) Development of an analytical reversed-phase high-performance liquid chromatography-electrospray ionization mass spectrometry method for characterization of recombinant antibodies. J Chromatogr A 1053:299–305
15. Dillon TM, Bondarenko PV, Rehder DS, Pipes GD, Kleemann GR, Ricci MS (2006) Optimization of a reversed-phase high-performance liquid chromatography/mass spectrometry method for characterizing recombinant antibody heterogeneity and stability. J Chromatogr A 1120:112–120
16. Gadgil HS, Pipes GD, Dillon TM, Treuheit MJ, Bondarenko PV (2006) Improving mass accuracy of high performance liquid chromatography/electrospray ionization time-of-flight mass spectrometry of intact antibodies. J Am Soc Mass Spectrom 17:867–872
17. Zhang Z, Pan H, Chen X (2009) Mass spectrometry for structural characterization of therapeutic antibodies. Mass Spectrom Rev 28: 147–176
18. Akesson P, Moritz L, Truedsson M, Christensson B, Pawel-Rammingen U (2006) IdeS, a highly specific immunoglobulin G (IgG)-cleaving enzyme from Streptococcus pyogenes, is inhibited by specific IgG antibodies generated during infection. Infect Immun 74:497–503
19. Chevreux G, Tilly N, Bihoreau N (2011) Fast analysis of recombinant monoclonal antibodies using IdeS proteolytic digestion and electrospray mass spectrometry. Anal Biochem 415:212–214
20. Vincents B, Pawel-Rammingen U, Bjorck L, Abrahamson M (2004) Enzymatic characterization of the streptococcal endopeptidase, IdeS, reveals that it is a cysteine protease with strict specificity for IgG cleavage due to exosite binding. Biochemistry 43:15540–15549
21. Pawel-Rammingen U, Bjorck L (2003) IdeS and SpeB: immunoglobulin-degrading cysteine proteinases of streptococcus pyogenes. Curr Opin Microbiol 6:50–55
22. Pawel-Rammingen U, Johansson BP, Bjorck L (2002) IdeS, a novel streptococcal cysteine proteinase with unique specificity for immunoglobulin G. EMBO J 21:1607–1615
23. Wenig K, Chatwell L, Pawel-Rammingen U, Bjorck L, Huber R, Sondermann P (2004) Structure of the streptococcal endopeptidase IdeS, a cysteine proteinase with strict specificity for IgG. Proc Natl Acad Sci U S A 101: 17371–17376
24. Ren D, Pipes GD, Hambly DM, Bondarenko PV, Treuheit MJ, Brems DN et al (2007) Reversed-phase liquid chromatography of immunoglobulin G molecules and their fragments with the diphenyl column. J Chromatogr A 1175:63–68
25. Sinha S, Pipes G, Topp EM, Bondarenko PV, Treuheit MJ, Gadgil HS (2008) Comparison of LC and LC/MS methods for quantifying N-glycosylation in recombinant IgGs. J Am Soc Mass Spectrom 19:1643–1654

26. Pipes GD, Campbell P, Bondarenko PV, Kerwin BA, Treuheit MJ, Gadgil HS (2010) Middle-down fragmentation for the identification and quantitation of site-specific methionine oxidation in an IgG1 molecule. J Pharm Sci 99:4469–4476
27. Damen CW, Chen W, Chakraborty AB, van Oosterhout M, Mazzeo JR, Gebler JC et al (2009) Electrospray ionization quadrupole ion-mobility time-of-flight mass spectrometry as a tool to distinguish the lot-to-lot heterogeneity in N-glycosylation profile of the therapeutic monoclonal antibody trastuzumab. J Am Soc Mass Spectrom 20:2021–2033
28. Inouye K, Ohnaka S (2001) Pepsin digestion of a mouse monoclonal antibody of IgG1 class formed F(ab')(2) fragments in which the light chains as well as the heavy chains were truncated. J Biochem Biophys Methods 48:23–32
29. Lau H, Pace D, Yan B, McGrath T, Smallwood S, Patel K et al (2010) Investigation of degradation processes in IgG1 monoclonal antibodies by limited proteolysis coupled with weak cation-exchange HPLC. J Chromatogr B Analyt Technol Biomed Life Sci 878:868–876
30. Volkin DB, Mach H, Middaugh CR (1997) Degradative covalent reactions important to protein stability. Mol Biotechnol 8:105–122
31. Kamoda S, Ishikawa R, Kakehi K (2006) Capillary electrophoresis with laser-induced fluorescence detection for detailed studies on N-linked oligosaccharide profile of therapeutic recombinant monoclonal antibodies. J Chromatogr A 1133:332–339

Chapter 8

Therapeutic Antibody Glycosylation Analysis: A Contract Research Organization Perspective in the Frame of Batch Release or Comparability Support

Arnaud Delobel, Fabrice Cantais, Anicet Catrain, Erell Dereux, and Géry Van Vyncht

Abstract

Glycosylation of the Fc moiety of a monoclonal antibody is a heterogeneous posttranslational process considered as a critical quality attribute of the purified drug substance due to its major impact on safety and efficacy (i.e., immunogenicity, CDC or ADCC effector functions, etc.).

Glycosylation should thus be addressed for batch-to-batch comparability and for drug substance characterization, in terms of identity and/or purity testing.

We present below a set of efficient, performing and complementary analytical tests that can be used alone or in combination, depending on the information needed and available laboratory instrumentation.

The results obtained using these techniques for "global" glycosylation profile, *N*-glycans profiling, monosaccharides, and sialic acids determination are presented for the Trastuzumab (Herceptin)-humanized mAb produced in CHO.

Key words Monoclonal antibody, Glycosylation profile, *N*-Glycans, Monosaccharides, Sialic acids, ESI/MS, CE/LIF, UPLC/Fluo

1 Introduction

Robust and state-of-the-art analytical techniques should be used in the frame of batch release or comparability support of a monoclonal antibody manufacturing process. Although it is much less complex than for other glycoproteins, monoclonal antibody glycosylation is one of the critical quality attributes (1, 2) that should be addressed for the manufactured drug substance due to its major impact on safety and efficacy, i.e., immunogenicity and CDC or ADCC effector functions (galactosylation or fucosylation, respectively (3, 4)), anti-inflammatory activity (sialylation) (5, 6), and pharmacokinetics (high mannose) (7).

Even if numerous different methods are used for the quality control of a mAb, providing complementary results as regards drug

Alain Beck (ed.), *Glycosylation Engineering of Biopharmaceuticals: Methods and Protocols*, Methods in Molecular Biology, vol. 988, DOI 10.1007/978-1-62703-327-5_8, © Springer Science+Business Media New York 2013

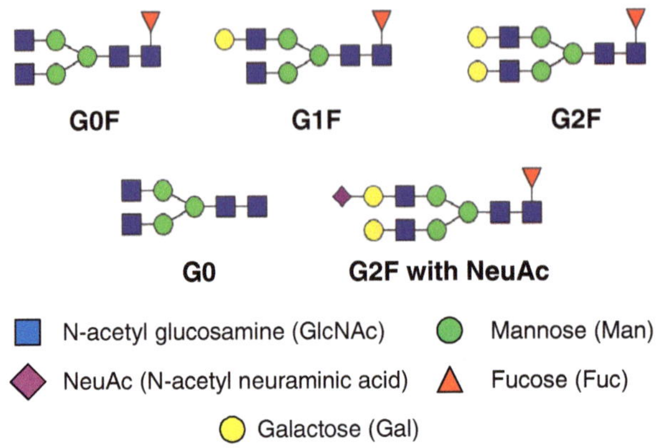

Fig. 1 Schematic representation of glycans structures commonly found on therapeutic monoclonal antibodies

substance purity (SEC-HPLC, IEX-HPLC, (CE-)IEF, SDS …) or potency (Bio-Assays, binding assays), N-glycosylation should also be addressed separately, using dedicated methods (2).

The control strategy established at Quality Assistance is based first on the study of the global glycosylation pattern of the reduced "intact" mAb by high-resolution ESI/MS. This test also enables the identification of the drug substance.

Depending on the information needed, this approach can be followed by the PNGase release of the *N*-glycans and profiling by CE/LIF or UPLC/Fluo (to be compared to the *N*-glycans profiling by MALDI/TOF after permethylation).

Finally, monosaccharides contents can be determined by UPLC/Fluo or CE/LIF, and sialic acids by UPLC/Fluo.

Examples of *N*-glycans commonly found on therapeutic monoclonal antibodies are presented in Fig. 1.

The methods described in this chapter are not an exhaustive list of the methods available for the characterization of monoclonal antibodies. They correspond to the package that our group recommends for the characterization and release testing of therapeutic monoclonal antibodies. These methods can be validated according to ICH guidelines.

1.1 Intact Mass Measurement of Intact Monoclonal Antibodies and Related Substructures by Liquid Chromatography Hyphenated to High-Resolution Mass Spectrometry

This test is very easy to implement and can give valuable information in a minimum amount of time. That is why this is the first test that we would recommend to characterize monoclonal antibody glycosylation or to carry out batch-to-batch comparisons (8).

Salts and additives (surfactants …) that are contained in most formulation buffers are not compatible with electrospray ionization. They are usually removed by dialysis or filtration on a centrifugal filter, but this is time-consuming and not compatible with the analysis of numerous samples. The most efficient way to desalt the sample is to use an online cartridge. The cartridge used for this test contains a reversed-phase material that traps the protein while salts are eluted to waste using a highly aqueous mobile phase. The protein is then eluted to the mass spectrometer with a gradient of acetonitrile.

The mass spectrum obtained can be deconvoluted to yield the molecular mass of the intact protein and a global glycosylation pattern.

This method can be applied to intact or reduced monoclonal antibodies, and is compatible with the presence of surfactants such as Polysorbate 20 or 80 in the samples.

1.2 N-Glycosylation Profiling of Monoclonal Antibodies

When additional information is required, *N*-glycans profiling should be the method of choice. Although this test can be carried out by MALDI-TOF mass spectrometry (on native *N*-glycans or after permethylation), this technique suffers from higher variability compared to chromatographic methods. That is why we recommend the use of UHPLC in HILIC mode with fluorimetric detection or capillary electrophoresis with laser-induced fluorescence detection (CE/LIF).

In UHPLC/FLUO, *N*-glycans released by PNGase F are labeled with 2-AB (2-aminobenzamide), separated by UHPLC and detected by fluorescence (sample preparation is adapted from (9)). In CE/LIF, *N*-glycans released by PNGase F are labeled with APTS, separated by CE and detected by fluorescence (10).

Semi-quantitation of the different glycans is performed in % area. As the response originates from the 2-AB or APTS label, similar response factor is expected for the different glycans, even for the sialylated ones.

The choice between the two techniques will be led mainly by the availability of the equipment in the lab. However, it is commonly admitted that UHPLC is more robust than CE.

1.3 Monosaccharides Content of Monoclonal Antibodies

For batch-to-batch comparison, or to get information about the content of one specific monosaccharide, monosaccharides composition of a monoclonal antibody can be a useful test. For example, the presence of fucosylated glycans can have a great impact on the activity of the monoclonal antibody.

This test can be done in UHPLC with fluorimetric detection (adaptation of method described in (11) to UHPLC) or in CE/LIF (12, 13). In both cases, the first step is an acid hydrolysis of the protein to yield monosaccharides. This step is critical since the hydrolysis conditions usually have to be optimized to get the optimum yield for each monosaccharide. In most cases, several hydrolyses have to be performed in parallel.

After hydrolysis, monosaccharides are labeled with APTS (1-amynopyrene-3,6,6-trisulfonate) for CE/LIF analysis or with

2-AB (2-aminobenzamide) for UHPLC/FLUO, separated and detected by fluorescence. Quantification is done by external calibration.

Sialic acids are dealt with using another method, as they are destroyed using classical hydrolysis conditions. After mild acid hydrolysis, sialic acids are derivatized by DMB, separated by UHPLC and detected by fluorimetry (adaptation of method described in (14) to UHPLC). Quantification is done by external calibration using *N*-acetylneuraminic acid as the standard.

2 Materials

Prepare all solutions using ultrapure water (prepared by purifying deionized water to attain a sensitivity of 18 MΩ cm at 25°C) and analytical grade reagents. Whenever possible, the use of mass spectrometry-grade reagents is recommended.

2.1 Intact Mass Measurement of Intact Monoclonal Antibodies and Related Substructures by Liquid Chromatography Hyphenated to High-Resolution Mass Spectrometry

2.1.1 Apparatus

1. UPLC system (Autosampler and Binary solvent manager) (Acquity system, Waters).
2. Q-TOF mass spectrometer equipped with an electrospray ion source (Q-TOF premier, Waters).
3. LC Column (MassPREP™ Micro Desalting Column, length 5 mm, diameter 2.1 mm) (Waters).
4. Salt divert valve (Column selector, Waters).

2.1.2 UHPLC Preparation

1. Mobile phase A: 0.1% Formic acid (v/v) in water.
2. Mobile phase B: 0.1% Formic acid (v/v) in acetonitrile.
3. Weak wash solvent: 0.1% Formic acid (v/v) in water.
4. Strong wash solvent: acetonitrile/water (50/50: v/v).

2.1.3 Sample Preparation and Sample Reduction

1. 50 mM ammonium bicarbonate.
2. 300 mM Dithiothreitol (Sigma) in 50 mM ammonium bicarbonate.

2.1.4 Mass Spectrometer (Q-TOF) Calibration

1. Cesium iodide 2 mg/mL in 50/50 water/isopropanol (v/v).

2.2 N-Glycosylation Profiling of Monoclonal Antibodies

2.2.1 N-Glycosylation Profiling of Monoclonal Antibodies by UPLC/Fluo

Apparatus

1. UHPLC system (Autosampler and Binary solvent manager) (Acquity system, Waters).
2. Fluorimetric FLR detector for Acquity (Waters).
3. LC Column (UPLC BEH Glycan, length: 150 mm, ID: 2.1 mm, particle size: 1.7 μm) (Waters).

UPLC Preparation

1. Mobile phase A: 100 mM ammonium formate pH 4.5. Weigh 6.31 g of ammonium formate in a 1 L volumetric flask and bring to volume with purified water. Adjust the pH to 4.5 ± 0.05 with formic acid.
2. Mobile phase B: acetonitrile.
3. Strong wash solvent: acetonitrile/water (10/90: v/v).
4. Weak wash solvent: acetonitrile/water (90/10: v/v).
5. Seal wash solvent: acetonitrile/water (50/50: v/v).

Sample Preparation

1. Denaturation buffer: Transfer 250 μL of 10% SDS (w/v) (BioRad), 50 μL of mercapto-2-ethanol, and 50 μL of 25% ammoniac into a 5 mL volumetric flask and bring up to volume with H_2O (see Note 1).
2. Acetate/borate/methanol buffer: Weigh 4 g of sodium acetate trihydrate and 2 g of boric acid in a 100 mL volumetric flask and bring up to volume with methanol.
3. Derivatization buffer: Weigh 300 mg of anthranilamide and 200 mg of sodium cyanoborohydride in a 10 mL inactinic volumetric flask and bring up to volume with acetate/borate/methanol buffer (see Note 2).
4. 5% Nonidet P40: Dilute 5 mL of Nonidet P40 with 95 mL of H_2O.
5. 95% Acetonitrile: acetonitrile/H_2O (95/5 v/v).
6. 20% Acetonitrile: acetonitrile/H_2O (20/80 v/v).
7. 50% Acetonitrile: acetonitrile/H_2O (50/50 v/v).

Standard

1. SA-MAb4 Glycan reference panel (QA Bio, reference: SA-MAB4). If required, other standards can be used (high mannose, sialylated glycans). 2-AB-labeled glycans standards can also be obtained from some suppliers.

2.2.2 N-Glycosylation Profiling of Monoclonal Antibodies by CE/LIF

This method uses a commercial kit from Beckman Coulter referenced as "Carbohydrate Labeling and Analysis Chemistry" kit. The kits contain:

- An N-CHO-coated capillary.
- Glucose ladder standard (G20).

- Quantitation/mobility marker (Maltose) (G22).
- Carbohydrate separation buffer (25 mM acetate buffer pH = 4.75 + 0.4% PEO).
- Labeling dye (APTS) (L6).
- APTS-M (monosaccharide grade).
- Labeling dye solvent (15% acetic acid) (L3).

Apparatus

1. CE system (PA800 plus, Beckman Coulter).
2. Laser Induced Fluorescence detector, $\lambda_{excitation} = 488$ nm, $\lambda_{emission} = 520$ nm (Beckman Coulter).
3. CE column (N-CHO coated capillary, length: 50.2 cm, diameter: 50 μm, effective length to detector: 40 cm) (Beckman Coulter).

Sample Preparation

1. Ethanol.
2. 0.5 M HCl.
3. 50 mM ammonium bicarbonate.
4. 0.2% Rapigest (v/v): Dissolve the content of one vial (1 mg; Waters, art. no 186001861) in 500 μL of 25 mM ammonium bicarbonate buffer.
5. 105 mM Dithiothreitol (Sigma) in water.
6. 95 mM Iodoacetamide (Sigma) in water.
7. 1 U/μL PNGase F solution: 100 units of recombinant *N*-Glycosidase F (Roche, art. no 11365185001) are dissolved in 100 μL of water.
8. Maltose quantitative control/mobility marker: Dissolve the vial G22 with 500 μL of water. This makes a solution containing 500 nmol of maltose or concentration of 1 nmol/μL.
9. Glucose ladder standard: Weigh and dissolve 5 mg of G20 in 80 μL of water in a 1.5 mL tube. Sonicate if necessary. Aliquot 2 μL portions of the glucose ladder standard solution and dry them in a speedvac under vacuum.
10. APTS labeling reagent: To a vial L6, add 300 μL of L3. Vortex until complete dissolution.
11. 1 M sodium cyanoborohydride/THF.

Standard

1. SA-MAb4 Glycan reference panel (QA Bio, reference: SA-MAB4).

Buffer Trays

1. Fill the vials with the appropriate volumes of each reagent and put the vials in the appropriate position.
   (a) 1,500 μL of water per H_2O vial (4 vials): Position A1 and A4 on the inlet and outlet trays.

(b) 1,500 μL of carbohydrate separation buffer per Gel-R vial (1 vial): Position C1 on the inlet and outlet trays.

(c) 1,300 μL of carbohydrate separation buffer per Gel-S vial (2 vial): Position B1 on the inlet tray.

(d) 800 μL of water per waste vial (1 vial): Position B1 on the outlet tray.

2.3 Monosaccharides Content of Monoclonal Antibodies

2.3.1 Monosaccharides Content of Monoclonal Antibodies by UHPLC/Fluo (See Note 1)

Apparatus

1. UHPLC system (Autosampler and Binary solvent manager) (Acquity system, Waters).
2. Fluorimetric FLR detector for Acquity (Waters).
3. LC Column (UPLC BEH C18, length: 150 mm, ID: 2.1 mm, particle size: 1.7 μm) (Waters) (see Note 3).

UHPLC Preparation

1. Glacial acetic acid.
2. Mobile phase A: 50 mM sodium acetate pH 4.3. Weigh 13.6 g of sodium acetate trihydrate in a 2 L volumetric flask and bring to volume with purified water. Adjust the pH to 4.3 with acetic acid glacial.
3. Mobile phase B: methanol.
4. Strong wash solvent: mobile phase A.
5. Weak wash solvent: mobile phase A.

Sample Preparation

1. 40% trifluoroacetic acid (TFA) (v/v).
2. 37% chlorhydric acid (v/v).
3. Dilution solvent (1% anhydrous sodium acetate): Dissolve 25.5 g of sodium acetate trihydrate with 1,500 mL of water.
4. Internal standard solution (ISS) (xylose concentration ≈15 μM): Into a 100 mL volumetric flask, accurately weigh 45 mg of xylose, dissolve, and bring to volume with 1.7% sodium acetate. In a 1 L volumetric flask, introduce 5 mL of the previous solution; bring to volume with 1.7% sodium acetate.
5. Monosaccharides stock solutions (SS): Into a 200 mL volumetric flask, accurately weigh 12 mg of fucose (≈365 μM), 25 mg of galactose (≈694 μM), 27 mg of glucose (≈749 μM), 32 mg of galactosamine·HCl (≈742 μM), 36 mg of mannose (≈999 μM), 58 mg of glucosamine·HCl (1,345 μM). Dissolve and bring to volume with ISS.
6. Standard solutions: In a 100 mL volumetric flask, introduce 2 mL of each of the monosaccharides stock solution and bring

to volume with ISS. Monosaccharide concentrations of standard solutions are fucose (≈7 μM), galactose (≈14 μM), glucose (≈15 μM), galactosamine·HCl (≈15 μM), mannose (≈20 μM), and glucosamine·HCl (≈27 μM).

7. Acetate/borate/methanol buffer: Weigh 4 g of sodium acetate trihydrate and 2 g of boric acid in a 100 mL volumetric flask and bring up to volume with methanol.
8. Derivatization buffer: Weigh 300 mg of anthranilic acid and 300 mg of sodium cyanoborohydride in a 10 mL inactinic volumetric flask and bring up to volume with acetate/borate/methanol buffer (see Note 4).

2.3.2 Monosaccharides Content of Monoclonal Antibodies by CE/LIF (See Note 1)

This method uses a part of the commercial kit from Beckman Coulter referenced as "Carbohydrate Labeling and Analysis Chemistry" kit.

Apparatus

1. CE system (PA800 plus, Beckman Coulter).
2. Laser Induced Fluorescence detector, $\lambda_{excitation} = 488$ nm, $\lambda_{emission} = 520$ nm (Beckman Coulter).
3. CE column (capillary, length: 31 cm, diameter: 45 μm, effective length to detector: 21 cm) (Beckman Coulter).

Sample Preparation

1. 4 N TFA.
2. Derivatization solution: APTS-M labeling reagent (1-amynopyrene-3,6,6-trisulfonate, "monosaccharides grade"): To a vial of 20 mg of APTS-M, add 200 μL of 0.9 M citric acid. Vortex until complete dissolution.
3. 1 M sodium cyanoborohydride/THF.
4. Acetylation solution: 25 mM sodium bicarbonate. Weigh 23.3 mg of sodium bicarbonate, and add 10 mL of water. Adjust the pH to 9.5 with 50% NaOH.
5. Acetic anhydride.

Standard Solutions

1. Monosaccharide stock solutions: Prepare 1 mg/mL stock solutions (SS) of the different monosaccharide standard solution in PBS 1× (glucose, fucose, mannose, galactose, *N*-acetyl galactosamine). See Note 5.
2. Monosaccharide standard solution: Mix 5 μL of *N*-acetylgalactosamine SS, 24 μL of mannose SS, 16 μL of glucose SS, 8 μL of fucose SS, and 20 μL of galactose SS. Add 63 μL of 4 N TFA. Vortex.

Buffer Trays

1. Separation buffer (240 mM boric acid): Weigh 1.48 g of boric acid and add 80 mL of water. Mix until complete dissolution and adjust the pH to 9 with 50% NaOH.
2. 0.1 N HCl.
3. Water.
4. 1 N NaOH.

2.3.3 *Sialic Acid Content*

Apparatus

1. UHPLC system (Autosampler and Binary solvent manager) (Acquity system, Waters).
2. Fluorimetric FLR detector for Acquity (Waters).
3. LC Column (UPLC BEH C18, length: 100 mm, ID: 2.1 mm, particle size: 1.7 μm) (Waters).

UHPLC Preparation

1. Mobile phase A: methanol/acetonitrile/water (7/9/84 v/v/v).
2. Mobile phase B: methanol/acetonitrile/water (7/25/68 v/v/v).
3. Strong wash solvent: methanol/acetonitrile/water (7/9/84 v/v/v).
4. Weak wash solvent: methanol/acetonitrile/water (7/25/68 v/v/v).

Sample Preparation

1. 1.4 M acetic acid: Dilute 800 μL of acetic acid in a 10 mL volumetric flask and bring to volume with purified water.
2. 0.75 M mercaptoethanol in 1.4 M acetic acid: Dilute 526 μL of mercaptoethanol in a 10 mL inactinic volumetric flask and bring to volume with 1.4 M acetic acid.
3. Derivatization buffer: Weigh 9.1 mg of sodium hydrosulfite and 1.59 mg of DMB (1.2-diamino-4.5-methylenoxybenzene) in an Eppendorf tube and transfer 1,000 μL of 0.75 M mercaptoethanol in 1.4 M acetic acid. Vortex until complete dissolution (see Note 6).
4. 2 M acetic acid: Dilute 1,143 μL of acetic acid in a 10 mL volumetric flask and bring up to volume with purified water.

Standard

1. *N*-Acetylneuraminic acid standard solution (5 μg/mL = 16.2 nmol/mL): Weigh 1 mg of *N*-acetylneuraminic acid in a 20 mL volumetric flask and bring to volume with purified water. Transfer 500 μL of this solution in a 5 mL volumetric flask and bring to volume with purified water.
2. Sialic acid reference panel (Ludger).

3 Methods

3.1 Intact Mass Measurement of Intact Monoclonal Antibodies and Related Substructures by Liquid Chromatography Hyphenated to High-Resolution Mass Spectrometry

3.1.1 Mass Spectrometer (Q-TOF) Calibration

The analyses are performed in positive electrospray ion mode. The TOF analyzer is calibrated before the analysis by infusion of a solution of 2 mg/mL cesium iodide in H_2O/isopropanol 50/50 (v/v), on *m/z* acquisition range: 600–5,000. External multi-point calibration is based on singly charged CsI ions.

3.1.2 Sample Preparation

1. Dilute the monoclonal antibody to 1 mg/mL with 50 mM ammonium bicarbonate.
2. Add 10 μL of 300 mM DTT to 140 μL of 1 mg/mL monoclonal antibody and incubate for 30 min at 40°C for sample reduction (see Note 7).

3.1.3 UHPLC Separation and Mass Spectrometry Analysis

1. Inject 20 μL of sample. Start by running in isocratic mode at 95% of mobile phase A and 5% of mobile phase B during 0.5 min, sent to waste to avoid contamination of the ion source with the salts. Then run a gradient of mobile phase A and mobile phase B, from 5 to 90% of mobile phase B at a flow rate of 0.2 mL/min during 1.5 min. To finish, apply twice a gradient from 5 to 90% of mobile phase B at a flow rate of 0.5 mL/min during 0.6 min and return to initial conditions (95% of mobile phase A) during 0.5 min (see Notes 8 and 9).
2. Hyphenation with mass spectrometer (positive electrospray ion mode): Record MS spectra from *m/z* 600 to 5,000. Source conditions should be optimized for optimum sensitivity.
3. Deconvolute the resulting mass spectra using MaxEnt1 algorithm (MassLynx Software, Waters) (see Note 10). The molecular masses of the antibody for native sample analysis and the light and heavy chains for reduced sample analysis are obtained. Moreover, a global glycosylation pattern can also be determined on the heavy chain, looking for mass differences between the peaks observed. Examples of MS spectra of intact and reduced Trastuzumab antibody and corresponding deconvoluted MS spectra are presented in Figs. 2, 3, 4, and 5. From the reduced antibody analysis, three biantennary N-linked oligosaccharides are identified on Trastuzumab monoclonal antibody: G0F, G1F, and G2F. From the intact monoclonal antibody, the glycosylation pattern is determined and is constituted by different associations of N-linked oligosaccharides (G0F/G0F, G1F/G1F, and/or G0F/G2F and G2F/G2F) (see Note 11).

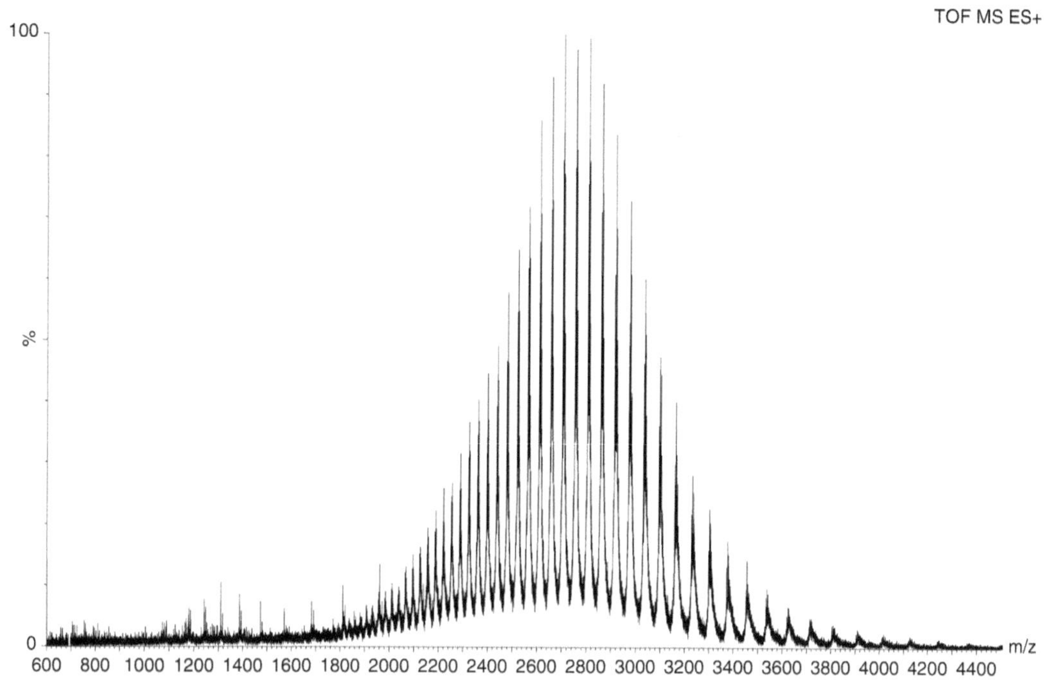

Fig. 2 MS spectrum of intact Trastuzumab monoclonal antibody

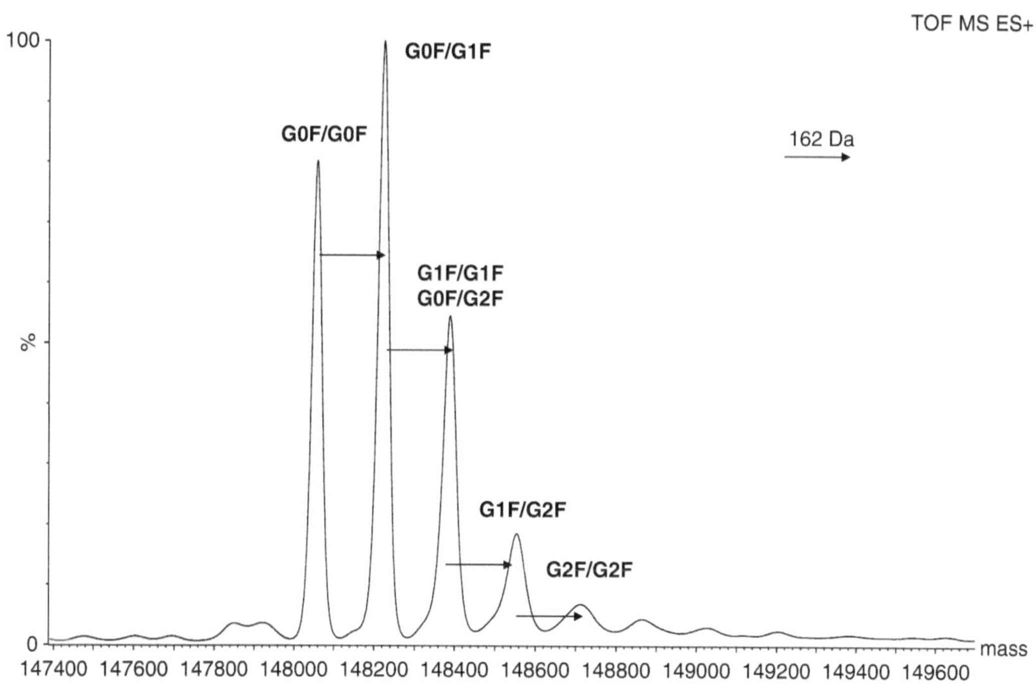

Fig. 3 Deconvoluted MS spectrum of intact Trastuzumab monoclonal antibody

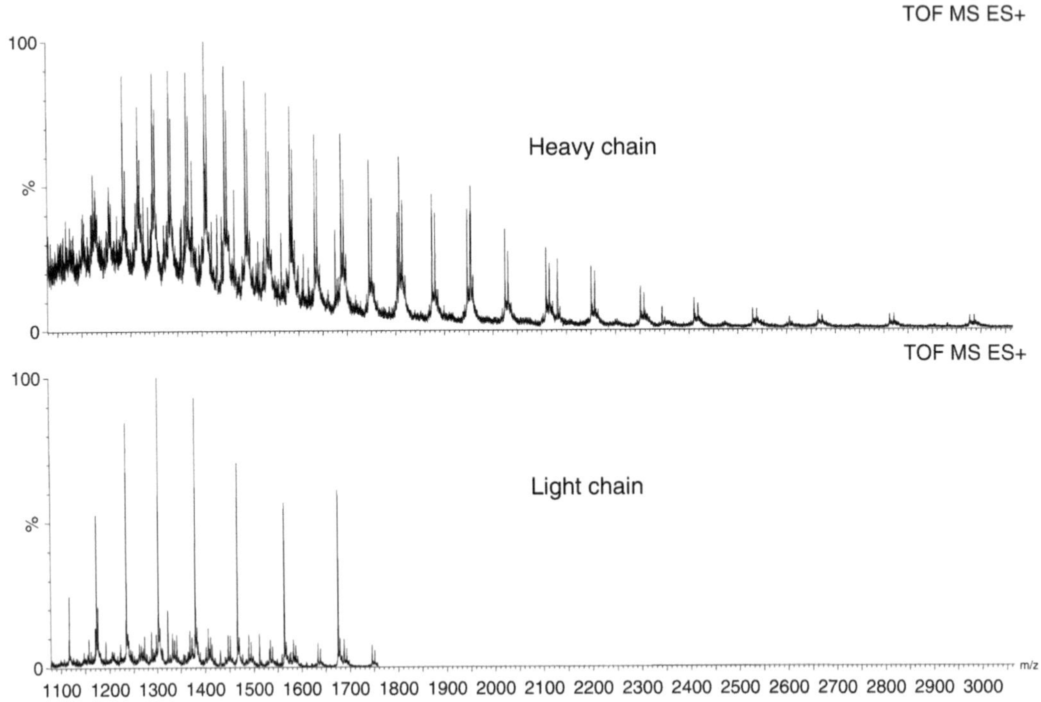

Fig. 4 MS spectrum of reduced Trastuzumab monoclonal antibody

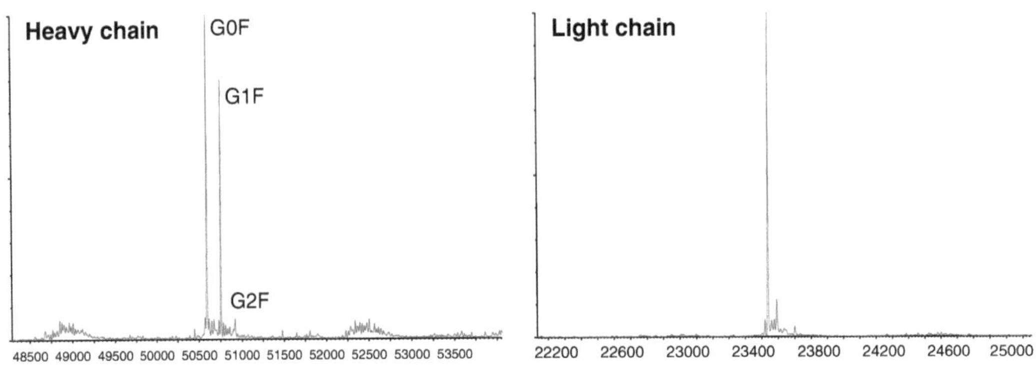

Fig. 5 Deconvoluted MS spectra of reduced Trastuzumab monoclonal antibody

3.2 N-Glycosylation Profiling of Monoclonal Antibodies

3.2.1 N-Glycosylation Profiling of Monoclonal Antibodies by UPLC/Fluo

Sample Preparation

1. Sample: Use 10 μL of a monoclonal antibody solution diluted at 5 mg/mL in formulation buffer and follow all the steps starting with the denaturation step.
2. Blank: 10 μL of formulation buffer are used as blank solution.
3. Standard solution: Use 10 μL of SA-MAb4 Glycan reference panel at a concentration of 0.1 mg/mL in water (mixture of glycans commonly found on monoclonal antibodies).

4. Denaturation step (blank and samples only): Add 30 μL of denaturation buffer, vortex, incubate for 3 min at 100°C, and then allow to cool down to room temperature. Centrifuge at 14,000 × *g* for 30 s in order to recover the entire sample and add 5 μL of 5% Nonidet P40. Vortex.
5. Digestion step (blank and samples only): Add 4 μL of PNGase F (1 U/μL) to each vial, vortex, and incubate overnight at 37°C.
6. Derivatization step (see Note 12): Add 2 μL of acetic acid and vortex. Add 100 μL of derivatization buffer, vortex, and incubate at 80°C for 1 h. Let the sample cool down to room temperature and centrifuge for 30 s at 14,000 × *g* in order to recover the entire sample. Dilute with 1,000 μL of 95% acetonitrile. Vortex (see Note 13).
7. Pre-purification on Oasis HLB 1 mL 30 mg SPE columns (Waters) by gravity flow: Prime the Oasis HLB 1 mL column with 2 × 1,000 μL of 95% acetonitrile, and then equilibrate with 3 × 1,000 μL of 95% acetonitrile. Add the diluted solution and allow to flow through the column. Wash with 2 × 1,000 μL of 95% acetonitrile and discard the prime, equilibration, and wash fractions. Elute oligosaccharides with 2 × 500 μL of 20% acetonitrile (see Note 14).
8. Concentration step: Dry the samples using a SpeedVac and add 50 μL of 50% acetonitrile (see Note 15).

UHPLC Separation

1. Inject 5 μL of sample on the UPLC column (see Note 16). The injected mode used is partial loop with needle overfill (3 μL). The detector parameters are the following: $\lambda_{excitation}$ = 330 nm, $\lambda_{emission}$ = 420 nm, and gain = 200 (see Note 17). The column temperature is set at 60°C. Samples are eluted by a linear gradient from 75 to 60% of mobile phase B during 46.5 min at a flow rate of 0.5 mL/min, from 46.5 to 48 min at 60% of mobile phase B at a flow rate of 0.5 mL/min, from 48 to 49 min at 0% of mobile phase B at a flow rate of 0.25 mL/min, and from 50 to 63 min at 75% of mobile phase B (initial condition) at a flow rate of 0.5 mL/min.
2. All the peaks detected are integrated on the LC profile. The identification of *N*-glycans is achieved by comparing the LC profile of the antibody glycans standard (see the identified *N*-glycans in Fig. 1) to the LC profile of glycosylated monoclonal antibodies (see Note 18). Calculate the relative retention time and the % peak area of each *N*-glycan (individual and mean values). The limit of quantification of the method is estimated at 0.5% of the standard solution (250 μg/mL). Examples of chromatogram of SA-Mab4 Glycan reference standard and *N*-oligosaccharides in Trastuzumab monoclonal antibody sample are presented in Figs. 6 and 7.

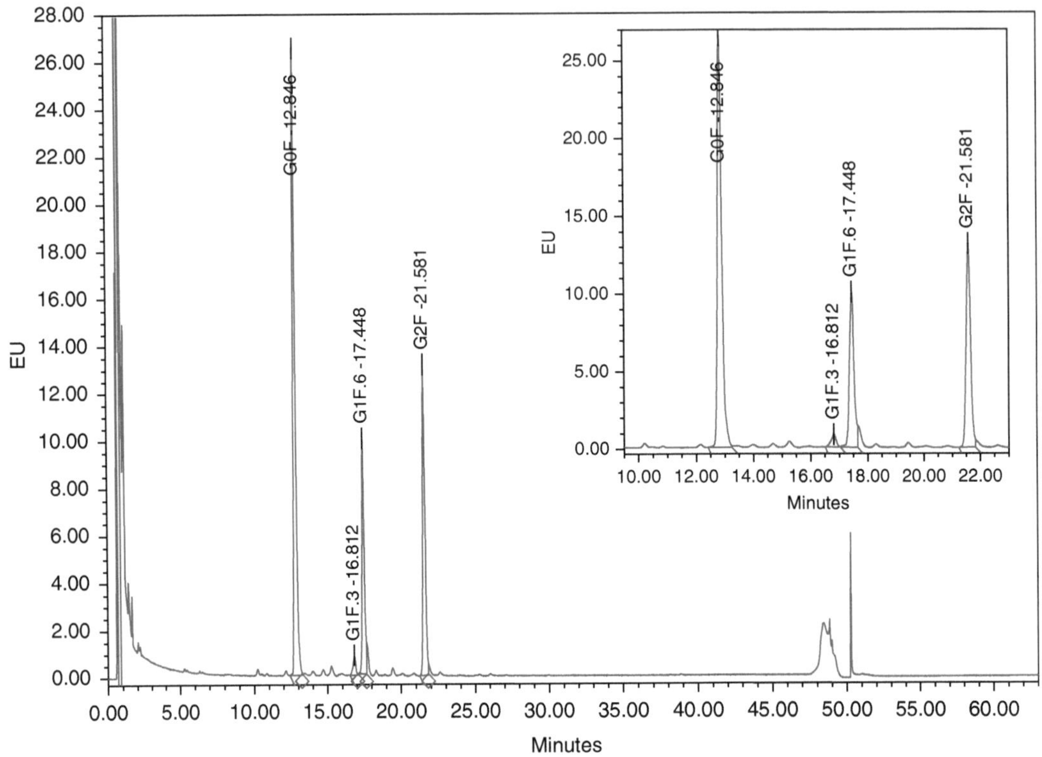

Fig. 6 Chromatogram of 2-AB-labeled *N*-glycans of the SA-MAb4 glycan reference panel

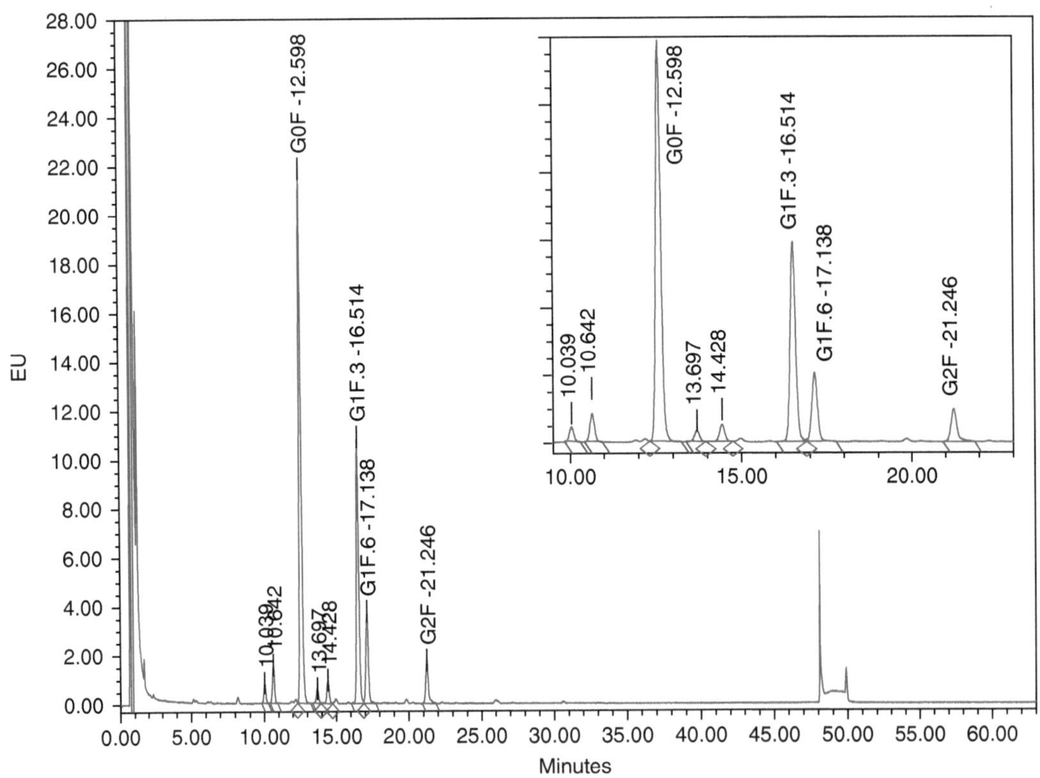

Fig. 7 Chromatogram of 2-AB-labeled *N*-glycans in Trastuzumab monoclonal antibody sample

3.2.2 N-Glycosylation Profiling of Monoclonal Antibodies by CE/LIF

Sample Preparation

1. Sample and reference solutions must be diluted in 50 mM ammonium bicarbonate buffer to reach a final concentration of ~10 mg/mL. For samples with a concentration below 10 mg/mL, the desalting protocol can be adapted accordingly.
2. Sample desalting: This step can be adapted to take into account the initial sample concentration. Sample desalting is achieved using "Amicon Ultra-4/Ultracel–10 k" filters (Millipore) according to the following protocol: Transfer 1,000 μL of sample solution diluted at 10 mg/mL onto the Amicon filter and centrifuge for 10 min at 4,000 rpm. Check that the remaining volume on the filter is lower than 200 μL (otherwise, centrifuge for 5 more minutes), add 2,000 μL of 50 mM ammonium bicarbonate buffer, and centrifuge for 15 min at 4,000 rpm. Check that the remaining volume on the filter is lower than 200 μL (otherwise, centrifuge for 5 more minutes), add 2,000 μL of 50 mM ammonium bicarbonate buffer, and centrifuge for 15 min at 4,000 rpm. Check that the remaining volume on the filter is lower than 200 μL (otherwise, centrifuge for 5 more minutes). Transfer the remaining solution into a 1 mL volumetric flask and bring to volume with 50 mM ammonium bicarbonate buffer. In a 1.5 mL low bind tube, add 25 μL of a 10 mg/mL desalted protein and evaporate overnight under vacuum at room temperature.
3. Denaturation: Dissolve the dried sample in 50 μL of 0.2% Rapigest. Incubate at 37°C for 10 min. Add 2.5 μL of 105 mM DTT and vortex (final concentration of DTT: 5 mM). Incubate at 60°C for 45 min. Let the sample cool to room temperature, add 10 μL of 95 mM IAA, and vortex (final concentration of IAA: 15 mM). Incubate at room temperature for 30 min, protected from light (see Note 19).
4. Digestion step: Add 4 μL of PNGase F (1 U/μL) to each vial, vortex, and incubate overnight at 37°C (see Note 20).
5. Glycan isolation step: Add 10 μL of 0.5 M HCl and incubate at 37°C for 30 min. Centrifuge the sample at 16,000×*g* for 4 min. Transfer the supernatant into a 1.5 mL low bind vial and add about 150 μL of cold ethanol (or three times the actual reaction mixture volume). Vortex the mixture and place it on ice for complete protein precipitation for 10 min. Centrifuge the samples at 16,000×*g* for 10 min. Withdraw and save the supernatant. Transfer the supernatant into a 0.5 mL low bind tube. Add 4 μL of maltose quantitative control/mobility marker. Dry down the sample in a SpeedVac (see Note 21).
6. Derivatization step (for sample and glucose ladder solution): Add 5 μL of 1 M sodium cyanoborohydride/THF to the dried oligosaccharide sample. Add 15 μL of APTS labeling reagent to the sample. Vortex the mixture for 1 min and centrifuge at

5,000 rpm for 10 s. Incubate the reaction mixture at 60°C for 90 min in the dark. Add 30 μL of water to the sample to stop the reaction, and 80 μL of water to the glucose ladder standard. Take an aliquot of 10 μL of the sample solution and add 190 μL of water (final solution 250 μg/mL). Take an aliquot of 5 μL of the glucose ladder standard solution and add 195 μL of water. The samples are ready for CE analysis (see Notes 22 and 23).

CE Separation

1. Inject during 8 s in hydrodynamic mode (0.5 psi). The temperature of the cartridge is set at 20 °C. During the conditioning, prerinse a new capillary for 10 min at 30 psi pressure with DDI water and then rinse for 10 min at 30 psi with carbohydrate separation buffer prior to the first run. Prior to separation step, rinse the system by pressure (30 psi) during 3 min (inlet and outlet position in B1). Separate during 20 min, in reverse polarity at 30 kV (inlet and outlet position in gel separation C1) (see Notes 24–29).
2. All the peaks detected are integrated on the LC profile. The identification of *N*-glycans is achieved by comparing the LC profile of the antibody glycan standard (see the identified *N*-glycans in Fig. 1) to the LC profile of glycosylated monoclonal antibodies. Calculate the relative retention time and the % peak area of each *N*-glycan (individual and mean values). The glucose standard solution gives an idea of the size of the oligosaccharides. The limit of quantification of the method is estimated at 0.2% of the standard solution (250 μg/mL). Examples of electropherograms of glucose standard solution, Mab4 glycan reference standard, and Trastuzumab monoclonal antibody are presented in Figs. 8, 9, and 10.

3.3 Monosaccharides Content of Monoclonal Antibodies

3.3.1 Monosaccharides Content of Monoclonal Antibodies by UPLC/Fluo (See Note 30)

Sample Preparation

1. Desalt the solutions using one of the two protocols (depending on the quantity of sample available). The formulation buffer is used as blank. The advantage of the second protocol (large amount of sample available) is the possibility to carry out several hydrolyses using only one sample desalting step.

   Using 50 μg of sample, desalting is achieved using "Amicon Ultra-0.5/Ultracel–30 k" filters (Millipore) according to the following protocol. Transfer 500 μL of sample solution diluted at 0.1 mg/mL in water onto the Amicon filter and centrifuge for 15 min at 14,000 × *g*. Add 400 μL of water onto the filter and centrifuge for 15 min at 14,000 × *g*. Repeat this step once. Remove the filter device from the centrifuge tube and place it upside down in a clean centrifuge tube. Centrifuge for 2 min at 1,000 × *g* to transfer the sample from the device to the tube. Add 30 μL of water to the tube. The tube is ready for hydrolysis.

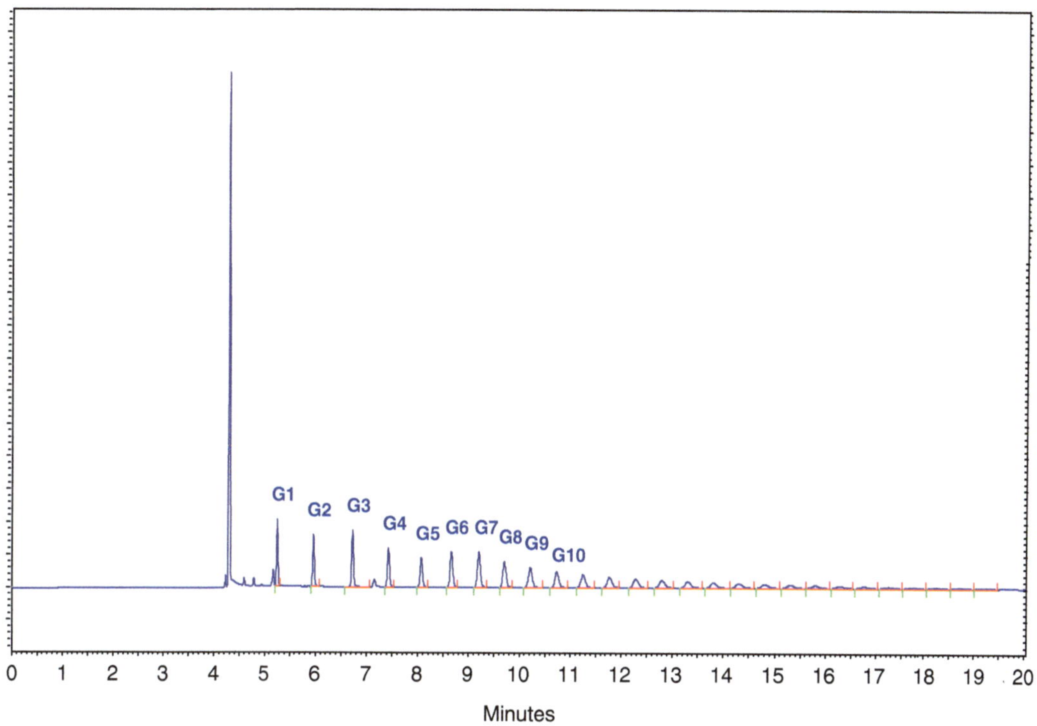

Fig. 8 Electropherogram of APTS-labeled glucose standard solution

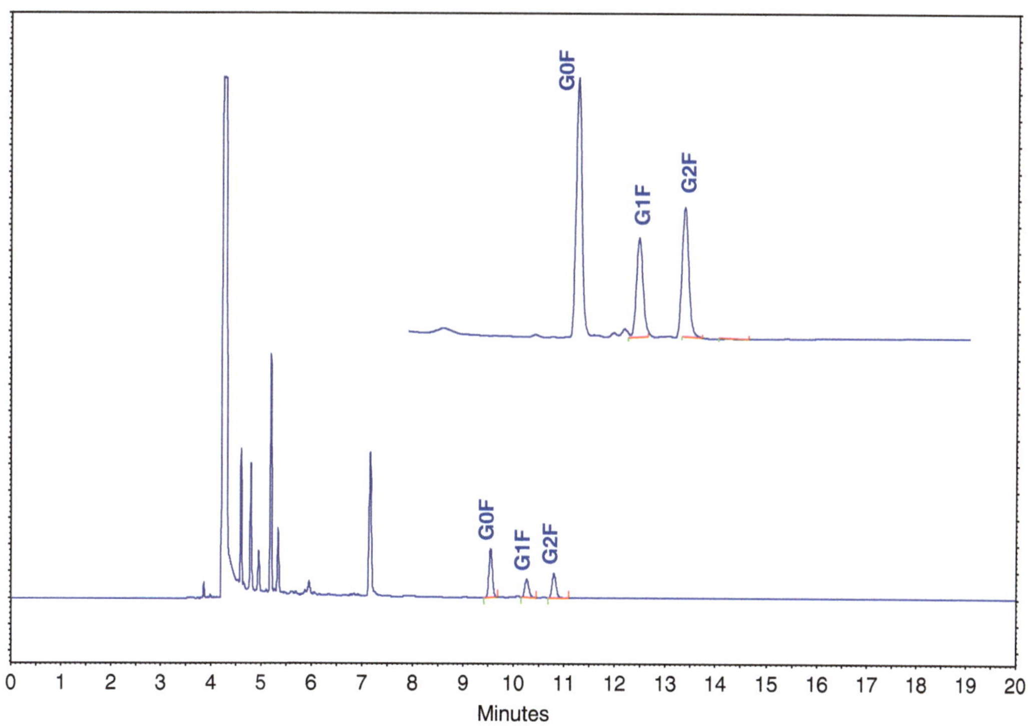

Fig. 9 Electropherogram of APTS-labeled *N*-glycans of SA-mAB4 reference panel

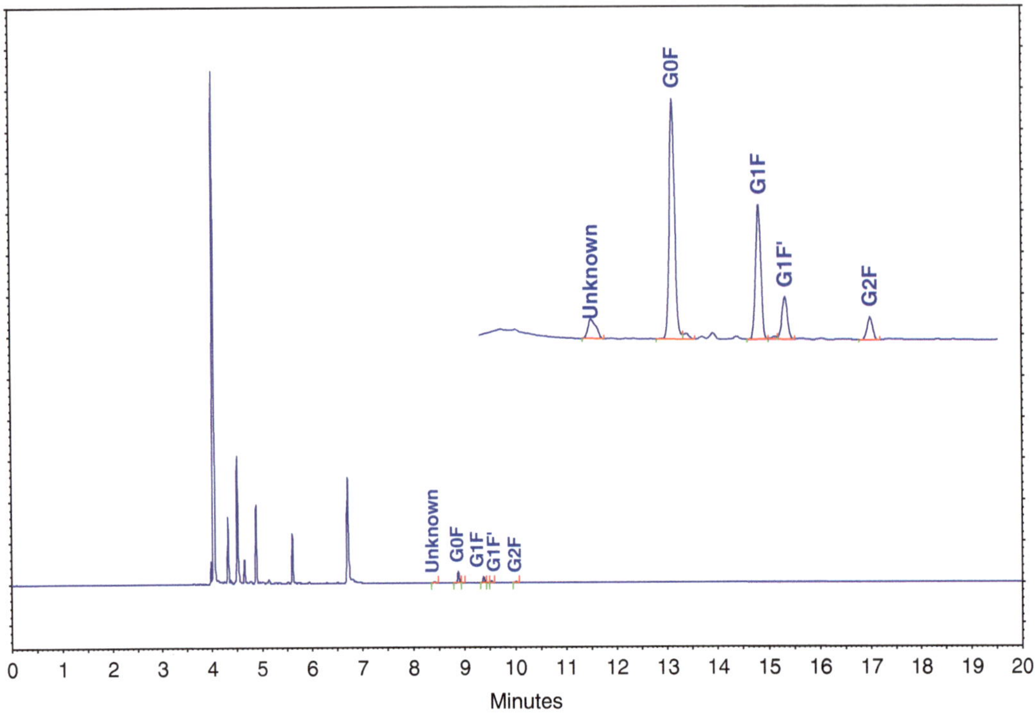

Fig. 10 Electropherogram of APTS-labeled *N*-glycans in Trastuzumab monoclonal antibody sample

Using 1 mg of sample, desalting is achieved using "Amicon Ultra-4/Ultracel–10 k" filters (Millipore) according to the following protocol. Transfer 1,000 μL of sample solution diluted at 1 mg/mL in water onto the Amicon filter and centrifuge for 10 min at 4,000 rpm. Check that the remaining volume on the filter is lower than 200 μL (otherwise, centrifuge for 5 more minutes), add 2,000 μL of water, and centrifuge for 15 min at 4,000 rpm. Repeat this step once. Transfer the remaining solution into a 1 mL volumetric flask and bring to volume with water. For each independent hydrolysis to carry out: transfer 50 μL of this solution into a 1.5 mL tube.

2. Hydrolyze the desalted solutions with the appropriate protocol. Prepare it in triplicate.
3. For the determination of fucose, add 50 μL of 40% TFA and incubate for 3 h at 80°C.
4. For the determination of glucosamine, galactose, and mannose, add 50 μL of 37% hydrochloric acid and incubate for 3 h at 80°C.
5. For the determination of glucosamine only, add 50 μL of 37% chlorhydric acid and incubate for 3 h at 100°C.

6. Evaporate to dryness with a SpeedVac. In each tube, add 100 μL of ISS and vortex until complete dissolution.
7. Derivatization: The solutions to label are blank, standard solutions and samples. In an HPLC vial, introduce 50 μL of solution. Add 100 μL of derivatization buffer. Vortex. Incubate at 80°C for 45 min. Add 850 μL of mobile phase A. Vortex.

UHPLC Separation

1. Inject 5 μL of sample on the UPLC column. The detector parameters are the following: $\lambda_{excitation} = 230$ nm, $\lambda_{emission} = 425$ nm, and gain = 1. The column temperature is set at 40°C. Samples are eluted by a linear gradient from 1.5 to 8% of mobile phase B during 25 min at a flow rate of 0.3 mL/min, then from 8 to 16% of mobile phase B during 6 min at a flow rate of 0.3 mL/min, from 16 to 80% of mobile phase B during 3 min at a flow rate of 0.3 mL/min, at 80% of mobile phase B during 3 min, from 80 to 1.5% during 1 min at a flow rate of 0.3 mL/min, at 1.5% of mobile phase B during 2 min at a flow rate of 0.3 mL/min, 20 min at a flow rate of 0.4 mL/min, and during 3 min at a flow rate of 0.3 mL/min. The total run time is 63 min.
2. It is possible to predict the number of residues of each monosaccharide in a human monoclonal antibody. Because of the composition of the main oligosaccharides (G0F, G1F, and G2F), the structure of the antibody (two heavy chains, two light chains linked by disulfide bonds), and the number of N-glycosylation sites (one site per heavy chain), the mean number of residues expected per mole of protein are two for fucose, eight for *N*-acetylglucosamine (see Note 31), six for mannose, and zero to four for galactose (see Note 32). Galactosamine and glucose are not expected on most monoclonal antibodies. The losses due to the sample preparation are compensated by the use of xylose as internal standard (the monosaccharide areas obtained are corrected by the area of xylose). Examples of chromatograms of monosaccharides standard solution and monosaccharides from Trastuzumab monoclonal antibody sample are presented in Figs. 11 and 12.

3.3.2 Monosaccharides Content of Monoclonal Antibodies by CE/LIF (See Note 33)

Sample Preparation

1. Desalt the sample solutions using "Amicon Ultra-0.5/Ultracel–10 k" filters (Millipore) according to the following protocol. The formulation buffer is used as blank.

   Transfer 200 μL of sample solution diluted at 0.5 mg/mL in PBS 1× onto the Amicon filter and centrifuge for 15 min at 14,000 × *g*. Add 200 μL of PBS 1× onto the filter and centrifuge for 15 min at 14,000 × *g*. Repeat this step once. Remove the filter device from the centrifuge tube and place it upside down in a clean centrifuge tube. Centrifuge for 2 min at 1,000 × *g* to transfer the sample from the device to the tube.

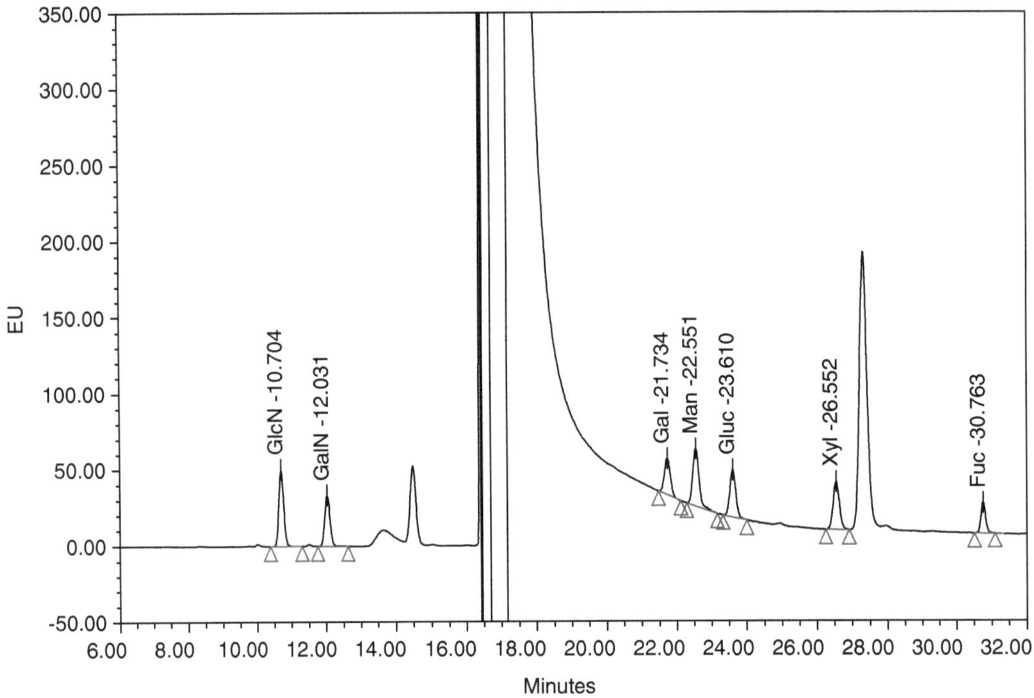

Fig. 11 Chromatogram of 2-AA-labeled monosaccharides standard solution

Divide the final solution into two equal solutions. The tube is ready for hydrolysis.

2. Standard solution hydrolysis: Incubate the standard solution at 100°C during 5 h. Let the sample cool down to room temperature. Evaporate 12 μL of the hydrolyzed solution to dryness with a SpeedVac (see Note 34).
3. Sample hydrolysis: Add 20 μL of 4 N TFA and incubate the standard solution at 100°C during 5 h. Let the sample cool down to room temperature. Evaporate 4 μL of the hydrolyzed solution to dryness with a SpeedVac (see Note 35).
4. Derivatization: To hydrolyzed samples or standards, add 2 μL of APTS-M solution and 1 μL of 1 M sodium cyanoborohydride/THF. Vortex, spin down, and incubate for 2 h at 55°C. Add 200 μL of water to stop the reaction. Dilute the solution tenfold with water.

CE Separation

1. Inject during 5 s in hydrodynamic mode (0.5 psi). The temperature of the cartridge is set at 20 °C. During the conditioning, prerinse a new capillary for 5 min at 30 psi pressure with HCl 0.1 N, for 5 min at 30 psi with water, for 10 min with NaOH 1 N, and for 5 min at 30 psi with water and rinse with separation buffer during 10 min prior to the first run.

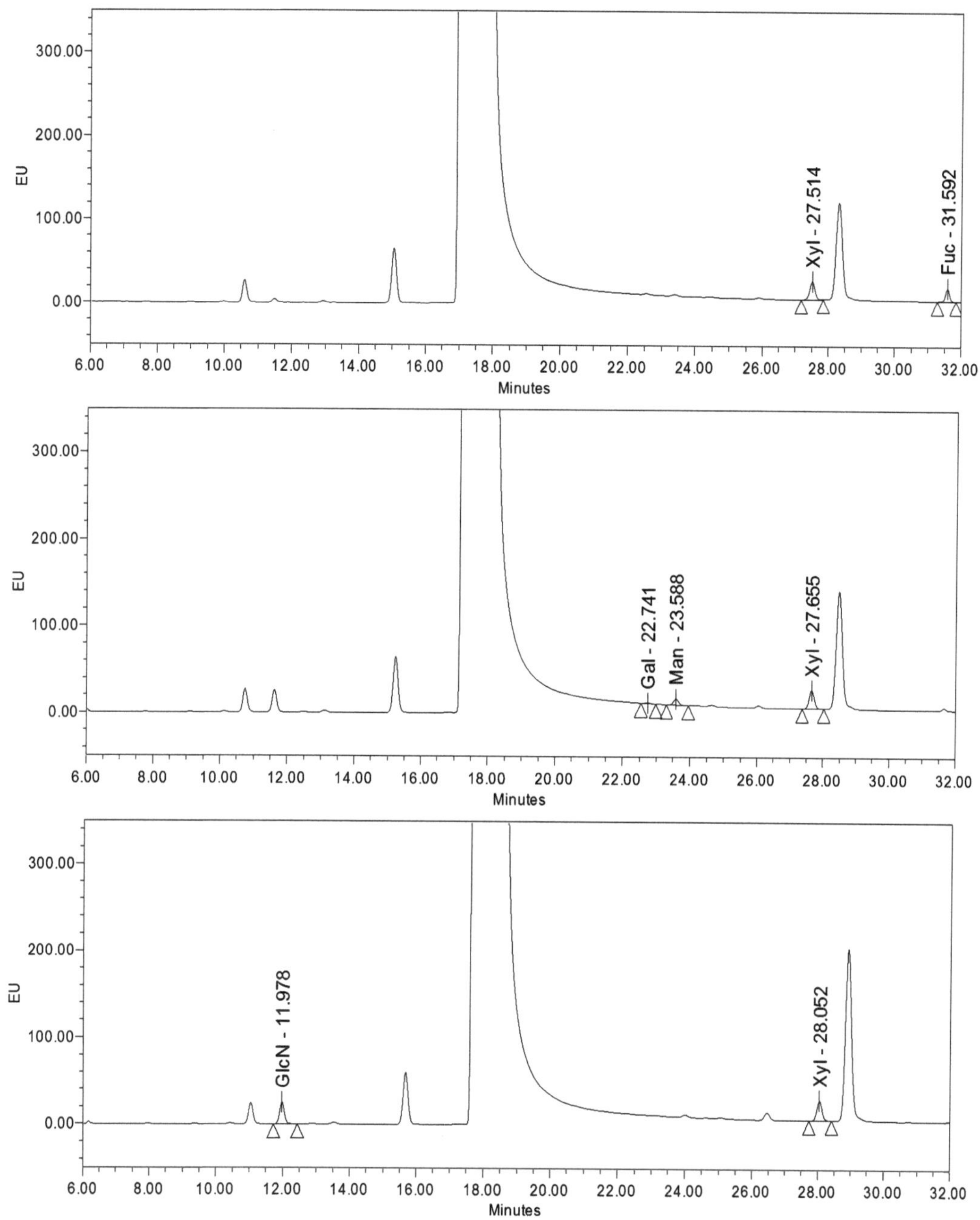

Fig. 12 Chromatogram of 2-AA-labeled monosaccharides in Trastuzumab monoclonal antibody sample (*upper trace*: fucose quantification; *middle trace*: mannose and galactosamine quantification; *lower trace*: glucosamine quantification)

Separate during 20 min, in normal polarity, at 20 kV (inlet and outlet position in separation buffer).

2. *N*-acetylglucosamine, mannose, fucose, galactose, and glucose are quantified by external calibration. As glucose is not expected

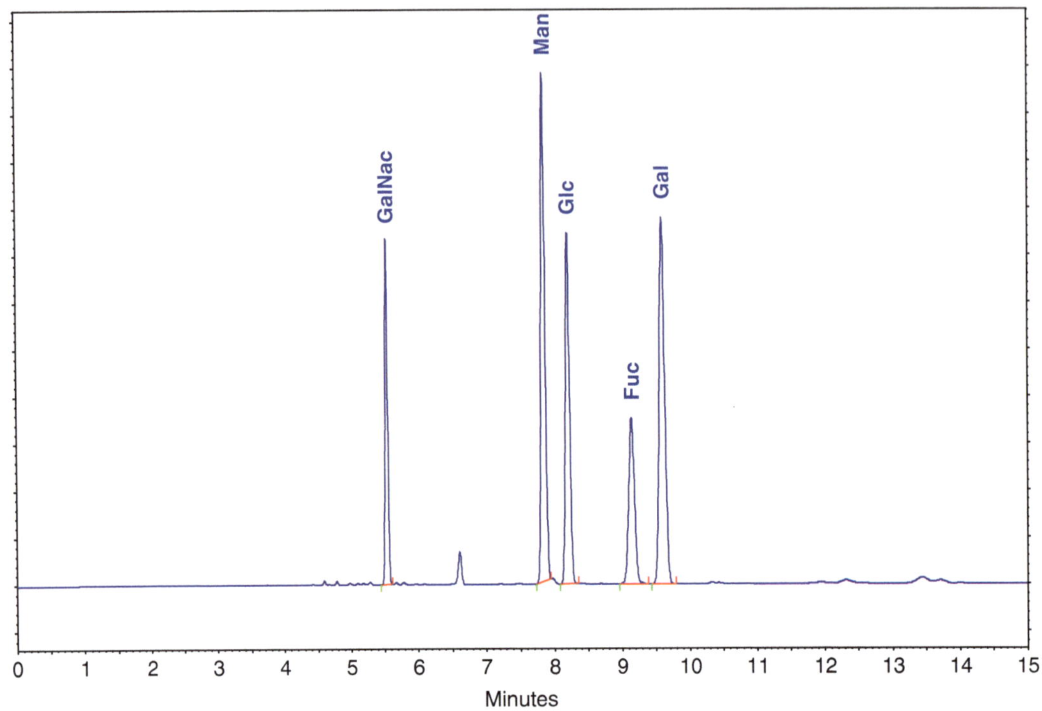

Fig. 13 Electropherogram of APTS-labeled monosaccharides standard solution

in most monoclonal antibodies, it can be used as internal standard to compensate for variability due to sample preparation. Examples of electropherograms of monosaccharides standard solution and monosaccharides in Trastuzumab monoclonal antibody sample are presented in Figs. 13 and 14.

3.3.3 Sialic Acids Content

Sample Preparation

1. Sample solution: Dilute the sample to 2 mg/mL with water. 5 μL of this solution are used.
2. Hydrolysis step: Dry the samples or the blank using a SpeedVac. Add 100 μL of 2 M acetic acid, vortex, and incubate for 2 h at 80°C. Then allow to cool down to room temperature, vortex, and transfer 20 μL of the hydrolyzed sample into an Eppendorf tube.
3. Derivatization step (see Notes 36 and 37): Add 80 μL of the derivatization solution to 20 μL of the standard, blank, or sample solution; vortex; and incubate for 3 h at 50°C. Let the sample cool down to room temperature. Dilute the sample with 1,135 μL of purified water and transfer into an HPLC vial.

UHPLC Separation (see Note 38)

1. Inject 20 μL of sample on the UPLC column (full loop injection). The detector parameters are the following: $\lambda_{excitation} = 373$ nm, $\lambda_{emission} = 448$ nm, and gain = 10. The column temperature is set

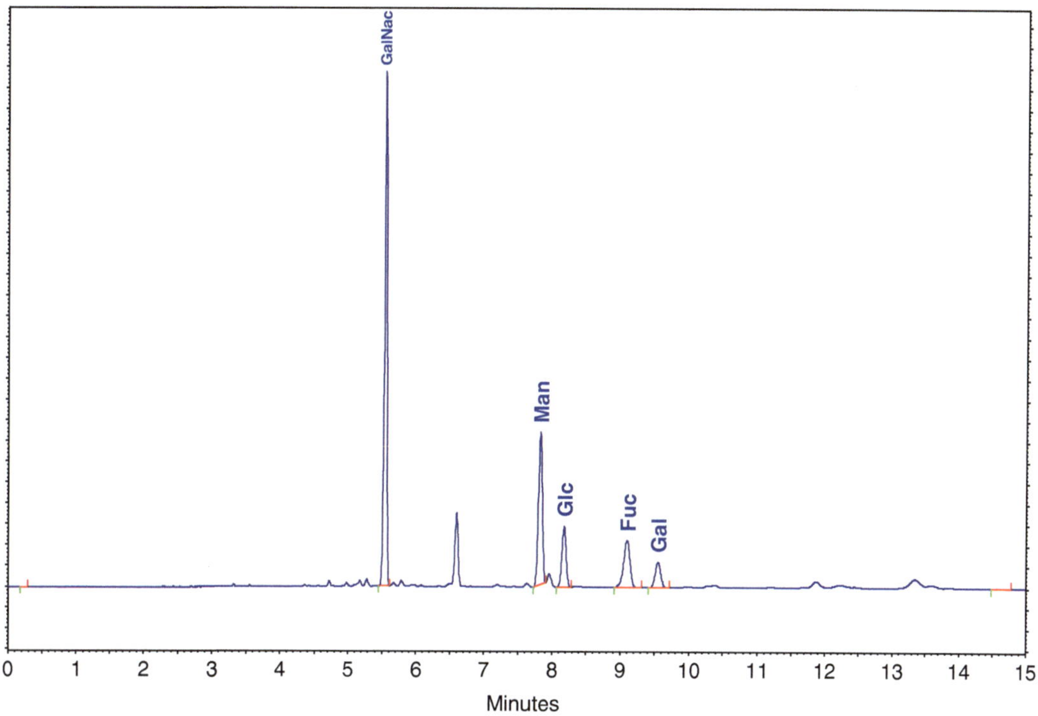

Fig. 14 Electropherogram of APTS-labeled monosaccharides in Trastuzumab monoclonal antibody sample

at 30°C. Samples are eluted at 0% of mobile phase B during 5 min, then by a linear gradient from 0 to 100% of mobile phase B during 1 min, at 100% of mobile phase B during 1 min, from 100 to 0% of mobile phase B during 1 min, and at 0% of mobile phase B during 1 min at a flow rate of 0.2 mL/min. The total run time is 9 min.

2. All the peaks detected are integrated on the LC profile. The identification of sialic acids is achieved by comparing the LC profile of the sialic acid reference panel (see the identified sialic acids in Fig. 13) (see Note 39). The *N*-acetylneuraminic acid content (mean result of two preparations) of each sample is calculated out by external calibration (Fig. 15). The limit of quantification of the method is estimated at 0.1% of the standard solution (5 μg/mL) (v/v). Quantification of other sialic acids can also be carried out by external calibration using *N*-acetylneuraminic acid standard. Examples of chromatograms of sialic acid reference panel, *N*-acetylneuraminic acid standard solution, and Trastuzumab monoclonal antibody sample are presented in Figs. 16 and 17. No sialic acid is detected in Trastuzumab monoclonal antibody sample (see Note 40).

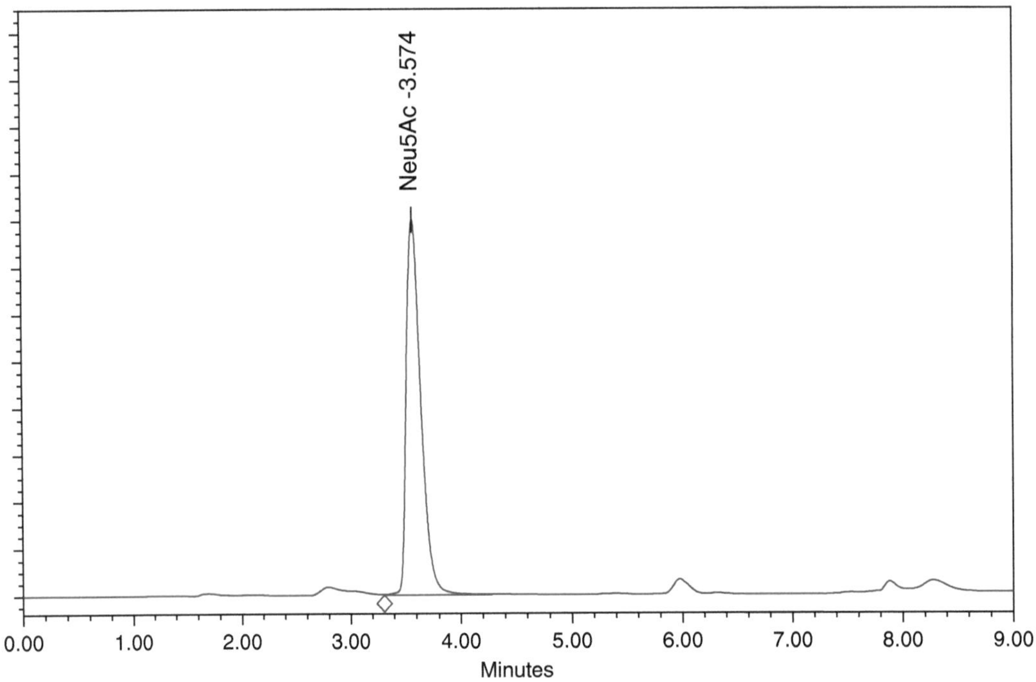

Fig. 15 Chromatogram of DMB-labeled *N*-acetylneuraminic acid standard

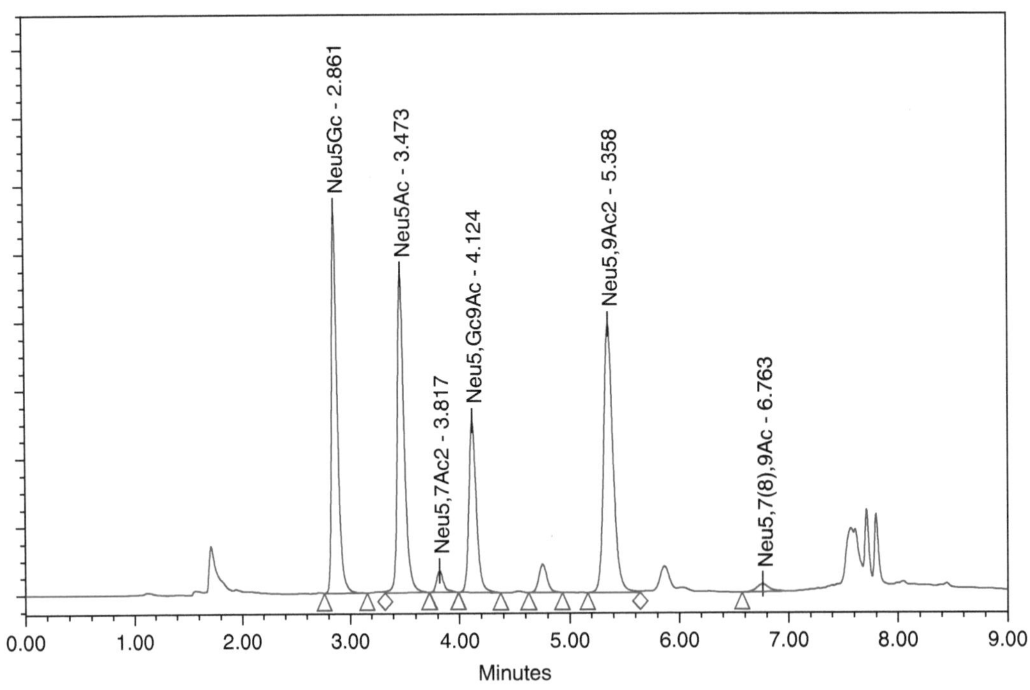

Fig. 16 Chromatogram of DMB-labeled sialic acids reference panel

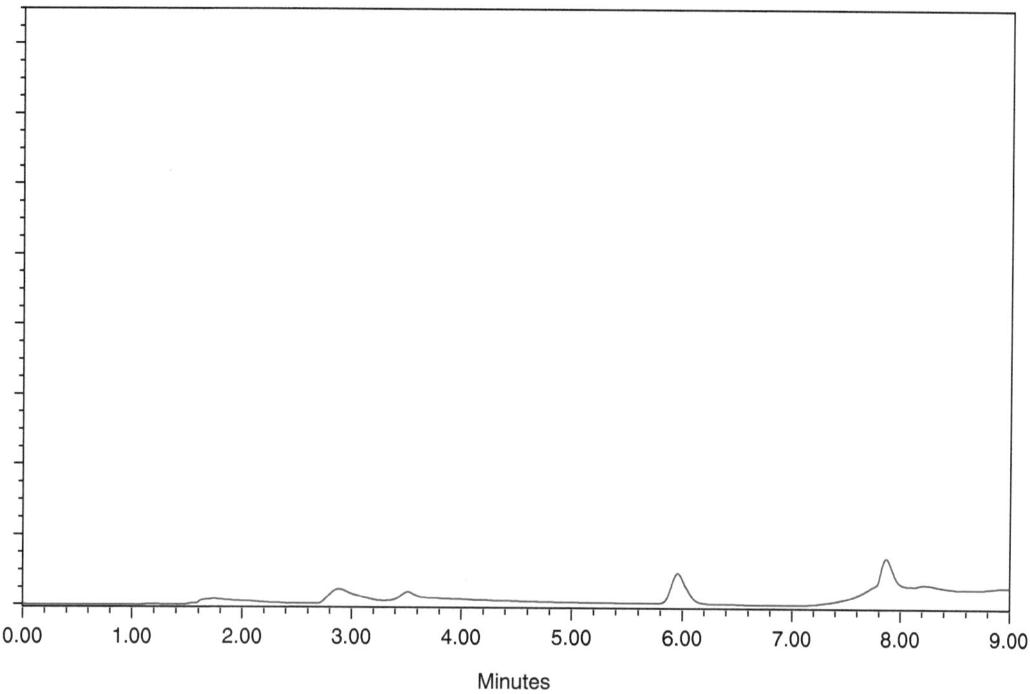

Fig. 17 Chromatogram of DMB-labeled sialic acids in Trastuzumab sample

4 Notes

1. The denaturation buffer is stable for 1 day when stored at room temperature. It has to be prepared on the day of the denaturation step.
2. The derivatization buffer is stable for 1 week when stored at room temperature and protected from light.
3. The column length is critical: With 50 mm columns, the peaks of the galactose, mannose, and glucose are not well separated from the labeling reagent peak. With 100 mm columns, the ribose (internal standard) has the same retention time as a peak system.
4. The derivatization buffer is stable for 1 day when stored at room temperature.
5. As *N*-acetylglucosamine is not commonly found in monoclonal antibodies, it was not included in the standard solution. If required, it can anyway be added in the solution as this monosaccharide is well separated from other monosaccharides.
6. Derivatization buffer is stable for 1 h. If the derivatization solution precipitates a few minutes after the complete dissolution of

the reductant, it is necessary to prepare it again. The consequence of the precipitation appears on the chromatographic profile: peaks of sialic acids have lower intensity than the ones obtained by using clear derivatization solution.

7. Some monoclonal antibodies may be unstable after reduction. In this case, the reduction step should be adapted to avoid sample precipitation (modification of reaction temperature, addition of a surfactant, modification of the reaction buffer ...).
8. The column temperature is set at 80°C to improve resolution between heavy and light chains.
9. The separation is done in two steps. The first step desalts the sample and the second one separates the sample and (for reduced samples) the heavy and the light chains. The run time of gradient step can be optimized in order to perform the separation between the light and heavy chains. To finish the run, a cleanup step is applied at a flow rate of 0.5 min with a gradient step (from 95 to 10% of mobile phase A) during 0.6 min, repeated twice, and then a return to initial condition (95% of mobile phase A) during 0.5 min.
10. MaxEnt1 typical deconvolution parameters:
    (a) m/z range corresponds to the five most intense charge states of the antibody.
    (b) The resolution is set up at 1 Da/channel.
    (c) The peak width at half height is determined on the major charge state.
    (d) The iteration *must* be completed until convergence for accurate molecular mass determination; however, some artifacts may appear: in that case, to get a correct view of the mass profile, it is recommended to stop the iteration as soon as an acceptable profile is obtained.
11. It is not possible to distinguish between the two isomers of G1F by mass spectrometry. Additional analyses are therefore necessary, such as *N*-glycans profiling.
12. The derivatization step must be done avoiding exposure to direct light.
13. At the end of this step, the theoretical concentration in *N*-glycans is 0.667 pmol/μL. Calculations are based on the molecular weight and on the number of glycosylation sites of the molecule of interest. For example: An antibody has a molecular weight of 150,000 g/mol: including one glycosylation on each heavy chain.
14. The final concentration in *N*-glycans is ≈13.3 pmol/μL.
15. Samples have to be stored at +4°C to be stable. When they are stored at +20°C and analyzed after a few days, peaks are split on the chromatographic profile.

16. A maximum of 8 μL has to be injected into the column; otherwise peak splitting is observed. Sensitivity can be improved by increasing the injection volume (up to 8 μL), by increasing the initial volume of the sample solution, or by decreasing the final volume (concentration step).
17. If the peak response is saturated, decrease the gain in the parameters of the fluorimetric detector.
18. This method is MS compatible. MS can therefore be used for peak identification. However the use of a very sensitive mass spectrometer is required, and adaptations to the method may be required to get enough sensitivity (for example, increase of the concentration of the sample injected).
19. Some proteins, including IgG, tend to precipitate during the denaturation step. This critical step can be optimized for each sample (use of Rapigest, SDS, β-mercaptoethanol, and denaturation at 37, 70, or 100°C).
20. If required, add Nonidet P40 (2%) to improve the PNGase digestion.
21. It is important to have a completely dried sample before labeling. Due to the reaction of sodium cyanoborohydride with water, avoid moisture and store this product under dry conditions. Use a dry needle to dispense this chemical. A mild labeling condition can be necessary to avoid losing the sialic acid residues from the oligosaccharides chain. Reaction mixture can be left at room temperature overnight and in the dark to do so. Tris buffer is not compatible with this protocol as it will undergo reductive amination and will compete with the carbohydrate for the APTS labeling.
22. The use of APTS as the fluorophore and laser-induced fluorescence (argon ion laser at 488 nm) by CE allows extremely high sensitivity and excellent resolution of components.
23. The efficiency of the labeling reaction is dependent on temperature and the amount of oligosaccharides. The protocol is optimized for labeling 5 nmol or less of total oligosaccharides. The stoichiometry of labeling reaction is one APTS molecule per molecule of oligosaccharide. The binding efficiency is 99%.
24. Prior to using the instrument, turn on the laser and let it stabilize for 15 min. Turn it off when not in use for long periods of time.
25. When the capillary is not in use or after finishing with all experiments, it is recommended to run a shutdown method to rinse the capillary and to leave the ends immersed in water to keep the capillary from drying out.
26. Prior to using the capillary for the first time, it is recommended to condition it (rinse it for 10 min at 30 psi with H_2O and then

rinse for 10 min at 30 psi with the carbohydrate separation gel buffer).

27. During the separation, salts migrate from one vial of gel buffer to the other, changing the composition of the buffer. Therefore, it is recommended to use fresh buffer after 20 runs.
28. Water is used to clean the capillary tips after the sample introduction step to prevent sample carryover. Replace the vials with fresh water if rinse water becomes contaminated.
29. The injection time can be adapted for a better sensitivity.
30. The hydrolysis step is critical in this method: Hydrolysis conditions shoud be strong enough to release completely amino sugars, but sufficiently mild not to degrade neutral sugars. Different conditions are provided depending on the sugar of interest. These different conditions are usually carried out in parallel. However, for optimal quantitative results on a given protein, kinetic experiments should be carried out to find out the best hydrolysis conditions.
31. The quantification of the monosaccharides is done, after fluorimetric detection, by external calibration and using an internal standard (xylose).
32. During the hydrolysis, *N*-acetylglucosamine and *N*-acetylgalactosamine are converted into glucosamine and galactosamine, respectively.
33. The hydrolysis step is critical in this method: Hydrolysis conditions should be strong enough to release completely amino sugars, but sufficiently mild not to degrade neutral sugars. A single hydrolysis condition is described, which is a good compromise for the efficient release of both neutral and amino sugars. However, for optimal quantitative results on a given protein, kinetic experiments should be carried out to find out the best hydrolysis conditions.
34. The dried hydrolyzed standard solutions are stocked at −20°C before use.
35. In order to improve the recovery of some monosaccharides, a re-acetylation step may be necessary: Add to hydrolyzed dried sample 5 μL of 25 mM sodium bicarbonate pH 9.5 and 2 μL of acetic anhydride and incubate for 30 min at room temperature. Evaporate to dryness with a SpeedVac.
36. During the derivatization step, avoid exposure to direct light.
37. At the end of the derivatization, a sample dilution can be done so as the final concentration is not higher than the concentration of *N*-acetyl-neuraminic standard solution.
38. Be careful with the composition of the mobile phase. The chromatographic profile depends strongly on the proportion

of each solvent. The mobile phases must be degassed by sonication, but not filtered or degassed under vacuum.

39. This method is MS compatible. MS hyphenation can therefore be used for peak identification.

40. Interfering peaks are usually present at the retention time of the NeuAc in the blank sample. If they are significant (higher than the peak in the LOQ solution), remove their area from the samples to get the exact amount of each sialic acid in the sample. An oscillation of the baseline is usually observed for the solutions with a low concentration in NeuAc (for example the LOQ solution).

References

1. CMC Biotech Working Group (2009) A-Mab case: a case study in bioprocess development. Version 2.1, pp 251–253.
2. Beck A, Wagner-Rousset E, Bussat MC, Lokteff M, Klinguer-Hamour C, Haeuw JF, Goetsch L, Wurch T, Van Dorsselaer A, Corvaïa N (2008) Trends in glycosylation, glycoanalysis and glycoengineering of therapeutic antibodies and Fc-fusion proteins. Curr Pharm Biotechnol 9: 482–501
3. Raju TS (2008) Terminal sugars of Fc glycans influence antibody effector functions of IgGs. Curr Opin Immunol 20:471–478
4. Gramer M, Eckblad J, Donahue R, Brown J, Shultz C, Vickerman K, Priem P, van den Bremer E, Gerritsen J, van Berkel P (2011) Modulation of antibody galactosylation through feeding of uridine, manganese chloride and galactose. Biotechnol Bioeng 108:1591–1602
5. Kaneko Y, Nimmerjahn F, Ravetch J (2006) Anti-inflammatory activity of immunoglobulin g resulting from Fc sialylation. Science 313: 670–673
6. Ghaderi D, Taylor R, Padler-Karavani V, Diaz S, Varki A (2010) Implications of the presence of *N*-glycolylneuraminic acid in recombinant therapeutic glycoproteins. Nat Biotechnol 8: 863–869
7. Goetze A, Liu Y, Zhang Z, Shah B, Lee E, Bondarenko P, Flynn G (2011) High mannose glycans on the Fc region of therapeutic IgG antibodies increase serum clearance in humans. Glycobiology 21:949–959
8. Xie H, Chakraborty A, Ahn J, Yu Y, Dakshinamoorthy D, Gilar M, Chen W, Skilton SJ, Mazzeo J (2010) Rapid comparison of a candidate biosimilar to an innovator monoclonal antibody with advanced liquid chromatography and mass spectrometry technologies. MAbs 2:379–394
9. Anumula KR, Dhume ST (1998) High resolution and high sensitivity methods for oligosaccharide mapping and characterization by normal phase high performance liquid chromatography following derivatization with highly fluorescent anthranilic acid. Glycobiology 8:685–694
10. Gennaro LA, Salas-Solano O (2008) On-Line CE-LIF-MS technology for the direct characterization of N-linked glycans from therapeutic antibodies. Anal Chem 80:3838–3845
11. Saddic GN, Dhume ST, Anumula KR (2008) Carbohydrate composition analysis of glycoproteins by HPLC using highly fluorescent anthranilic acid (AA) tag. Methods Mol Biol 446:215–229
12. Chen FA, Dobashi TS, Evangelista RA (1998) Quantitative analysis of sugar constituents of glycoproteins by capillary electrophoresis. Glycobiology 8:1045–1052
13. Račaitytė K, Kiessig S, Kálmán F (2005) Application of capillary zone electrophoresis and reversed-phase high-performance liquid chromatography in the biopharmaceutical industry for the quantitative analysis of the monosaccharides released from a highly glycosylated therapeutic protein. J Chromatogr A 1079:354–365
14. Hara S, Takemori Y, Yamaguchi M, Nakamura M, Ohkura Y (1987) Fluorometric high-performance liquid chromatography of N-acetyl and N-glycolylneuraminic acids and its application to their microdetermination in human and animal sera, glycoproteins, and glycolipids. Anal Biochem 164:138–145

Chapter 9

Mass Spectrometric Analysis of *O*-Linked Oligosaccharides from Various Recombinant Expression Systems

Diarmuid T. Kenny, Stefan Gaunitz, Catherine A. Hayes, Anki Gustafsson, Magnus Sjöblom, Jan Holgersson, and Niclas G. Karlsson

Abstract

Analysis of *O*-linked glycosylation is one of the main challenges during structural validation of recombinant glycoproteins. With methods available for *N*-linked glycosylation in regard to oligosaccharide analysis as well as glycopeptide mapping, there are still challenges for *O*-linked glycan analysis. Here, we present mass spectrometric methodology for *O*-linked oligosaccharides released by reductive β-elimination. Using LC-MS and LC-MS^2 with graphitized carbon columns, oligosaccharides are analyzed without derivatization. This approach provides a high-throughput method for screening during clonal selection, as well as product structure verification, without impairing sequencing ability. The protocols are exemplified by analysis of glycoproteins from mammalian cell cultures (CHO cells) as well as insect cells and yeast. The data shows that the method can be successfully applied to both neutral and acidic *O*-linked oligosaccharides, where sialic acid, hexuronic acid, and sulfate are common substituents. Further characterization of *O*-glycans can be achieved using permethylation. Permethylation of *O*-linked oligosaccharides followed by direct infusion into the mass spectrometer provide information about oligosaccharide composition, and subsequent MS^n experiments can be carried out to elucidate oligosaccharide structure including linkage information and sequence.

Key words Glycomics, *O*-Linked oligosaccharides, β-Elimination, Graphitized carbon LC-MS, Permethylation, Electrospray-MS

1 Introduction

The surface of cells is covered by carbohydrates, the structural complexity of which is vast. Glycans on cell surfaces can be protein or lipid bound. Cell surface glycans have, because of their structural complexity, been utilized by nature to act as binding partners, ligands, for an extensive number of carbohydrate-binding protein receptors. Such receptors, or lectins, can mediate many functions including, but not limited to, cell–cell/matrix interactions, antigen uptake, and intracellular sorting 1, 2). Also bacteria, viruses, and

Alain Beck (ed.), *Glycosylation Engineering of Biopharmaceuticals: Methods and Protocols*, Methods in Molecular Biology, vol. 988, DOI 10.1007/978-1-62703-327-5_9, © Springer Science+Business Media New York 2013

bacterial toxins attach to carbohydrate ligands as a first step during infection (3).

Mucins are important constituents of the protective mucus layer covering our mucosal surfaces as well as the glycocalyx layer of the cell surface (4, 5). They are characterized by a high degree of *O*-glycan substitution of their mucin domains, which are built up of short peptide repeats with abundant serines, threonines, and prolines (4, 5).

The total binding strength of a particular protein–carbohydrate interaction is dependent on several lectins binding multiple copies of a particular carbohydrate ligand in what is referred to as a multivalent interaction (6–8). Inhibitors of such interactions preferably should also carry several copies of the carbohydrate determinant. In an attempt to produce a new generation of inhibitors of protein–carbohydrate interactions, genetic engineering was used to establish mammalian, insect, and yeast cells secreting a mucin-type immunoglobulin fusion proteins with tailored *O*-glycan substitution (7). Host cells can be glyco-engineered to express multiple copies of a desired carbohydrate determinant—suitable for a particular target—on the mucin scaffold by co-expressing known glycosyltranferase sequences (7). Recombinant mucin-type fusion proteins carrying blood group A and B, sialyl-Le^x, Le^b, and other determinants on different *O*-glycan core structures have been engineered (9–14).

The structural complexity of glycans puts high demands on the analytical techniques used to resolve it. Mass spectrometry is a sensitive method by which information on saccharide composition and sequence can be obtained (15). Permethylation is often performed up-front the MS in order to increase ionization efficiency as well as to aid in interpretation of the fragmentation (16–18). However, this derivatization can destroy alkali-labile modifications and involves extra steps that extend the time of analysis as well as increase the likelihood for contamination. As an alternative, negative-ion MS and MS^n have been shown to complement the positive-ion mode MS of permethylated oligosaccharides (19, 20). If combined with a chromatographic method such as liquid chromatography using graphitized carbon (21), the structures of oligosaccharide isomers can be resolved. For *O*-linked oligosaccharides, the release using reductive β-elimination prevents reducing end derivatization. Hence, permethylation followed by MS and native LC-MS analysis are the suggested methods that are capable of giving reproducible results for glycomic studies (22). In this chapter we describe the mass spectrometric protocols used to analyze native as well as permethylated *O*-glycans released from recombinant mucin-type immunoglobulin fusion proteins expressed in mammalian, insect, and yeast cells. Tools used to interpret mass spectrometric data are presented and explained.

2 Materials

If not specified, all water is 18 MΩ/cm water from a MilliQ system (Millipore, Billerica, MA).

2.1 Affinity-Isolation of PSGL-1/mIgG$_{2b}$ Fusion Protein (IgG Column)

1. PSGL-1/mIgG$_{2b}$ in cell culture media.
2. Centrifuge, e.g., Beckman J6-MI with JS-4.2 rotor (Beckman Coulter, Brea, CA, USA).
3. 0.2 μm filter-top 500 mL TPP PES (TPP, Zurich, Switzerland) or Millipore PES (Millipore, Billerica, MA, USA).
4. Protease inhibitor cocktail (Sigma-Aldrich, St. Louis, MO, USA).
5. Sodium azide (71289, Sigma-Aldrich).
6. Goat anti-mouse IgG (whole molecule)—agarose, A6531 (Sigma-Aldrich).
7. Low pressure HPLC system, e.g., Bio-Rad BioLogic LP system (Hercules, CA, USA).
8. Washing buffer: Phosphate Buffer Saline (PBS, 0.01 M, 0.8% NaCl pH 7.3–7.4).
9. Elution buffer: 3 M sodium thiocyanate (251410, Sigma-Aldrich).
10. Corning 359096-well ELISA plates (Lowell, MA, USA).
11. Goat anti-mouse IgG Fc (M4280), goat anti-mouse IgG Fc-HRP (A2554), goat anti-mouse IgG F(ab')$_2$-HRP (A2304)—all from Sigma-Aldrich.
12. Mouse anti-PSGL-1 antibody CD162 (clone KPL-1) (BD PharMingen, San Diego, CA).
13. MWCO 12–14 kDa dialysis tubing (Spectra Por, Breda, The Netherlands).
14. Lyophilizer.

2.2 Affinity-Isolation of PSGL-1/mIgG$_{2b}$ Fusion Protein (Protein A Column)

1. WAVE single use bioreactors (GE Healthcare).
2. Cogent M Tangent Flow Filtration system (Millipore).
3. Microfiltration: Millistak+Pod deep filter 0.054 m^2 (Millipore).
4. Ultrafiltration: Pellicon 3 0.11 m^2 cassettes (Millipore).
5. Protease inhibitor Cocktail, P-8340 (Sigma-Aldrich).
6. Sodium azide.
7. HiTrap MabSelect SuRe columns (GE Healthcare).
8. Washing buffer: PBS (0.01 M, 0.8% NaCl pH 7.4–7.5).
9. Elution buffer: sodium citrate 0.1 M pH 3.
10. Corning 359096-well ELISA plates (Lowell, MA, USA).

11. Goat anti-mouse IgG Fc (M4280), goat anti-mouse IgG Fc-HRP (A2554), goat anti-mouse IgG F(ab')$_2$-HRP (A2304)—all from Sigma-Aldrich.
12. Mouse anti-PSGL-1 antibody CD162 (clone KPL-1) (BD PharMingen, San Diego, CA).
13. MWCO 12–14 kDa dialysis tubing (Spectra Por, Breda, The Netherlands).
14. Lyophilizer.

2.3 Gel Filtration of PSGL-1/mIgG$_{2b}$ Fusion Protein Purified with IgG Affinity Column

1. HiPrep Sephacryl 16/60 S-300 (GE Healthcare LifeSciences, Uppsala, Sweden).
2. High-pressure HPLC system, e.g., ÄKTA Explorer system (GE Healthcare LifeSciences).
3. Table top centrifuge.
4. 0.2 μm filter-top 500 mL TPP PES (TPP, Zurich, Switzerland).
5. Loading buffer: 3 M sodium thiocyanate (Sigma-Aldrich).
6. SYPRO Ruby stain and Pro-Q Emerald 300 glycoprotein gel stain (Invitrogen, Carlsbad, CA, USA).
7. Goat anti-mouse IgG Fc-HRP (A2554), goat anti-mouse IgG F(ab')$_2$-HRP (A2304)—all from Sigma-Aldrich.
8. Mouse anti-PSGL-1 antibody CD162 (clone KPL-1) (BD PharMingen, San Diego, CA).
9. MWCO 12–14 kDa dialysis tubing (Spectra Por, Breda, The Netherlands).
10. Lyophilizer.

2.4 Gel Filtration of PSGL-1/mIgG$_{2b}$ Fusion Protein Purified with Protein A Affinity Column

1. HiPrep 26/60 Sephacryl S-300 HR column (GE Healthcare LifeSciences, Uppsala, Sweden).
2. High-pressure HPLC system, e.g., ÄKTA Explorer system (GE Healthcare LifeSciences).
3. Tabletop centrifuge.
4. Filter-top (0.2 μm) 500 mL TPP PES (TPP, Zurich, Switzerland).
5. Sodium phosphate (0.1 M), pH 7.2, 0.5 M NaCl.
6. SYPRO Ruby stain and Pro-Q Emerald 300 glycoprotein gel stain (Invitrogen, Carlsbad, CA, USA).
7. Goat anti-mouse IgG Fc-HRP (A2554), goat anti-mouse IgG F(ab')$_2$-HRP (A2304)—all from Sigma-Aldrich.
8. Mouse anti-PSGL-1 antibody CD162 (clone KPL-1) (BD PharMingen, San Diego, CA).
9. MWCO 12–14 kDa dialysis tubing (Spectra Por, Breda, The Netherlands).
10. Lyophilizer.

2.5 Dot-Blotting of Glycoproteins

1. Glycoprotein sample dissolved in buffer.
2. Dot-blot apparatus (BioRad).
3. Vacuum pump or water suction device.
4. Blot papers (BioRad).
5. Immobilon-P transfer membrane (Millipore, Billerica, MA, USA).
6. Methanol.
7. Alcian blue stain solution: 25% ethanol, 10% acetic acid, 0.125% (w/v) alcian blue.
8. Alcian blue destaining solution: methanol.
9. Direct blue 71 (DB71) stain stock solution: 1% (w/v) DB71 (Bio-Rad) in MilliQ water.
10. DB71 destaining solution: 10% acetic acid, 40% ethanol in MilliQ water.
11. DB71 stain solution: 8 mL of DB71 stock solution in 100 mL of DB71 destaining solution, gel and blot shaker.

2.6 Release of Oligosaccharides

1. Glycoprotein samples (dissolved in 10 μL of water) or cut from glycoproteins blotted onto PVDF-membranes.
2. β-Elimination solution: 100 mM NaOH, 1 M $NaBH_4$.
3. Desalting solution: 1% glacial acetic acid in methanol.
4. AG 50W-X8 Cation Exchange Resin (Bio-Rad).
5. 1 M HCl.
6. P10 C18 Zip Tips (Millipore).

2.7 Packing of Graphitized Carbon Column

1. Fused silica capillary 250 μm i.d., 370 μm o.d. (Polymicro technologies, Phoenix, AZ).
2. 1/16 in. union (ZRU1.5, 0.25 mm bore, Vici AG, Schenkon, Switzerland).
3. Steel screen (0.5 μm or 1 μm pores Vici AG).
4. PEEK tubing 20–25 mm, 400 μm i.d. (Upchurch Scientific, Oak Harbor, WA).
5. Hypercarb (graphitized carbon particles) 5 μm Particle size (Thermo Scientific, Waltham, MA).
6. Vespel Ferrule (370 μm i.d, Vici AG).
7. Tetrahydrofuran (THF).
8. Methanol.
9. Ultrasonic bath.
10. Small magnetic bar (7 × 2 mm).
11. 2 mL borosilicate vial, 12 × 32 mm (Part no. VL-002, Western Fluid Engineering, Wildomer, CA, or equivalent).
12. Compressed air.

13. Binary HPLC pump (Hewlett Packar 100 series or equivalent).
14. T-split (Valco-T, Vici AG).
15. Acetonitrile (ACN).
16. MilliQ water.
17. Nanobaume (Western Fluid Engineering, Wildomar, CA).

2.8 Tuning of LC-MS

1. Maltopentaose solution (10 μg/mL) in 10 mM ammonium bicarbonate (40% ACN: 60% water).
2. Ion trap mass spectrometer with standard electrospray interface.
3. 250 μL syringe.
4. Syringe pump.
5. PEEK tubing 20–25 mm, 400 μm i.d. (Upchurch Scientific).

2.9 LC-MS of Released Oligosaccharides

1. ESI-LC-MS system: Thermo LCQ Deca-XP Ion Trap mass spectrometer, Agilent XCT or equivalent, equipped with a standard electrospray module.
2. Auto-sampler (HTC-PAL, CTC Analytics, Zwingen, Switzerland, or equivalent) with a 2 μL sample loop.
3. Graphitized carbon column: Packed in-house or alternatively a commercially available graphitized carbon column (e.g., Thermo Scientific 100 mm × 3 mm 5 μm Hypercarb column).
4. Mobile Phase A: 10 mM ammonium bicarbonate.
5. Mobile Phase B: 10 mM ammonium bicarbonate in 80% (v/v) ACN.

2.10 Interpretation of LC-MS Data of Released Oligosaccharides

1. Component identification: GlycoMod online tool (http://www.expasy.ch/tools/glycomod/).
2. Structure verification 1: GlycoWorkBench, downloadable java tool (http://www.ebi.ac.uk/eurocarb/gwb/home.action).
3. Structure verification 2: Database MS^2 matching and uploading of data, UniCarb-DB (http://unicarb-db.net/).

2.11 Permethylation of Oligosaccharides

1. Recommended minimal amount of oligosaccharides from 200 μg of glycoprotein.
2. Dimethylsulfoxide.
3. Iodomethane.
4. NaOH powder (see Note 1).
5. Magnetic stirrer.
6. Glass Pasteur pipet.
7. Water.
8. Pre-calibrated spoon.
9. Chloroform.

2.12 MS of Permethylated Oligosaccharides

1. LCQ ion-trap mass spectrometer (e.g., Thermo Finnigan, CA, USA).
2. Hamilton syringe.
3. Methanol, MS grade.
4. Water.
5. Acetic acid, MS grade.
6. Stachyos 1 mg/mL dissolved in methanol: water (S4001, Sigma-Aldrich).

2.13 Interpretation of Permethylated Oligosaccharides from MS

1. Component and structure verification 1: GlycoWorkBench, downloadable java tool (http://www.ebi.ac.uk/eurocarb/gwb/home.action).
2. Structure verification 2: ChemDraw Std 12.0 (CambridgeSoft, UK).

3 Methods

3.1 Isolation of PSGL-1/mIgG$_{2b}$ Fusion Protein (IgG Affinity Column)

1. Produce the PSGL-1/mIgG$_{2b}$ proteins in suspension cultures in Erlenmeyer flasks. The fusion protein should be excreted to the media for easy purification.
2. Remove cells from the suspension culture by centrifugation (5,020 × *g*) followed by filtration through a 0.2 μm filter.
3. Add protease inhibitor cocktail (1 mL/L medium and sodium azide (0.02% w/v)) to prevent degradation of the fusion protein during storage and downstream processing.
4. Purify the protein by an affinity goat anti-mouse IgG agarose column connected to a Bio-Rad BioLogic LP system running at low pressure.
5. Load the supernatant onto the column at a low flow rate (0.5–0.7 mL/min), wash with 12 column volumes of washing buffer, and elute protein with 24 column volumes of elution buffer. Both washing and elution are performed at a flow rate of 1 mL/min.
6. Collect fractions containing the fusion protein (determined with anti-IgG (Fc specific) Western blot) and pool and dialyze with MWCO 12–14 kDa dialysis tubing in a 5 L bottle.
7. The dialyze buffer should be changed from 5 mM PBS down to distilled water in a stepwise manner.
8. Collect the recovered protein after lyophilization.
9. Resuspend the sample in 1–2 mL of water.
10. The fusion protein concentration is determined in a sandwich ELISA (anti-IgG) (see Note 2).

3.2 Gel Filtration of PSGL-1/mIgG$_{2b}$ Fusion Protein Purified with IgG Affinity Column

1. Remove low-molecular-weight contaminating proteins by a final gel filtration step on a HiPrep 16/60 Sephacryl S-300 HR column coupled to an ÄKTA Explorer system.
2. Dissolve 1–2 mg of sample in sterile filtered 3 M sodium thiocyanate. Spin in tabletop centrifuge at 20,000×g for 15 min and load supernatant onto the column. The column is equilibrated in the same filtered buffer (to remove particles). Elute in 1.5 column volumes at 0.5 mL/min, with a 1.5 mL fraction size.
3. Fractions only containing the fusion protein as determined by Western blotting are collected and pooled.
4. The pooled sample is dialyzed and lyophilized in the same manner as previously.

3.3 Isolation of PSGL-1/mIgG$_{2b}$ Fusion Protein (Protein A Affinity Column)

1. Grow CHO cells in 10 L WAVE single-use bioreactors in which the PSGL-1/mIg$_{2b}$ is excreted to the media.
2. Reduce the 10 L of medium to less than 1 L using a Cogent M Tangent Flow Filtration system according to the manufacturer's recommendation.
3. Briefly, remove cells grown in serum-free media by filtration with a Millistak+Pod deep filter 0.054 m^2.
4. Concentrate supernatant and buffer exchange with a Pellicon 3 0.11 m^2 filter cassette.
5. Affinity purify the concentrated protein suspension with a HiTrap MabSelect SuRe column connected to an ÄKTA Explorer system (see Note 3).
6. Apply the sample in PBS 0.01 M loading buffer at 5 mL/min. Wash and elution volumes depend on the size of the column(s). The pre-packed columns can easily be connected in serial. 2×5 mL columns yield 500 mL 0.01 M PBS loading buffer and 300 mL sodium citrate 0.1 M pH 3 elution buffer.
7. Eluted fractions are neutralized by adding 0.3 mL 1 M Tris–HCl pH 9 per mL fraction volume.
8. Analyze eluted proteins with ELISA and Western blotting prior to pooling of selected fractions.
9. Dialyze pooled fractions directly in water followed by lyophilization.

3.4 Gel Filtration of PSGL-1/mIgG$_{2b}$ Fusion Protein Purified with Protein A Affinity Column

1. Remove low-molecular-weight proteins with gel filtration.
2. Reconstitute the sample in 0.1 M sodium phosphate pH 7.2 and 0.5 M NaCl and load onto a gel filtration column HiPrep 26/60 Sephacryl S-300 HR equilibrated with the same buffer. Elute in 1.5 column volumes at 1 mL/min, with a 5 mL fraction size.

3. Analyze eluted fractions with Western blotting prior to pooling.
4. Remove salt with dialysis in water followed by lyophilization.
5. Store dried glycoproteins at −80°C.

3.5 Dot Blotting of O-Linked Glycoproteins (See Note 4)

O-linked glycoproteins that have been isolated after affinity purification as described above or by other means can be blotted to PVDF membranes before release. This is adviced if the glycoproteins have been dissolved in organic buffers such as phosphate or Tris buffers (see Note 5). For this purpose, commercial dot blot apparatus and vacuum suction could be used.

1. Clean and dry dot blot apparatus.
2. Place gasket support plate in vacuum manifold.
3. Place sealing on top of gasket support plate.
4. Cut a PVDF membrane to 115 mm by 80 mm. Wet the membrane with methanol for 2 min and soak in MilliQ water for 2 min (always use forceps).
5. Soak a precut filter paper in MilliQ water for 2 min.
6. Place the wet filter paper on the gasket support plate.
7. Place the wetted PVDF membrane on top of the filter paper.
8. Assemble the dot blot according to the manufacturer's instructions.
9. Pipette 10 μL of prepared membrane samples into each well of the dot blot.
10. Add 50 μL of water to each well with sample to wash the sides of the wells.
11. Use a vacuum pump attached to the dot blot apparatus to pass the samples through the PVDF membrane.
12. Disassemble the dot blot apparatus and remove the PVDF membrane.
13. Membranes can be stained with Direct blue 71 (DB71) (protein) or Alcian blue (acidic sugars).
14. Wet the blot with wetting methanol.
15. Discard methanol and add 20 mL of Alcian blue stain solution or DB71 stain solution.
16. Shake the blots for 20 min.
17. Pour off the Alcian blue or DB71 stain.
18. Add 20 mL of destaining solution and shake for 5 × 2 min.
19. Dry the blot, either in air or in vacuum. The background will diminish as the blot dries.
20. The stained blot can be imaged in visible, reflected light. If faintly stained bands are not visible, definition can be improved by wetting the blot with wetting solution before imaging.

3.6 Release of Oligosaccharides

1. If glycoprotein is on a PVDF membrane: Cut out relevant protein bands from Alcian blue- or DB71-stained PVDF membranes, cut into small pieces and place each into well of a microtiter plate. Wet each band with ~20 μL methanol and remove excess methanol. Add 10 μL of water to each well.
2. If glycoprotein dissolved in water: Transfer 10 μL of the solution into a wall of a microtiter plate.
3. Add 10 μL β-elimination solutions to each well prepared freshly.
4. Cover tightly and incubate overnight (16 h) at 50°C.
5. Remove from oven, and add 1 μL glacial acetic acid to neutralize the reaction. Solutions should immediately fizz due to production of H_2 gas.
6. Prepare cation exchange columns:
   (a) Wash AG 50W-X8 resin five times in equal volume of methanol, and then five times in equal volume of water. Store at room temperature in an equal volume of methanol.
   (b) Pack columns by adding 25 μL of cation exchange suspension to Zip Tips (one per sample).
   (c) Place columns in 1.5 mL Eppendorf tubes and centrifuge for 10 s.
   (d) Wash columns with 2×60 μL methanol followed by 2×60 μL 1 M HCl, centrifuging for 10 s between each wash.
   (e) Remove the columns, discard the flow-through, and return to Eppendorf tubes.
   (f) Wash columns with 2×60 μL water, centrifuging for 10 s after each wash.
   (g) Remove the columns, discard the flow-through, and place columns in new labeled 1.5 mL Eppendorf tubes.
7. Pipette solutions from the neutralized samples to the corresponding columns. Centrifuge for 10 s.
8. Wash wells with 2×25 μL water, load onto columns, and centrifuge for 10 s after each wash.
9. Elute residual glycans from the columns by washing each with 30 μL water. Centrifuge for 10 s.
10. Remove cation exchange columns from Eppendorf tubes, and dry the eluted glycan solutions to completion in a speedivac at 45°C for approximately 60 min.
11. Remove white borate residues by addition of 50 μL desalting solution and drying to completion in a speedivac at 45°C for approximately 8 min. Repeat five times (or until no white residue is visible).

12. Samples can be stored dry for at least 1 week at 4°C, and re-suspended in water (10–50 μL) immediately prior to further analysis.

3.7 Packing of Graphitized Carbon Column (See Note 6)

1. Connect a fused silica capillary of 15–18 cm length (250 μm i.d, 370 μm o.d) to the 1/16 in. union containing the steel screen using 20–25 mm of PEEK tubing as a sleeve. Connect with a vespel ferrule a piece of fused silica to the opposite end of the union.
2. Prepare a slurry mixture of 15–20 mg of graphitized carbon in 1.2 mL methanol and sonicate in an ultrasonic bath for 5 min. Place the small magnet (7 × 2 mm) in the slurry container for stirring during packing.
3. Place the slurry container in the Nanobaume pressure unit. Prepare the unit according to the manufacturer's protocol. Once the pressure unit is set start the packing at approximately 50 bar and after 3–5 min increase to 120 bar. The packing takes about 40–45 min and the packing can be seen as the graphite particles are clearly visible through the fused silica column.
4. Turn off the pressure and allow the column to depressurize.
5. Connect to an HPLC pump via a T-split with a 50 cm × 50 μm fused silica capillary as splitter and wash with 80% ACN before connecting to the LC-MS. The LC is set to deliver a flow rate of 250 μL/min and the flow split in the T using a 50 cm × 50 μm of fused silica capillary as restrictor gives a flow through the 100 mm × 250 μm i.d column packed with 5 μm graphite particle of about 5–10 μL/min.

3.8 Tuning of LC-MS

1. Fill the syringe with the maltopentaose standard.
2. Connect the syringe to the PEEK tubing and insert the syringe in the pump.
3. Pump at 4–5 μL/min and perform the tuning on the $(M-H)^-$ ion of *m/z* 827 in negative-ion mode. Usually the auto-tuning function is sufficient but the temperature and gas flow may have to be optimized manually.

3.9 LC-MS of Released Oligosaccharides

Typical chromatograms and spectra of *O*-linked oligosaccharides from various glycoproteins expressed in various recombinant expression systems can be seen in Fig. 1, including mammalian, insect, and yeast cells.

The following outlines the details and conditions for oligosaccharide LC-MS analysis using a Thermo LCQ Deca-XP Ion Trap or other suitable mass spectrometer. Note that ion signal optimization may be required.

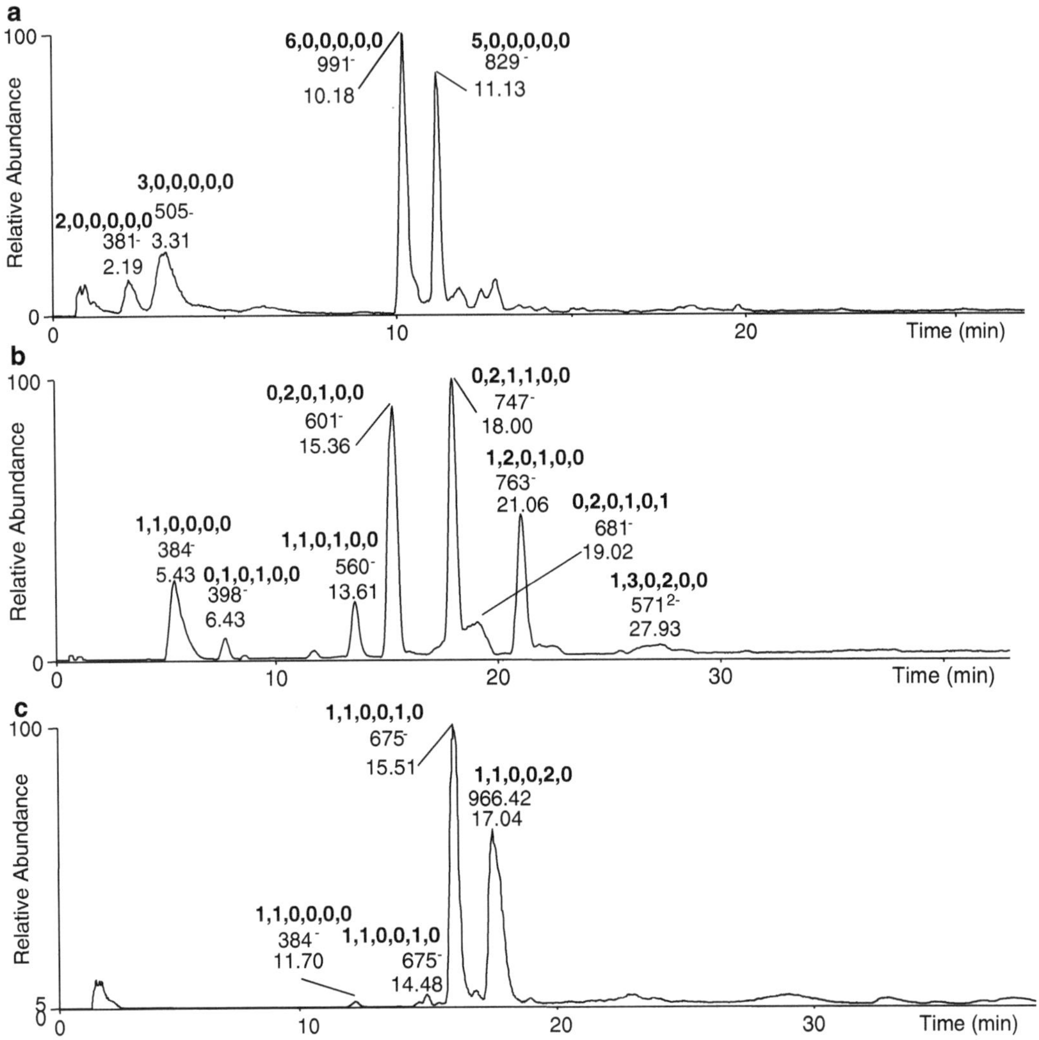

Fig. 1 LC-MS of *O*-linked oligosaccharides from the P-selectin glycoprotein ligand-1/mouse IgG_{2b} (PSGL-1/mIgG_{2b}) fusion protein produced in various expression systems. The base peak chromatogram of oligosaccharides released by β-elimination from PSGL-1/mIgG_{2b} produced in *Pichia pastoris* (**a**), Hi-5 insect cells (**b**), and CHO cells (**c**). The monosaccharide composition of the oligosaccharide alditol is indexed as (Hex, HexNAc, DeHex, HexA, Neu5Ac, Sulfate) containing one reducing end. The major structure in each chromatogram peak is indicated

1. Column details and preparation.
   (a) Column: Hypercarb.
   (b) Mobile phases: A: 10 mM ammonium bicarbonate, B: 10 mM ammonium bicarbonate in 80% ACN (v/v).
   (c) Column equilibration: with flow rate at 0.250 mL/min with splitting giving an on-column rate of 5–10 μL/min or with split-less system giving 7 μL/min.
   (d) Start with 100% A for 30 min, then 100% B for 10 min, and 100% A for 24 min.

2. LC conditions for separation of *O*-linked oligosaccharides:
   (a) Gradient: 0–7 min gradient increase from 0 to 5% B.
   (b) 7–46 min gradient increase from 5 to 35% B.
   (c) 46–46.1 min increase from 35 to 80% B.
   (d) 46–54 min steady at 80% B.
   (e) 54–54.1 min decrease from 80 to 0% B.
   (f) Stop time: 78 min.
   (g) Flow rate: 250 μL/min with splitting giving an on-column rate of 5–10 μL/min or with split-less system giving 7 μL/min.
   (h) Sample injection volume: 3 μL.
3. Typical MS conditions for oligosaccharide analysis:
   (a) Ionization mode: Negative electrospray.
   (b) Source voltage (kV): 3.50.
   (c) Source current (μA): 11.80.
   (d) Capillary voltage (V): –8.0.
   (e) Capillary temperature (°C): 280.

3.10 Interpretation of LC-MS Data of Released Oligosaccharides

The data from LC-MS contains both parent ion scanning and fragmentation data. The parent ion scanning gives information about oligosaccharide composition (Fig. 1, step 1 below, and Table 1). In order to generate sequence information, the fragmentation data needs to be interpreted (Fig. 2, steps 2–4 below, and Table 2). This procedure requires a combination of software interpretation as well as manual interpretation. In Table 3 is a list of common artifacts and interpretation problem in LC-MS of *O*-linked oligosaccharides.

1. Oligosaccharide composition identification:
   (a) Submit *m/z* values of peaks that are identified from the LC-MS trace to the GlyoMod Web site.
   (b) Fill in the relevant options as per sample information as follows (change parameters as necessary).
      - Ion mode and adducts: $(M-H)^-$.
      - *O*-linked oligosaccharides/reduced oligosaccharides.
      - Monosaccharide residues: underivatized.
      - Ranges of residues or possible if unsure.
   (c) Start GlycoMod. Output is a list of potential components that correspond to the input masses with links to GlycoSuiteDB if available. Alternatively, manual interpretation can be performed based on calculations of oligosaccharide residues (Table 1).

Table 1
Calculation of observed *m/z* values of *O*-linked oligosaccharide parent ions released by reductive β-elimination in negative-ion mode ((M – *x*H)*x*⁻ ions) and positive-ion mode after permethylation ((M + *x*Na)*x*⁺ ions)

Residue (common *O*-linked residues)	Underivatized monoisotopic	Permethylated monoisotopic[a]
DeHex-(Fuc-)	146.0579	174.0892
Hex-(Gal-)	162.0528	204.0998
HexA-(GlcA-)	176.0321	245.1263
HexNAc-(GlcNAc-, GalNAc-)	203.0794	361.1737
NeuAc-	291.0954	391.1842
NeuGc-	307.0903	174.0892
HSO_3–	79.9568	
H_2PO_3–	79.9663	
Reducing end	20.0262	62.0727

[a]Sulfation and phosphorylation usually not detected in the conditions used for permethylation in this chapter

$$\text{Negative ion parent ion} m / z = \frac{\text{Sum of residues} + \text{reducing end} - x * \text{H (1.0078 Da)}}{x}$$

$$\text{Positive ion parent ion} m / z = \frac{\text{Sum of residues} + \text{reducing end} + x * \text{Na (22.9898 Da)}}{x}$$

2. t l:
   (a) Extract the relevant MS^2 data for each of the identified components (m/z, intensity).
   (b) Using GlycoWorkBench draw putative structure(s) for the m/z value that is under investigation (e.g., from suggestions provided by the GlycoMod output from structures previously identified in GlycosuiteDB).
   (c) Input the peak list into GlycoWorkBench.
   (d) Using the FRAGMENT tool, perform a theoretical fragmentation of the putative structure(s).
   (e) Using the ANNOTATE tool, match the theoretical fragments to the MS^2 peak list. Repeat for all possible structures. Use the statistical data to identify the most likely structure.
3. Structure verification 2:
   (a) Make a list of possible structures and their corresponding MS^2 spectra.
   (b) Go to UniCarb-DB Web site and search for the relevant structures.

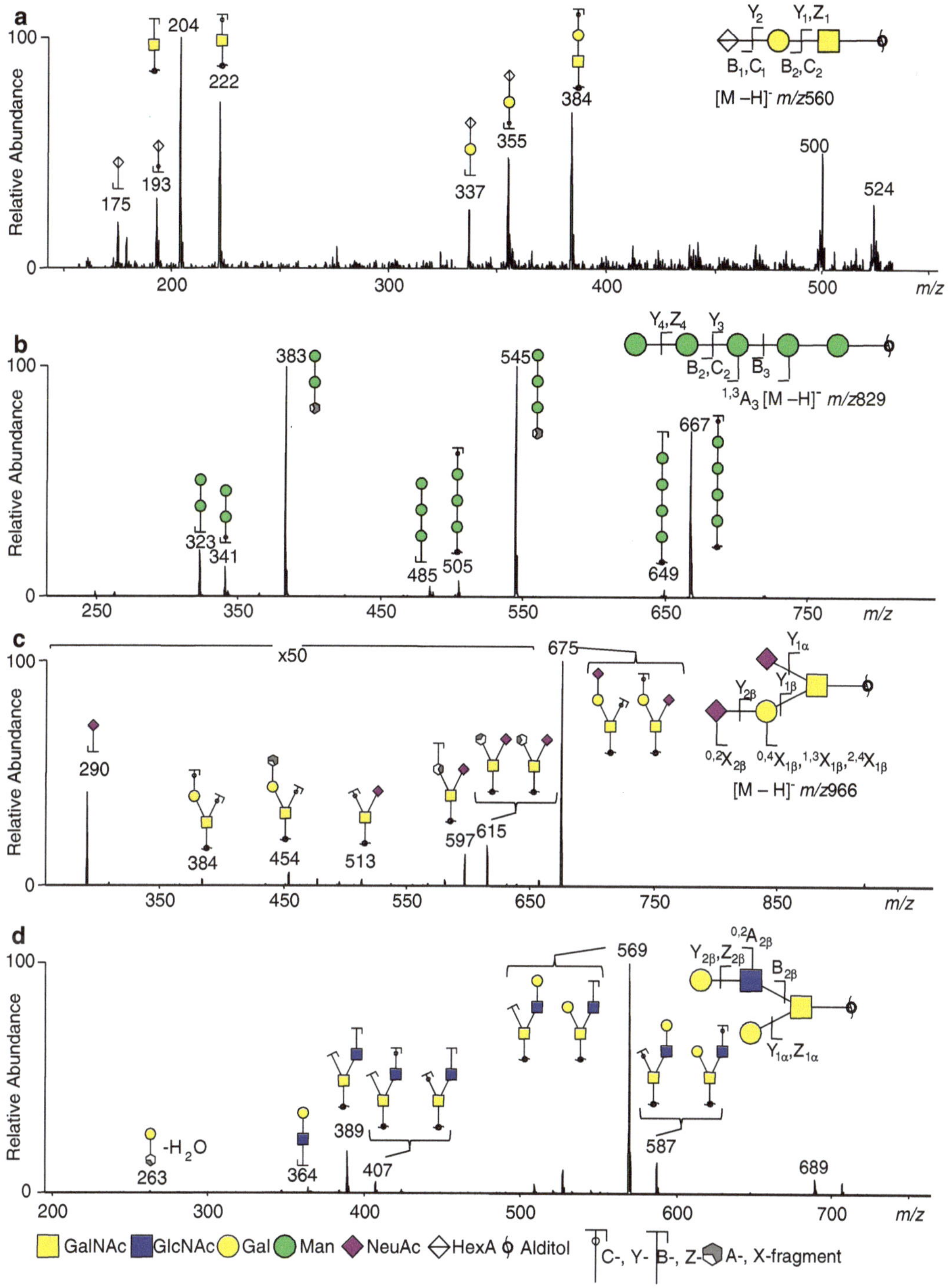

Fig. 2 LC-MS[2] of *O*-linked oligosaccharides showing typical fragmentation pathways. The MS[2] spectrum from a Hi-5 insect cell trisaccharide containing hexuronic acid (**a**) and a mannose pentasaccharide from *Pichia pastoris* (**b**) are shown. The fragmentation pattern of a disialylated core 1 *O*-glycan from CHO cells and a typical MS[2] spectrum of a neutral core 2 oligosaccharide from human MUC7 mucin are shown in (**c**) and (**d**), respectively

Table 2
Fragmentation pattern observed for native *O*-linked oligosaccharide alditols in negative-ion mode

Type	Most significant fragment	Other features
Neutral	Z-ion of the C-3 branch of GalNAcol	Diagnostic ions are also found for type 2 chains and Lewis type. Oligosaccharide sequence deduced mostly from Z-ions
Sialylated	Y-ion from loss of sialic acid (−291 Da)	Information on sialic acid location often lost. Spectra of singly sialylated species resemble their unsialylated counterpart. Only low-intense sequence ions of multiply sialylated oligosaccharides, especially of their $(M-H)^-$ ions
Sulfated	Y-ion from the residue furthest away from the sulfate	Oligosaccharide sequence deduced mostly from Y- and B-ions

Table 3
Common artifact and interpretation problems in LC-MS and LC-MS² of *O*-linked oligosaccharides

Name	Manifestation in the MS	Problems and explanations
Hexose ladders	Detected as $(M-H)^-$ ions of *m/z* 343, 505, 667, 829, 991,… or unreduced *m/z* 341, 503, 665, 827, 989,…	Isobaric with sulfated structures (e.g., $(M-H)^-$ of $Hex_1HexNAc_1HexNAcol_1Sulfate_1$ = *m/z* 667, $Hex_2HexNAc_1HexNAcol_1Sulfate_1$ = *m/z* 829). Contamination from water, blotting paper, powder from gloves
Adducts	Na^+ ($(M-H)^-+22$) K^+ ($(M-H)^-+36$)	Satellite peaks eluting at the same retention as a major ion. Increase the amount of cation exchange resin in the desalting step
Covalent reactions	De-*N*-acetylation ($(M-H)^--42$) *O*-acetylation ($(M-H)^-+42$)	Satellite peaks eluting at a different retention as a major ion. To long incubation with either reducing agent or during removal of borates
Peeling	Sugars without GalNAc at reducing end	Found for instance with $(M-H)^-$ of *m/z* 472 ($NeuAc_1Hexol_1$) in samples with high amount of core 1 and 2 structures
N-linked glycans	Also released during reductive β-elimination	Can be identified by its strange composition and isotopic distribution (a mixture of $-NH_2$ and –OH at reducing end)
Glycopeptides	Both *N*-linked and *O*-linked glycopeptides are generated in the release	*C*-terminal of the glycopeptides has been reduced to alcohol, preventing further β-elimination
Fucose–sialic acid problem	Two Fuc have similar nominal mass (292) as one NeuAc (291). $NeuAc_1Hex_1$ identical mass as $NeuGc_1Fuc_1$	Can often be resolved by its particular fragmentation in MS². The more sialic acid a molecule has the more charges in the MS (sialic acid introduces one more negative charge, but two fucoses none). Hence, more sialic acid will manifest as more sodiated adducts and more charges

(c) The MS^2 spectra of structures solved by LC-MS are available for download or real-time viewing from the Web site. Verify that the database MS^2 matches that for the structure under investigation.

4. Structure verification 3:

   (a) Manual interpretation using the table below (Table 2).

3.11 Permethylation of Oligosaccharides

The chemicals used in this protocol require proper safety precautions such as working in a fume hood, wearing gloves, protective eyewear, and lab coat. No chemicals should be poured into the sink, but be taken care of for proper disposal. Use only sterile glassware (see Note 7).

1. Dissolve the released and reduced oligosaccharides in 1.5 mL water and transfer to a Kimax (with Teflon-faced screw cap) glass tube and dry at 45°C under nitrogen gas.
2. To ensure complete dryness put vial in a vacuum desiccator connected to a vacuum pump for 30 min.
3. Add a magnetic bar, and dissolve sample in 1 mL (for 0.2–2 mg) dimethyl sulfoxide (a brief incubation in an ultrasonic may be needed to dissolve the glycoprotein).
4. Add 0.2 mL of iodomethane and 50 mg NaOH powder. A pre-calibrated spoon may be necessary to prevent rehydration of the NaOH powder from water in the air since it is very hygroscopic.
5. Stir the sample for 10 min in room temperature.
6. The reaction is stopped with 4 mL of water.
7. Add 2 mL of chloroform. Invert the tube several times to mix the phases. Spin tube at low speed (200 ×*g*, 5 min) to settle the phase separation again.
8. Remove the *upper* water phase gently with a Pasteur pipette. Add 4 mL water, invert tube, and spin again.
9. Remove water phase and repeat the procedure another two times.
10. In a final step add 4 mL water, mix, spin, and remove the *lower* (chloroform phase) to a clean tube.
11. Add 2 mL chloroform to the original tube, mix, spin, and transfer the lower phase to the new tube.
12. Dry the oligosaccharides under a stream of nitrogen gas at 45°C.
13. Finally dry in a vacuum desiccator connected to vacuum pump for 30 min.

3.12 MS of Permethylated Oligosaccharides

1. Dissolve samples in 200 μL methanol:water 1:1 (v/v) corresponding to an original concentration of 1 mg/mL of glycoprotein.
2. Usually the samples are strong enough to be diluted a further ten times prior to loading into the mass spectrometer.

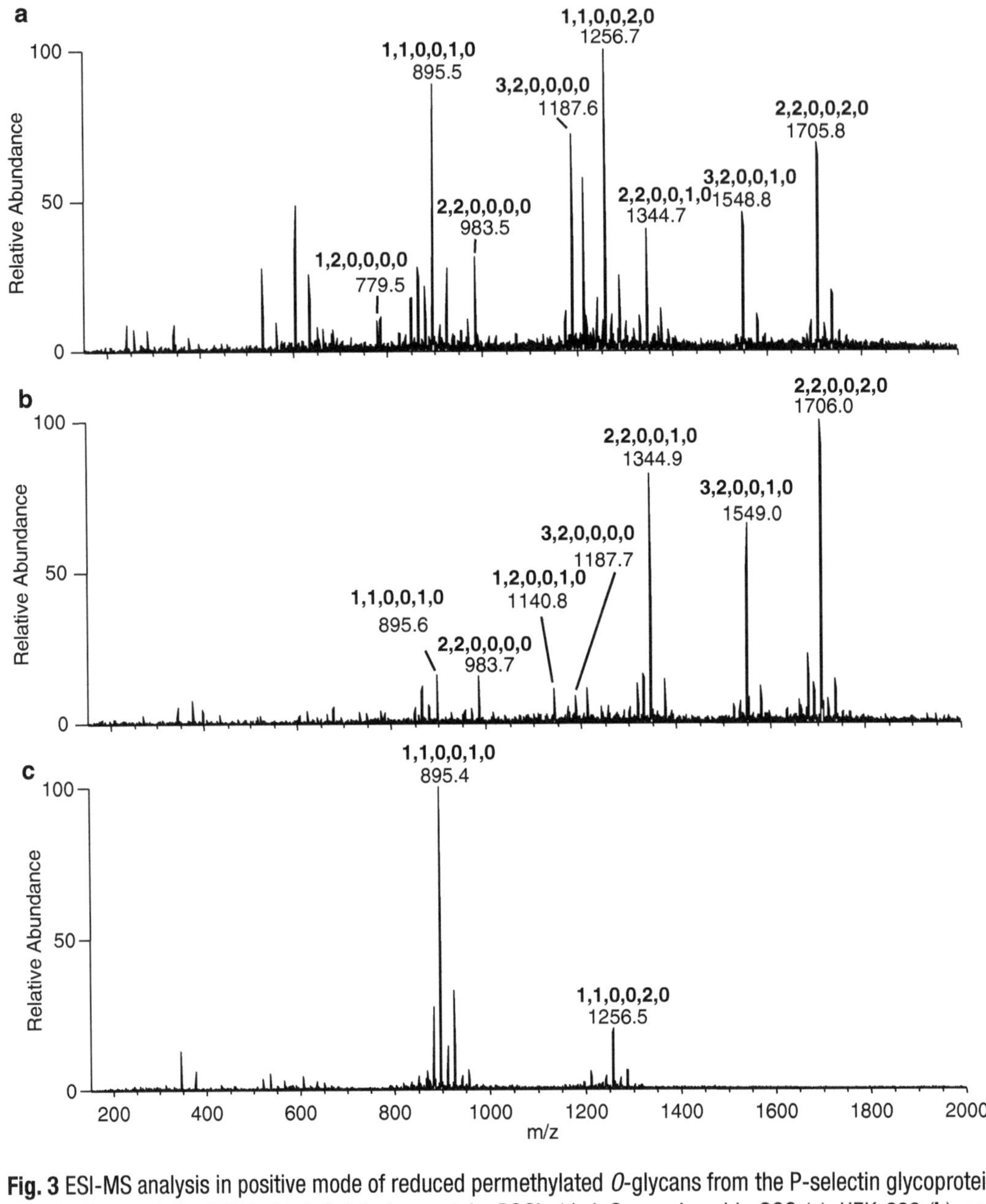

Fig. 3 ESI-MS analysis in positive mode of reduced permethylated *O*-glycans from the P-selectin glycoprotein ligand-1/mouse IgG_{2b} (PSGL-1/mIgG_{2b}) fusion protein. PSGL-1/mIgG_{2b} produced in COS (**a**), HEK-293 (**b**), and CHO (**c**) cells. The monosaccharide composition indexed as (Hex, HexNAc, DeHex, HexA, Neu5Ac, Sulfate), where HexNAc also includes reducing end GalNAc-ol, is indicated above each peak. The COS cells co-express the porcine α1,3-galactosyltransferase

3. Perform MS and MS^n of the $(M+Na)^+$ ions in positive mode in the ion-trap mass spectrometer (Figs. 3, 4, and 5 and Table 1).
4. Tune the instrument using stachyos (constant values: capillary voltage 4.50 V and capillary temperature: 200°C). Nitrogen is used as sheath gas.

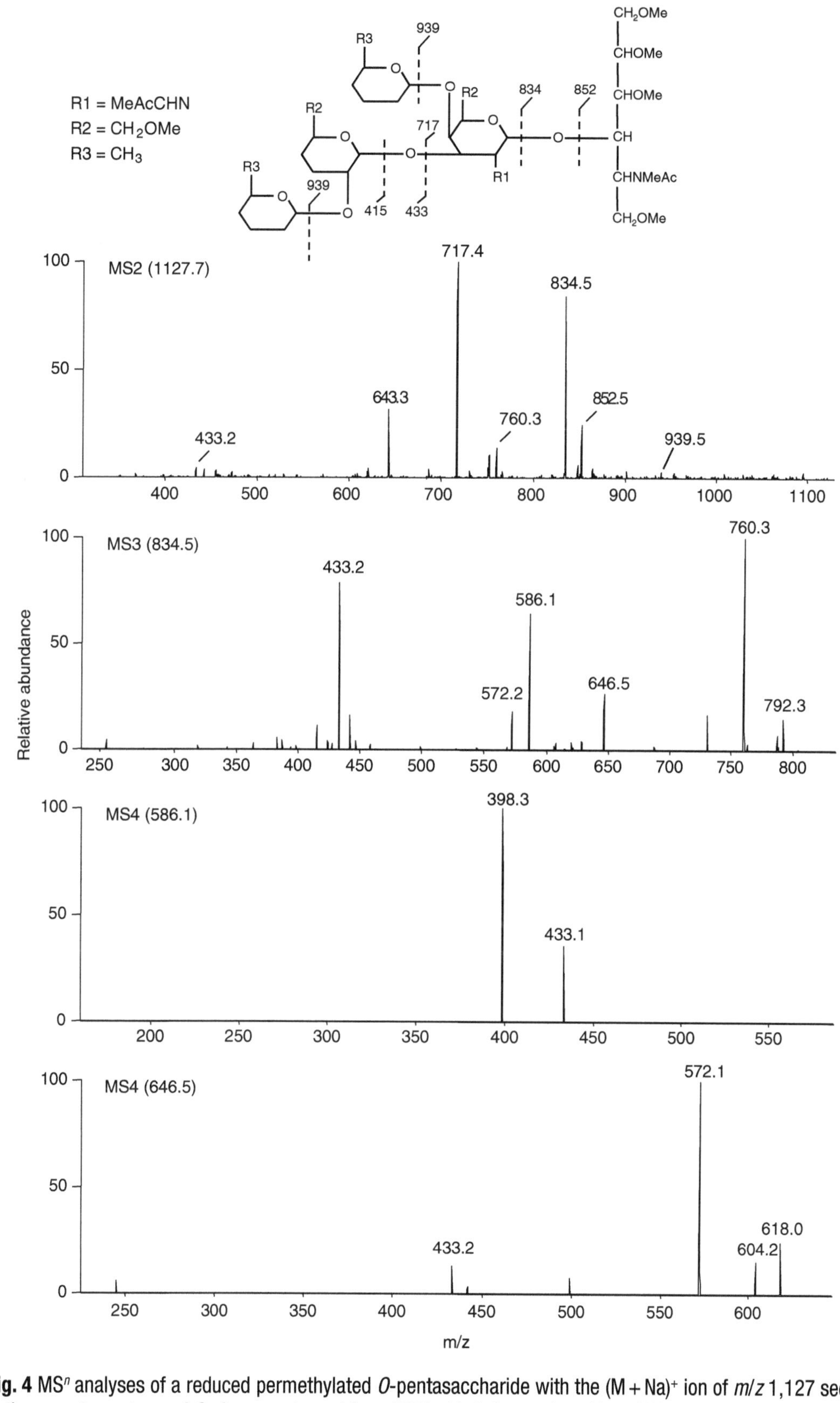

Fig. 4 MS^n analyses of a reduced permethylated *O*-pentasaccharide with the $(M+Na)^+$ ion of *m/z* 1,127 seen in the parent spectrum of *O*-glycans released from PSGL-1/mIgG$_{2b}$ produced in a CHO cell line expressing Le^b. ESI-MS$^{2-4}$ of peak *m/z* 1,127, together with the oligosaccharide sequence and its specific fragmentation as derived from the sequence ions identified in the daughter ion spectra

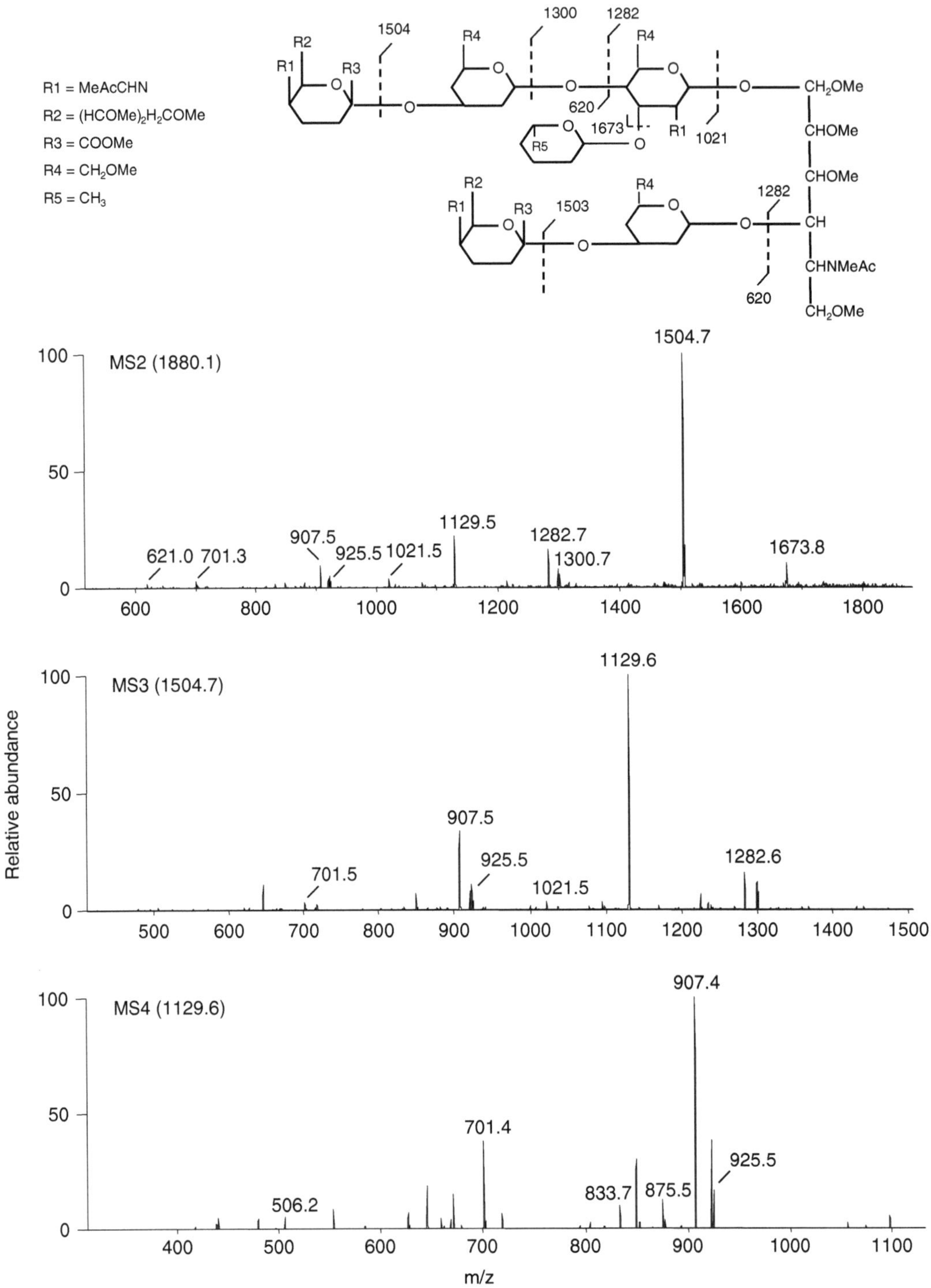

Fig. 5 MS^n analyses of reduced permethylated *O*-heptasaccharide with the $(M+Na)^+$ ion of *m/z* 1,880 seen in the parent spectrum of *O*-glycans released from PSGL-1/mIgG$_{2b}$ produced in a CHO cell line expressing sialyl-Le^x. ESI-MS$^{2-4}$ of peak *m/z* 1,880 together with the oligosaccharide sequence and its specific fragmentation as derived from the sequence ions identified in the daughter ion spectra

5. The samples are loaded manually (8 μL) with a Hamilton syringe and directly infused into the instrument at a flow rate of 5 μL/min.
6. MS^n was performed with a collision energy set to 50%.

3.13 Interpretation of Permethylated Oligosaccharides from MS

1. Export MS and MS^n chromatograms as ASCII text files. The GlycoWorkBench software can import the files and redraw the spectra.
2. GlycoWorkBench enables searches of *compositions* or *structures* with a given m/z value in CFG, Carbbank, or Glycosciences databases which serves as a good starting point to identify potential compositions and structures of the peaks in the ESI-MS mother spectra.
3. Structures drawn manually or imported from the databases can be subjected to in silico fragmentation which can be matched against the real peaks in the imported chromatograms.
4. Deduce structures from MS^n using information from branching (glycosidic bond cleavage) and linkage information (cross-ring cleavage) (see Note 8) (see example of MS^n in Figs. 4 and 5).
5. GlycoWorkBench usually does not produce all fragments observed in real spectra. Manual calculations of these fragment structures can be performed by drawing the parent structure in a molecule editing software such as ChemDraw where any fragmentation can be calculated by manually removing different parts of the drawn molecule.

4 Notes

1. NaOH must be dry, preferably prepared fresh from pellets with mortar and pestle.
2. Purity of protein can be assessed using Western blotting (anti-IgG Fc specific and anti-PSGL-1) and gel staining using Pro-Q Emerald 300 glycoprotein gel stain and SYPRO Ruby according to the manufacturer's recommendation.
3. The MabSelect SuRe columns carry alkali-tolerant recombinant protein A ligands in a highly cross-linked agarose matrix. Protein A has high binding affinity to the Fc region of IgG.
4. Oligosaccharides from glycoproteins separated by SDS-PAGE can also be electroblotted and stained by this same method.
5. Lower concentrations (<1%) of detergents such as SDS, CHAPS, Tween, and reducing agents such as mercaptoethanol and dithiothreitol as well as chaotropic agents such as urea, thiourea, and guanidium chloride (up to 8 M) could be present in the buffers as long as the blots are washed extensively with water after blotting and staining.

6. Columns for graphitized capillary LC-MS are also available from Thermo-Scientific.
7. Include a positive control such as oligosaccharides released from 1 mg pig gastric mucin available from Sigma-Aldrich.
8. A high number of matched fragments can reflect identification of the correct structure. However, a good understanding of the biological constraints of the source of the oligosaccharide is necessary. Identifying the structures derived from well-studied cells such as CHO cells is greatly facilitated compared to less studied cell lines, since the available glycosyltransferases in CHO cells are largely known.

Acknowledgements

This work was supported by the Swedish Research Council (621-2010-5322 to N.K. and K2011-65X-3031-01-6 to J.H.), the Swedish Foundation for International Research and Higher Education (to N. K.), the EU Marie Curie Program (PIRG-GA-2007-205302 to N.K.), the County Council of Västra Götaland (ALF to J.H.), and Recopharma AB. The mass spectrometers were obtained by grants from the Swedish Research Council (342-2004-4434).

Conflicts of interest. Jan Holgersson is founder, board member, and part-time CSO of Recopharma AB.

References

1. Gabius H-J (2009) Animal and human lectins. In: Gabius H-J (ed) The sugar code. Fundamentals of glycosciences. Wiley, Weinheim, pp 317–328
2. Gabius H-J et al (2011) From lectin structure to functional glycomics: principles of the sugar code. Trends Biochem Sci 36:298–313
3. Holgersson J, Gustafsson A, Gaunitz S (2009) Bacterial and viral lectins. In: Gabius H-J (ed) The sugar code. Fundamentals of glycosciences. Wiley, Weinheim, pp 279–300
4. Kufe DW (2009) Mucins in cancer: function, prognosis and therapy. Nat Rev Cancer 9: 874–885
5. Linden SK et al (2008) Mucins in the mucosal barrier to infection. Mucosal Immunol 1: 183–197
6. Coombs PJ et al (2010) Identification of novel contributions to high-affinity glycoprotein-receptor interactions using engineered ligands. J Mol Biol 396:685–696
7. Gustafsson A, Holgersson J (2006) A new generation of carbohydrate-based therapeutics: recombinant mucin-type fusion proteins as versatile inhibitors of protein-carbohydrate interactions. Expert Opin Drug Discov 1:161–178
8. Mammen M, Choi S, Whitesides G (1998) Polyvalent interactions in biological systems: implications for design and use of multivalent ligands and inhibitors. Angew Chem Int Ed 37:2754–2794
9. Gustafsson A et al (2006) Carbohydrate-dependent inhibition of Helicobacter pylori colonization using porcine milk. Glycobiology 16:1–10
10. Gustafsson A et al (2011) Pichia pastoris-produced mucin-type fusion proteins with multivalent O-glycan substitution as targeting molecules for mannose-specific receptors of the immune system. Glycobiology 21:107186
11. Holgersson J, Lofling J (2006) Glycosyltransferases involved in type 1 chain and Lewis antigen biosynthesis exhibit glycan and core chain specificity. Glycobiology 16:584–593
12. Liu J, Gustafsson A et al (2005) Anti-pig antibody adsorption efficacy of α-Gal carrying

recombinant P-selectin glycoprotein ligand-1/ immunoglobulin chimeras increases with core 2 β1, 6-N-acetylglucosaminyltransferase expression. Glycobiology 15:571–583
13. Lofling J, Holgersson J (2009) Core saccharide dependence of sialyl Lewis X biosynthesis. Glycoconj J 26:33–40
14. Lofling JC, Hauzenberger E, Holgersson J (2002) Absorption of anti-blood group A antibodies on P-selectin glycoprotein ligand-1/ immunoglobulin chimeras carrying blood group A determinants: core saccharide chain specificity of the Se and H gene encoded alpha1,2 fucosyltransferases in different host cells. Glycobiology 12:173–182
15. Zaia J (2004) Mass spectrometry of oligosaccharides. Mass Spectrom Rev 23:161–227
16. North SJ et al (2010) Curr Opin Struct Biol 19:498–506
17. Ashline D et al (2005) Congruent strategies for carbohydrate sequencing. 1. Mining structural details by MS^n. Anal Chem 77:6250–6262
18. Mechref Y, Kang P, Novotny MV (2009) Solid-phase permethylation for glycomic analysis. Methods Mol Biol 534:53–64
19. Sagi D et al (2002) Sequencing of tri- and tetraantennary N-glycans containing sialic acid by negative mode ESI QTOF tandem MS. J Am Soc Mass Spectrom 13:1138–1148
20. Harvey DJ (2005) Fragmentation of negative ions from carbohydrates: part 3. Fragmentation of hybrid and complex N-linked glycans. J Am Soc Mass Spectrom 16:647–659
21. Schulz BL, Packer NH, Karlsson NG (2002) Small-scale analysis of O-linked oligosaccharides from glycoproteins and mucins separated by gel electrophoresis. Anal Chem 74: 6088–6097
22. Wada Y et al (2010) Comparison of methods for profiling O-glycosylation: Human Proteome Organisation Human Disease Glycomics/Proteome Initiative multi-institutional study of IgA1. Mol Cell Proteomics 9:719–727

Chapter 10

Assessing Fc Glycan Heterogeneity of Therapeutic Recombinant Monoclonal Antibodies Using NP-HPLC

T. Shantha Raju

Abstract

Recombinant monoclonal antibodies (rMAbs) are becoming major human therapeutics to treat life-threatening diseases such as cancer. These rMAbs are produced using either in vitro cell culture processes or transgenic technology in animals or plants. Glycans present in the Fc region can affect functions of rMAbs. These Fc glycans are heterogeneous and impact binding of rMAbs to Fc gamma receptors (FcγRs) and C1q protein. As a result Fc glycans affect antibody-dependent cellular cytotoxicity and complement-dependent cytotoxicity of rMAbs. Thus understanding the glycan heterogeneity is necessary during the development of these rMAbs as human therapeutics. Because of their biological significance, understanding the glycan structure and their impact on the function of antibody molecules is also a regulatory requirement. Glycan mapping by NP-HPLC with fluorescence detection is a sensitive and reproducible method. Labeling of released glycans with anthranilic acid (AA) using reductive amination procedure improves sensitivity of detection. The NP-HPLC method resolves both neutral and sialylated glycans, thus enabling the user to obtain a broad heterogeneity profile of Fc glycans in a single run. Added advantage of the method is that the labeled glycans can be characterized using mass spectrometry and the method is also amenable for LC-MS analysis.

Key words Antibodies, IgGs, Monoclonal, CDC, ADCC, C1q, FcγR, Glycan, Mapping, PNGase F, Exoglycosidase, HPLC, NP-HPLC, Anthranilic acid, Fluorophore, Fluorescence, Reductive amination

1 Introduction

Protein glycosylation is a co- and posttranslational modification in which nascent polypeptide chain undergoes O- and N-glycosylation while processing through Golgi compartments (1). Since many different enzymes such as glycosyltransferases, glycosidases, nucleotide sugar synthetases, nucleotide sugar transporters, etc. are involved in protein glycosylation, the glycans are highly heterogeneous (2, 3). The glycosylation and the glycan microheterogeneity affect the physicochemical and biological functions of proteins (4). In addition glycosylation may also affect pharmacokinetic properties

Alain Beck (ed.), *Glycosylation Engineering of Biopharmaceuticals: Methods and Protocols*, Methods in Molecular Biology, vol. 988, DOI 10.1007/978-1-62703-327-5_10, © Springer Science+Business Media New York 2013

of proteins (5). For example glycoproteins containing terminal galactose residues have been shown to clear rapidly from serum compared to their sialylated counterparts (5, 6). This is because terminal galactose containing glycoproteins binds to asialoglycoprotein receptor present in the liver (5–7).

Immunoglobulins are serum glycoproteins that are involved in innate and adaptive immunity. Among immunoglobulins, IgGs are the major components that contain a conserved N-glycosylation site in the CH2 domain of the Fc. The monoclonal IgGs are often referred as monoclonal antibodies (MAbs). The N-glycans present in MAbs are highly heterogeneous and affect their binding to Fc receptors such as FcγRI, FcγRII, FcγRIII, etc. Accordingly, Fc glycans affect the antibody-dependent cellular cytotoxicity (ADCC) and antibody-dependent cellular phagocytosis (ADCP) of MAbs (8). In addition, Fc glycans may also affect MAbs binding to C1q protein, thus affecting the complement-dependent cytotoxicity (CDC) of MAbs (9). Since C1q is the first component of the complement cascade, binding to C1q protein is required for CDC activity of MAbs (8, 9).

More than 20 recombinant monoclonal antibodies (rMAbs) are currently in the market as human therapeutics to treat various diseases (10). Several hundred more rMAbs are in preclinical and clinical development (8–10). The majority of these rMAbs are currently produced using in vitro cell culture processes (11). These cell culture processes use mammalian cells such as Chinese hamster ovary or mouse hybridoma-derived cell lines (12). Since the antibody glycosylation varies with species, the glycosylation profiles of rMAbs produced in CHO cells are slightly different than the rMAbs produced using mouse hybridoma-derived cell lines (13). Since changes in glycosylation affect the antibody functions, it is necessary to characterize the glycosylation profiles of rMAbs. For example changes in terminal galactose content affect the CDC activity of Rituxan. In addition an increase in sialic acid content is reported to result in decreased ADCC activity of rMAbs but increase the anti-inflammatory activity (8–10). Further, the terminal galactose and sialic acid content of rMAbs varies with cell culture conditions and hence differs from lot to lot. As a result rMAb activity may also vary from lot to lot. Hence, it is necessary to monitor changes in glycosylation profiles of rMAbs. For this it is necessary to implement highly sensitive and reliable methods to analyze Fc glycans of rMAbs.

Methods for the analysis of glycans are complex processes that involve the release of glycans using either enzymatic or chemical methods, followed by purification to remove protein and salts, and then analysis to identify the glycans (14). An example of an enzymatic method involves the release of glycans with peptide-N-glycosidase F (PNGase F) under native or denaturing conditions. One chemical method involves treating pure glycoproteins with hydrazine under anhydrous conditions followed by removal of excess hydrazine. Since glycans do not contain chromophoric groups detection is a problem and hence some methods contain an additional labeling step to incorporate chromophoric or fluorophoric groups. In addition, the glycans are highly heterogeneous; thus method development is tedious. Several methods have been described in the literature to analyze glycans from glycoproteins including MAbs and rMAbs. These methods can be broadly categorized as electrophoretic methods, HPLC methods, and mass spectrometric methods. Among the electrophoretic methods, capillary electrophoresis with laser-induced fluorescent detection (CE–LIF) and fluorophore-assisted carbohydrate electrophoresis (FACE) methods are sensitive to detect microheterogeneity. However, the labeling reagents used in these methods such as APTS and ANTS are expensive and unstable. In addition, there is a run-to-run variation observed in CE–LIF and also lot-to-lot variation in capillary affects the analysis. For FACE analysis the specially prepared gels need to be obtained from one specific vendor and these gels are very expensive (15). Among the HPLC methods, high-performance anion exchange chromatography with pulsed amperometric detection (HPAEC-PAD) has been used for glycan analysis (16). The advantage of HPAEC-PAD method is that it does not require derivatization but requires a dedicated HPLC instrument. However, it has several disadvantages including high noise, unstable baseline, and high pH buffers. In addition, amino acids, peptides, and other oxidizable organic compounds interfere during the glycan analysis using HPAEC-PAD (17).

Glycan analysis using NP-HPLC coupled with fluorescent detection improves sensitivity and resolution (18, 19). Among the several fluorescent-labeling reagents described in the literature, anthranilic acid (AA) is a very sensitive labeling reagent that can be used under aqueous conditions for glycan analysis and is very stable (20). AA labeling is specific for reducing sugars and does not need fractionation of glycans from peptides or proteins prior to derivatization (19). In addition AA-labeling reagents are inexpensive and available from multiple vendors.

Fc glycans can be released from MAbs or rMAbs by treating with PNGase F under native conditions at or around neutral pH (21–25). In this way releasing Fc glycans from MAbs or rMAbs is much easier than releasing glycans from other glycoproteins that contain surface-exposed glycans that require denaturing and reduction followed by carboxymethylation of disulfide bonds. This chapter describes releasing N-glycans from MAbs using PNGase F followed by labeling of the released glycans with AA and NP-HPLC analysis using Agilent 1100 instrument. Although the method does not require a dedicated HPLC instrument, users are advised to use one instrument so that there is no need to change the instrument settings. Further, using the NP-HPLC method it is possible to obtain information on variations in terminal galactose and sialic acid contents of different lots or rMAbs.

2 Materials

2.1 Release of N-Glycans from IgGs

Prepare all buffer (20 mM Tris–HCl, pH 7.5) solutions in deionized water and use water bath preset at 37°C for incubation. Human IgG was obtained from Sigma and PNGase F was from either QaBio or ProZyme.

2.2 Derivatization

Use high-quality glacial acetic acid, sodium acetate trihydrate, and boric acid, preferably from Fluka or Aldrich. Preheat the heating block to 80°C for derivatization.

2.3 Purification of Derivatized Glycans

Use NAP-5 columns from GE Healthcare and deionized water for elution and for washings of the NAP-5 columns.

2.4 Chromatography

1. Use any HPLC equipped with fluorescence detector (Agilent 1100 system with Chemstation software was used in this study). The system should contain thermostated column compartment.

Use polymeric-amine-bonded HPLC column, 4.6×250 mm (Astec, cat. no. 56403) for mapping. Use proper autosampler vials (Agilent). Eluant A was 2% (v/v) acetic acid and 1% (v/v) tetrahydrofuran (inhibited) in acetonitrile (all HPLC grade). Eluant B was 5% (v/v) acetic acid, 1% (v/v) tetrahydrofuran (inhibited), and 3% (v/v) triethylamine (all HPLC grade) in Milli-Q-filtered water.

3 Methods

3.1 Preparation of the Glycans Using Human IgG as an Example

1. Dilute the human IgG ~1 mg/mL in 20 mM Tris–HCl buffer pH 7.5.
2. Label one Eppendorf tube for each blank and sample.
3. Place 50 μL of the buffer to blank tube and 50 μL of human IgG to the sample tube.
4. Add 1 μL PNGase F to each vial, gently mix by vortexing, and centrifuge to collect all the liquid in the bottom of the tube.
5. Incubate the tubes in a heating block that is set at 37°C for overnight digestion (~12 h of digestion is sufficient for MAbs).

3.2 Derivatization

1. Allow the blank and the sample tubes to cool to room temperature. Briefly centrifuge to collect the liquid at the bottom of the tubes (see Note 1).

2. Add 1 μL of 20% acetic acid to each vial and mix.
3. Prepare 100 mL of acetate–borate–methanol solution by dissolving 4 g of sodium acetate and 2 g of boric acid in 100 mL methanol (see Note 2).
4. Prepare anthranilic acid reagent by dissolving ~30 mg of anthranilic acid (Aldrich) and ~20 mg sodium cyanoborohydride in 1 mL of acetate–borate–methanol solution (see Note 3).
5. Add 100 μL of anthranilic acid reagent to each tube. Cap the tubes tightly and mix (see Note 5).
6. Heat the vials for 1 h in a heating block that is set at 80°C. Allow the vials to come to room temperature before proceeding.

3.3 Purification

1. Pre-equilibrate the NAP-5 columns with Tris buffer.
2. Briefly centrifuge the Eppendorf tubes containing AA-labeled glycans to collect all the liquids at the bottom of the tubes.
3. Add 350 μL of Tris buffer to the tubes and mix well.
4. Load ~500 mL of the content from each tube onto separate NAP-5 columns.
5. Allow the liquid to enter the columns and do not collect this.
6. Add 1.0 mL of Tris buffer to each NAP-5 column and collect the eluants from the columns using separate new Eppendorf tubes that are labeled appropriately as blank and sample.
7. Label two autosample vials as blank and sample.
8. Transfer the contents from Eppendorf collection tubes to autosample vials.

3.4 Chromatography

1. NP-HPLC runs are arranged as blank and sample.
2. For glycan mapping, prepare eluants A and B.
3. Set up the NP-HPLC gradient program as:

   0 min, B = 30%

   2 min, B = 30%

   82 min, B = 95% (linear increase)

   97 min, B = 95% (wash)

   97.1 min, B = 30%

   112.1 min, B = 30% (equilibration)
4. Set column temperature at 50°C.
5. Flow rate of 1.0 mL/min.
6. Detection at .ex = 360 nm and .em = 425 nm.
7. Fluorescence detector settings: Peak width response time: >0.4 min (8 s, slow); PMT gain: 12.

3.5 Acceptance Criteria for Glycan Mapping

NP-HPLC system and glycan mapping profile are acceptable when the oligosaccharide map of human IgG is comparable to the representative chromatogram shown in Fig. 1 (corresponding structure of glycans is shown in Fig. 2). Human IgG glycan map can be routinely used as a check for system suitability. In the event of abnormal human IgG glycan map, the chromatography run should be stopped and further investigation should be made on the system prior to the sample analysis. The retention times of G0, G1, G2, G2S1, and G2S2 peaks should have a standard deviation (SD) of less than 0.2 min and RSD of less than 5% from run to run.

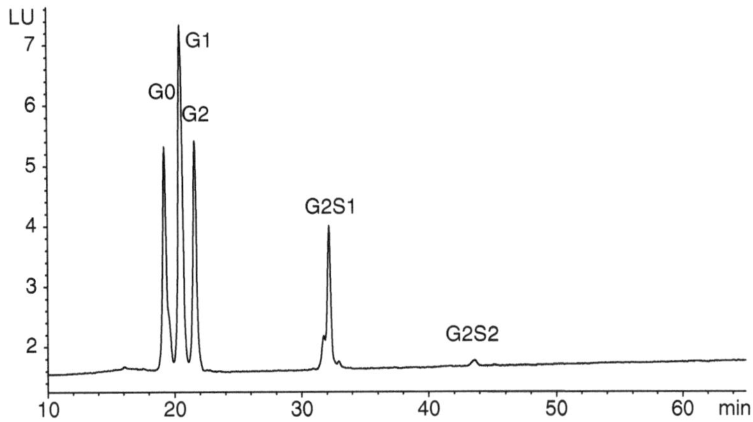

Fig. 1 NP-HPLC mapping of glycans released from human IgG. Corresponding structure of glycans is shown in Fig. 2

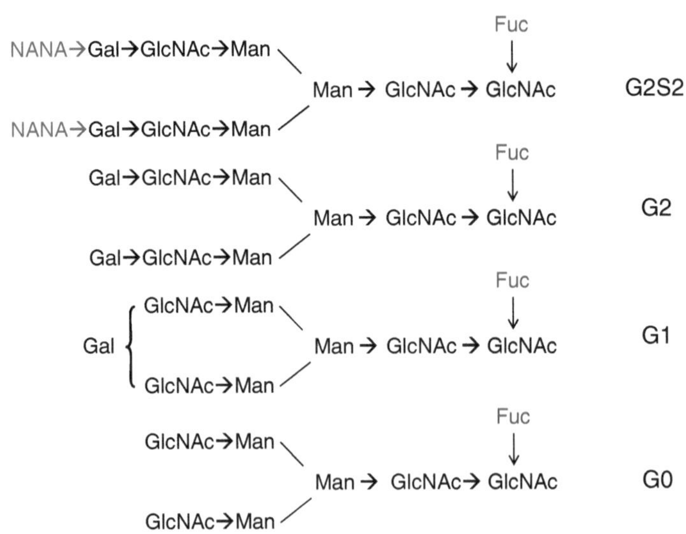

Fig. 2 Structure of major glycans found in rMAb

Alternatively, a reference standard of rMAb can be used to check the system suitability. Test results of the samples are acceptable when the oligosaccharide maps of the experimental samples are comparable to the reference standard.

4 Notes

1. During digestion of MAbs or rMAbs with PNGase F to release N-glycans some evaporation of liquid content occurs and it collects as moisture (water droplets) at the top of the tube. Hence, it is necessary to centrifuge the tubes prior to opening to avoid any loss of glycans.
2. Store the acetate–borate–methanol solution in brown bottles or in amber vials at room temperature and the reagent is stable up to a year.
3. Anthranilic acid reagent is stable for several weeks in the dark at room temperature. However, it is recommended that this reagent is prepared fresh. Sodium cyanoborohydride is poisonous and hygroscopic. Transfer of AA and sodium cyanoborohydride should be made in a chemical hood. Limit the exposure of these chemicals to air during weighing.
4. Highly accurate weighing of anthranilic acid and sodium cyanoborohydride is not a necessity. It has been observed that slight variations in the amounts of anthranilic acid and sodium cyanoborohydride do not affect the labeling of released glycans (26).
5. It is necessary to cap the Eppendorf tubes tightly so that they do not open during heating at 80°C for derivatization. If necessary use parafilm to secure the tubes tightly before incubating in the heating block.
6. After digestion of MAbs or rMAbs with PNGase F it is not necessary to remove the protein from released glycans prior to labeling. This is an advantage of AA labeling compared to other labeling reaction where it is necessary to remove the protein by ethanol precipitation followed by drying using speed-vac to remove ethanol and water. This is needed for labeling of glycans with APTS, ANTS, 2-AB, etc (19, 27).
7. Polymeric amine column is very rugged and can be used for several months (or even years). However, it is advisable to replace the column after 500 runs or whenever the peaks are becoming noticeably broad in the glycan mapping (26, 27).
8. Figures 3 and 4 show the comparison of NP-HPLC profiles of glycans released from rMAbs produced using CHO and SP2/0 cells, respectively. Although these rMAbs contain the same

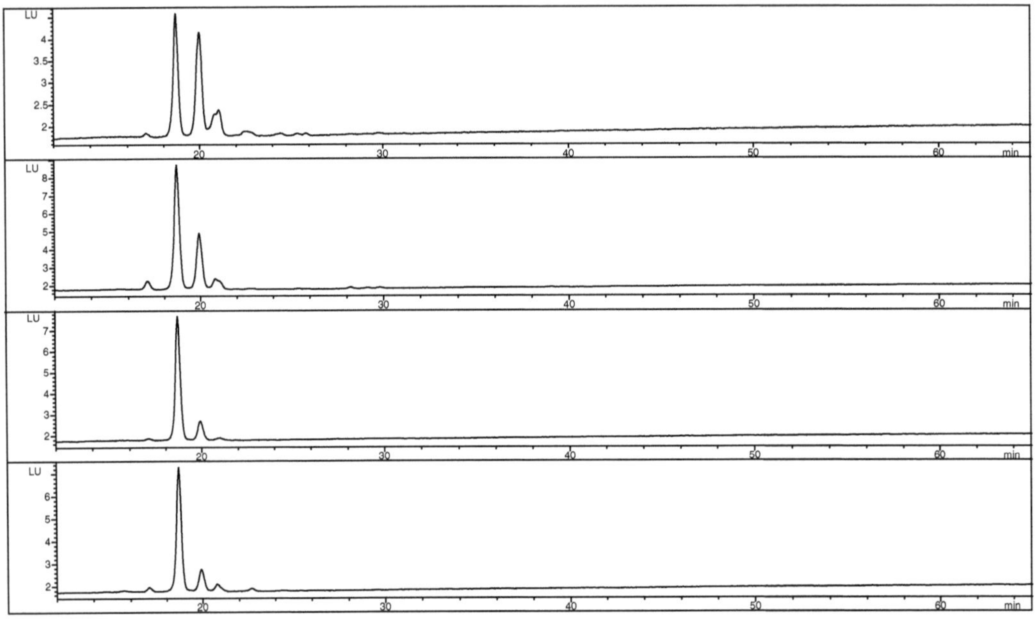

Fig. 3 Comparison of NP-HPLC profiles of glycans released from rMAbs produced using CHO cells

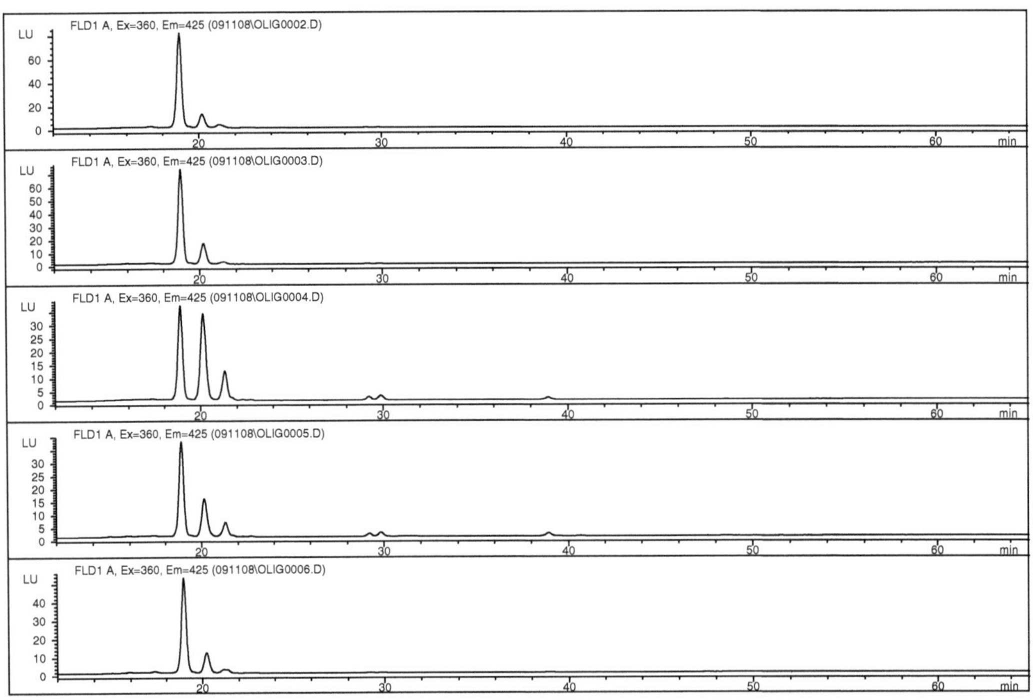

Fig. 4 Comparison of NP-HPLC profiles of glycans released from rMAbs produced using SP2/0 cells

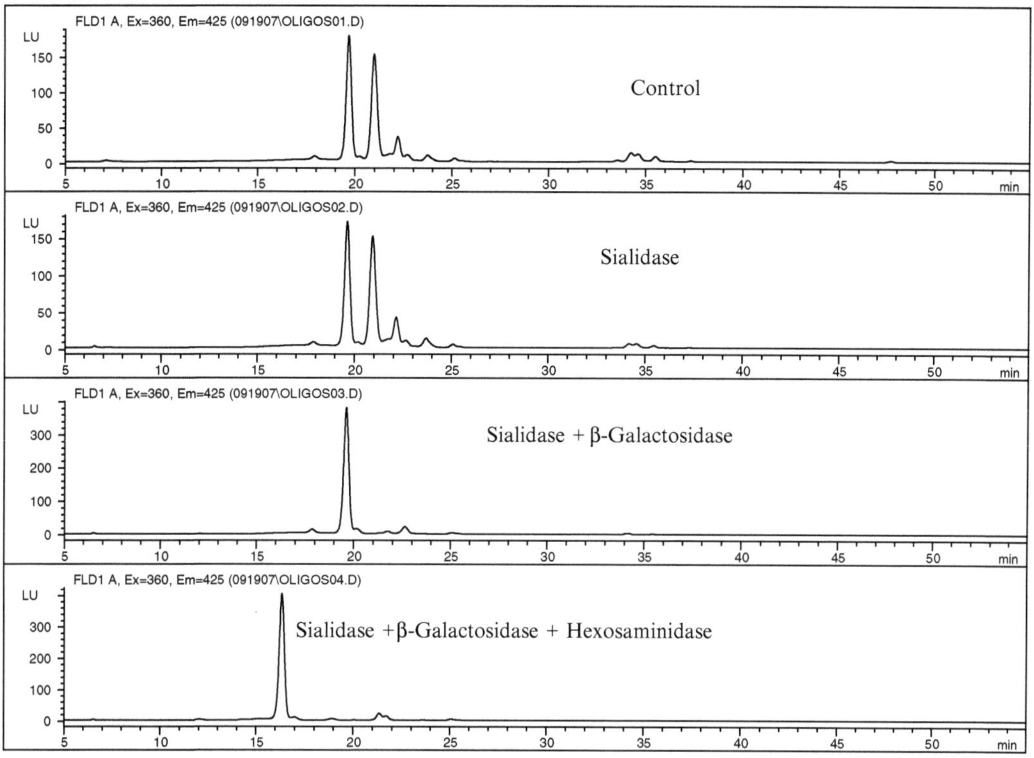

Fig. 5 Sequential exoglycosidase digestion of Fc glycans of rMAb

constant region of heavy chains, differences in their glycan profile are observed. These differences are due to species specific and also due to cell culture variations (13).

9. Figure 5 shows the glycan mapping of rMAb produced in SP2/0 that was subjected to sequential exoglycosidase digestion demonstrating the use of NP-HPLC mapping method to enzymatic sequencing of glycans (26, 27).
10. Figure 6 shows the glycan mapping of rMAbs that were produced under varying cell culture conditions. The glycan maps show minor differences in the ratio of G0, G1, G2, G2S1, and G2S2 glycans. This data demonstrates that the glycan mapping method can be used to monitor changes in glycan profiles of rMAbs during cell line development (26).
11. Figures 7 and 8 show the reproducibility of NP-HPLC maps using human IgG as an example. The peak retention times and peak area are highly reproducible (14, 26).
12. The NP-HPLC method is very rugged and minor changes in the procedure such as little variations in eluants and other reagents do not affect the reproducibility of the glycan map (13, 26).

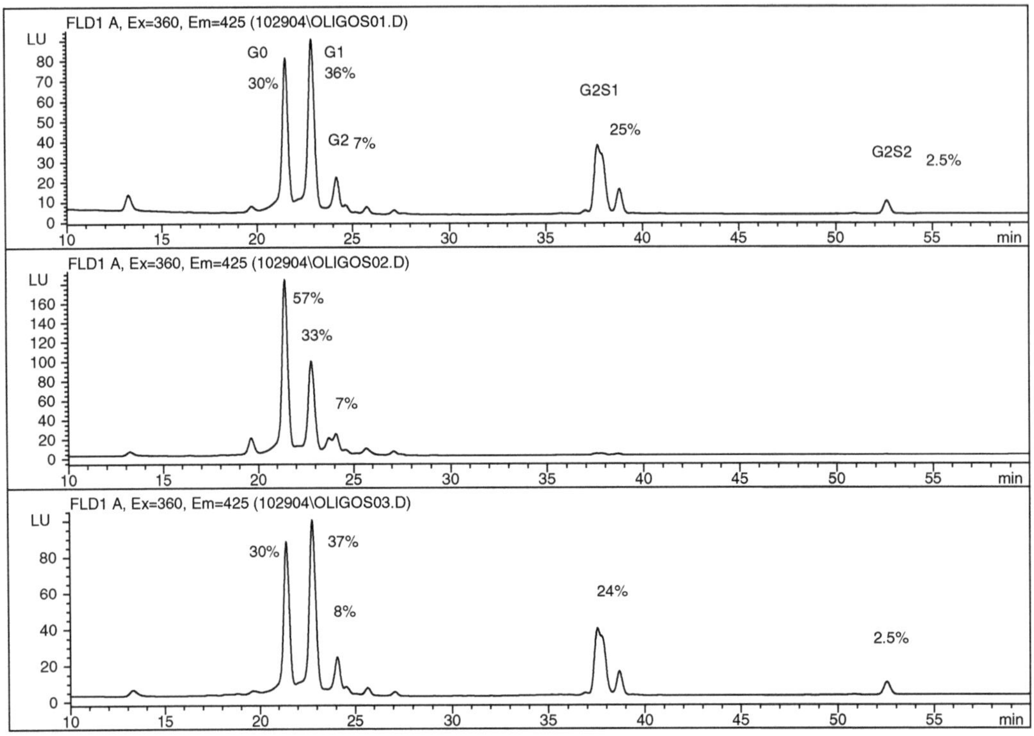

Fig. 6 Use of NP-HPLC to detect sialic acid variations during cell culture

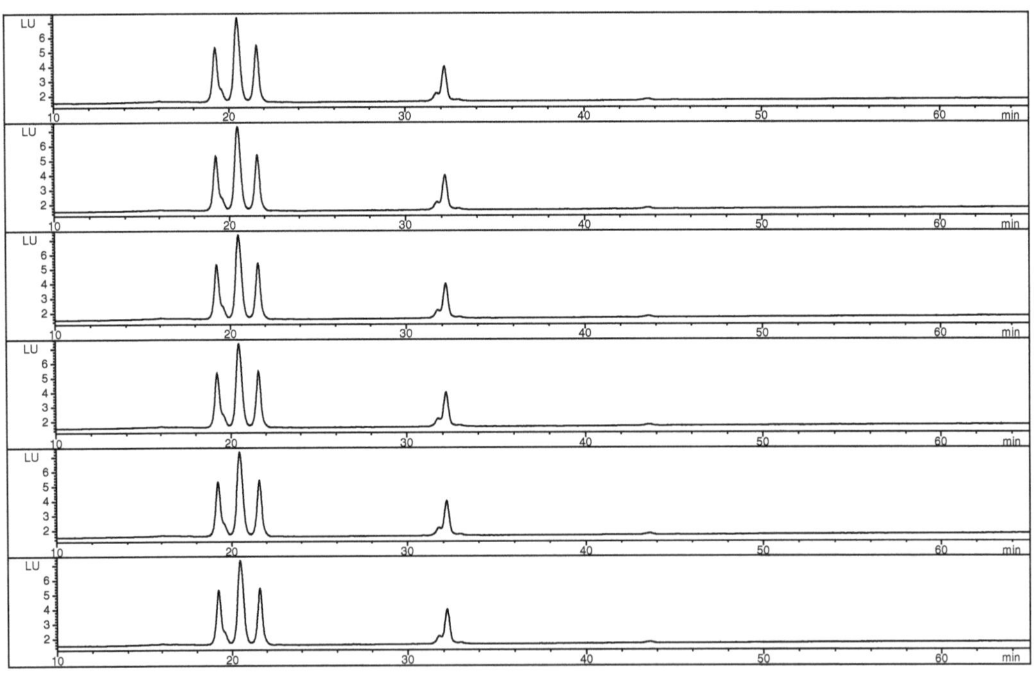

Fig. 7 Stacked chromatograms to show reproducibility of glycan maps using human IgG as an example

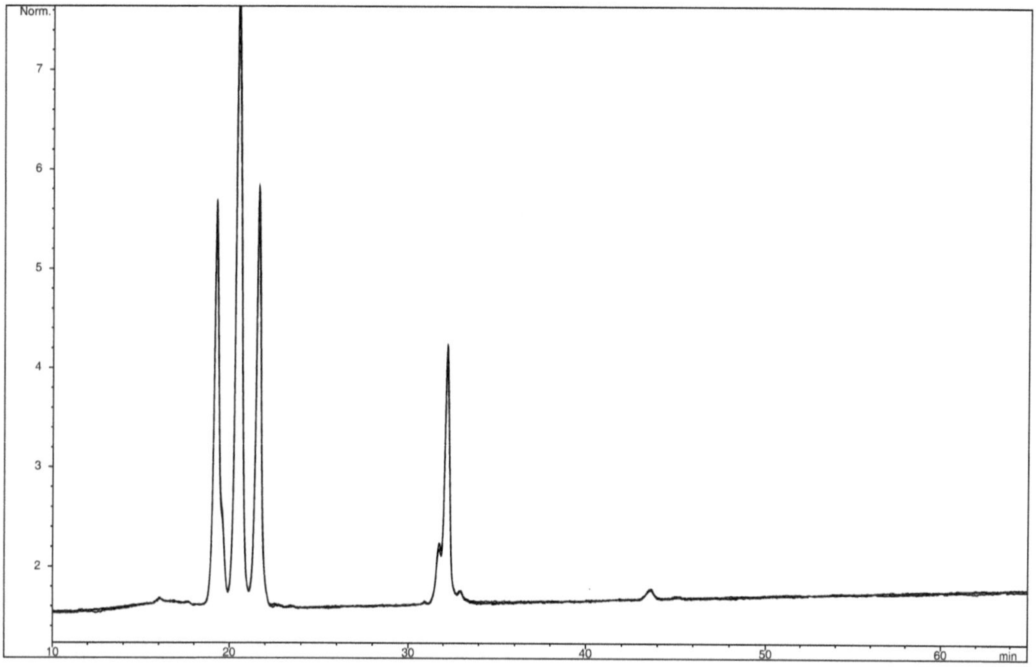

Fig. 8 Overlaid chromatograms to show reproducibility of glycan maps using human IgG as an example ($n = 10$ runs)

References

1. Kornfeld R, Kornfeld S (1985) Assembly of asparagine-linked oligosaccharides. Annu Rev Biochem 54:631–664
2. Varki A (1993) Biological roles of oligosaccharides: all of the theories are correct. Glycobiology 3:97–130
3. Kobata A (1996) A journey to the world of glycobiology. Glycoconj J 17:443–464
4. Dwek RA (1996) Toward understanding the function of sugars. Chem Rev 96:683–720
5. Ashwell G, Harford J (1982) Carbohydrate-specific receptors of the liver. Annu Rev Biochem 51:531–554
6. Stockert RJ, Morell AG, Ashwell G (1991) Structural characteristics and regulation of the asialoglycoprotein receptor. Targeted Diagn Ther 4:41–64
7. Raju TS, Lerner L, O'Connor JV (1996) Glycopinion: biological significance and methods for the analysis of complex carbohydrates of recombinant glycoproteins. Biotechnol Appl Biochem 24(Pt 3):191–194
8. Jefferis R (2005) Glycosylation of recombinant antibody therapeutics. Biotechnol Prog 21: 11–16
9. Raju TS (2008) Terminal sugars of Fc glycans influence antibody effector functions of IgGs. Curr Opin Immunol 20(4):471–478
10. Raju TS (2003) Glycosylation variations with expression systems and their impact on biological activity of therapeutic immunoglobulins. BioProcess Int 1(4):44–53
11. Presta LG (2002) Engineering antibodies for therapy. Curr Pharm Biotechnol 3(3):237–256
12. Raju TS, Scallon BJ (2006) Glycosylation in the Fc domain of IgG increases resistance to proteolytic cleavage by papain. Biochem Biophys Res Commun 341(3):797–803
13. Raju TS, Briggs JB, Borge SM, Jones AJ (2000) Species-specific variation in glycosylation of IgG: evidence for the species-specific sialylation and branch-specific galactosylation and importance for engineering recombinant glycoprotein therapeutics. Glycobiology 10(5):477–486
14. Raju TS (2000) Analysis of glycoconjugates. Anal Biochem 283(2):123–124
15. Raju TS (2000) Electrophoretic methods for the analysis of N-linked oligosaccharides. Anal Biochem 283(2):125–132

16. Townsend RR (1995) Analysis of glycoconjugates using high-pH anion-exchange chromatography. In: El Rassi Z (ed) Carbohydrate analysis: high-performance liquid chromatography and capillary electrophoresis. Elsevier, New York, pp 181–209
17. Davies MJ, Hounsell EF (1998) HPLC and HPAEC of oligosaccharides and glycopeptides. In: Hounsell EF (ed) Glycoanalysis protocols, Methods in molecular biology. Humana Press, Totowa, NJ
18. Anumula KR (2006) Advances in fluorescence derivatization methods for high-performance liquid chromatographic analysis of glycoprotein carbohydrates. Anal Biochem 350: 1–23
19. Anumula KR, Dhume ST (1998) High resolution and high sensitivity methods for oligosaccharide mapping and characterization by normal phase high performance liquid chromatography following derivatization with highly fluorescent anthranilic acid. Glycobiology 8:685–694
20. Naka R, Kamoda S, Ishizuka A, Kinoshita M, Kakehi K (2006) Analysis of total N-glycans in cell membrane fractions of cancer cells using a combination of serotonin affinity chromatography and normal phase chromatography. J Proteome Res 5:88–97
21. Patel T, Bruce J, Merry A, Bigge C, Wormald M, Jacques A, Parekh R (1993) Use of hydrazine to release in intact and unreduced form both N- and O-linked oligosaccharides from glycoproteins. Biochemistry 32:679–693
22. Dhume ST, Anumula KR (1998) Evaluation of various fluorescent tags for oligosaccharide characterization: a comparative study with anthranilic acid. Glycobiology 8, Abstract no. 34
23. Anumula KR (1994) Quantitative determination of monosaccharides in glycoproteins by high performance liquid chromatography with highly sensitive fluorescence detection. Anal Biochem 220:275–283
24. Anumula KR (1995) Novel fluorescent methods for quantitative determination of monosaccharides and sialic acids in glycoproteins by reversed phase high performance liquid chromatography. In: Atassi MZ, Apella E (eds) Methods in protein structure analysis. Plenum Press, New York, pp 195–206
25. Anumula KR (1997) Highly sensitive pre-column derivatization procedures for quantitative determination of monosaccharides, sialic acids and amino sugars alcohols of glycoproteins by reversed phase HPLC. In: Townsend RR, Hotchkiss AT (eds) Techniques in glycobiology. Marcel Dekkar, New York, pp 349–357
26. Raju TS, Jordan RE (2012) Galactosylation variations in marketed therapeutic antibodies. MAbs 4(3):385–391
27. Naso MF, Tam SH, Scallon BJ, Raju TS (2010) Engineering host cell lines to reduce terminal sialylation of secreted antibodies. MAbs 2(5): 519–527

Chapter 11

Application of Capillary Electrophoresis in Glycoprotein Analysis

Richard R. Rustandi, Carrie Anderson, and Melissa Hamm

Abstract

Capillary electrophoresis (CE) is a versatile analytical method used to characterize glycoproteins. We have used several modes of CE separation such as CE-SDS gel, imaged capillary isoelectric focusing (icIEF), and capillary zone electrophoresis (CZE) to study therapeutic glycoprotein products. CE-SDS gel is applied to characterize the glycan occupancy and number of glycosylation sites, and icIEF is used to study the charge heterogeneities due to sialic acids in glycoproteins. To further characterize the glycoprotein, removal of N-linked glycans is necessary and a CZE technique is employed to analyze each glycan moiety. Examples from a monoclonal antibody, erythropoietin, and granulocyte colony-stimulating factor are presented here to demonstrate the utility of these CE modes. The details of sample preparation and separation conditions for each CE mode are described in this chapter.

Key words Capillary electrophoresis, CE-SDS gel, icIEF, Capillary zone electrophoresis, mAb, EPO, GCSF, N-glycan, O-glycan, Ampholytes, APTS

1 Introduction

A glycoprotein is a protein that has different types of carbohydrate moieties attached to one of its amino acids. The carbohydrate groups are linked either through an amide nitrogen on an asparagine group (N-linked glycans) or through an oxygen on a serine/threonine group (O-linked glycans). The oligosaccharides can contain several types of neutral monosaccharides such as galactose, mannose, N-acetylglucosamine, fucose, and negatively charged glycans such as sialic acid. The evaluation of the glycoforms can be done intact (with the glycans still attached to the protein) or by removing glycans from the protein by either enzymatic digestion or chemical hydrolysis. There are several methods available to analyze these glycoforms such as HPLC with fluorescent labels, pulsed amperometric detector (PAD), charged aerosol detector (CAD), SDS-PAGE, MS, NMR, and CE (1–3).

Alain Beck (ed.), *Glycosylation Engineering of Biopharmaceuticals: Methods and Protocols*, Methods in Molecular Biology, vol. 988, DOI 10.1007/978-1-62703-327-5_11, © Springer Science+Business Media New York 2013

In last several years, CE has emerged as an indispensable tool to analyze glycoproteins due to its high resolving power, quantitative analysis, fast analysis, ability to be automated, and more importantly commercial equipment is much more robust. There are several modes of CE separation which include CE-SDS gel, micellar electrokinetic chromatography (MEKC), imaged capillary isoelectric focusing (icIEF), and capillary zone electrophoresis (CZE) to name a few. CE-SDS gel and MEKC are useful to characterize glycoproteins containing neutral oligosaccharides and in particular, CE-SDS is a powerful method to evaluate N-linked or O-linked glycan occupancy. For glycoproteins containing charged oligosaccharides such as sialic acids, icIEF, and CZE methods of separation are more useful. Both methods are complementary to each other; however, icIEF methods are quicker and easier to develop as compared to CZE. The icIEF method is based on total charge separation while CZE is based on charge-to-mass ratio separation. Here we provide examples of methods used to analyze intact glycoproteins using CE-SDS gel, icIEF, and CZE for released oligosaccharides from glycoproteins.

2 Materials

2.1 CE-SDS Gel

1. Beckman PA800 CE system (Fullerton, CA).
2. Barefuse silica capillary 50 μm ID and 365 μm OD $L_T = 30$ cm $L_D = 20$ cm (Beckman Coulter or MicroSolv).
3. Photodiode array detector at 220 nm and reference at 350 nm.
4. SDS Gel Buffer (Beckman Coulter).
5. 0.1 N NaOH (Beckman Coulter or MicroSolv).
6. 0.1 N HCl (Beckman Coulter or MicroSolv).
7. Tris buffer 1 M, pH 8.0 (Sigma).
8. SDS solution, 10% (w/v) (Fisher Scientific).
9. β-Mercaptoethanol (βME) (Acros) (see Note 1).
10. Protein desalting spin column, PN 89849 (Pierce).

2.2 Imaged Capillary Isoelectric Electrophoresis

1. IcIEF instrument, iCE 280 IEF Analyzer with PrinCE MicroInjector (ProteinSimple).
2. FC coated fused silica cIEF cartridge capillary 5 cm long, 100 μm ID, 200 μm OD (ProteinSimple).
3. 1% Methylcellulose solution (ProteinSimple).
4. Pharmalyte Ampholyte, pH 3–10 (GE Healthcare).
5. Pharmalyte Ampholyte, pH 4–6.5 (GE Healthcare).
6. Pharmalyte Ampholyte, pH 5–8 (GE Healthcare).

7. Pharmalyte Ampholyte, pH 8–10.5 (GE Healthcare).
8. o-Phosphoric acid, 85% A.C.S Grade (Fisher Scientific).
9. 1 M NaOH (Fluka).
10. p*I* Marker 3.59 (ProteinSimple).
11. p*I* Marker 4.65 (ProteinSimple).
12. p*I* Marker 7.60 (Sigma-Aldrich).
13. p*I* Marker 9.0 (Sigma-Aldrich).
14. p*I* Marker 9.22 (ProteinSimple).
15. Sialidase A (Glyko).
16. Reaction Buffer, 5× for sialidase treatment (Glyko).

2.3 CZE for N-Glycan Analysis

1. Beckman PA800 CE system (Fullerton, CA).
2. Polyvinyl Alcohol-Coated (PVA) capillary 50 μm ID and 365 μm OD; $L_T = 50$ cm, $L_D = 40$ cm (Beckman Coulter or Agilent).
3. Laser-induced fluorescence (LIF) detector with an Ar laser; $\lambda_{ex} = 488$ nm, $\lambda_{em} = 520$ nm.
4. Carbohydrate separation buffer (Beckman Coulter).
5. 8-Aminopyrene-1,3,6-trisulfonate (APTS) high purity (Fluka or Invitrogen) (see Note 2).
6. Labeling APTS dye solvent (Beckman Coulter).
7. Maltose quantitation marker (Beckman or Supelco).
8. Sodium cyanoborohydride 1 M in THF (Aldrich).
9. PNgase (New England Biolabs).
10. Phosphate buffer saline (PBS).
11. 10% (w/v) SDS solution (Fisher Scientific).
12. βME (Aldrich).
13. Nonidet NP-40 (Pierce).
14. Ethanol, 200 Absolute Proof (Aldrich).
15. Protein desalting columns, PN 89849 (Pierce).

3 Methods

3.1 CE-SDS Gel

This CE mode of separation is based on the molecular mass of the protein migrating through a replaceable sieving gel matrix in the capillary. The method was developed originally to replace the traditional, labor-intensive SDS-PAGE slab gel. CE-SDS has the ability to be automated, quantitative, and robust, and provides better resolution compared to slab gels. The high resolving power of CE-SDS gel is a major advantage in analyzing glycoproteins, especially in

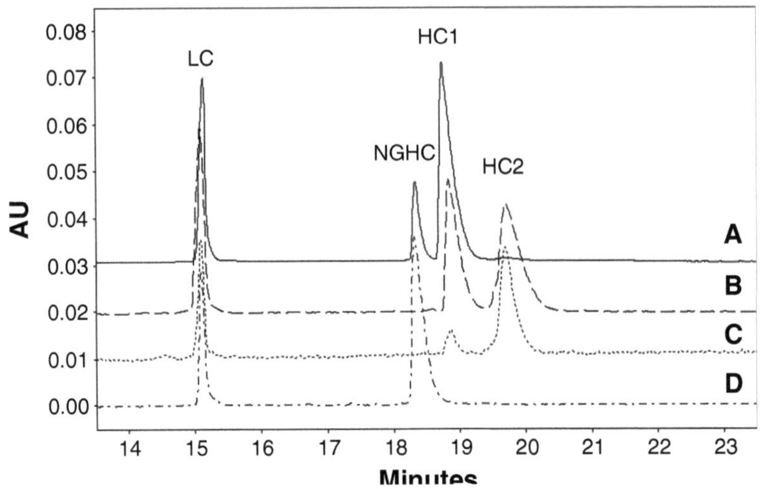

Fig. 1 Electropherograms of reduced mAb produced under various cell culture conditions. (*A*) NGHC is produced in significant amount (12%); (*B*) cell culture produces mAb with the normal one glycosylation site (HC1) and two glycosylation sites in HC (HC2); (*C*) significant HC2 material is produced in this sample; (*D*) PNgase-treated sample in (*C*) shows that both peaks are converted into NGHC confirming HC2 contains two glycosylation sites. PNGase-treatment is described in Subheading 3.3

determining the glycan occupancy (4–6). For example, the method is capable of separating the glycosylated and non-glycosylated forms of the protein. This method can be applied to determine occupancy for both N-linked (see Figs. 1 and 2) or O-linked (Fig. 3) glycans.

CE-SDS gel employs SDS to denature and bind to protein in a constant weight ratio (1.4 g SDS per 1 g protein). This leads to a constant charge-to-mass ratio for all proteins providing the mechanism of separation based on hydrodynamic size. Proteins traveling in the sieving matrix with highest MW migrate the slowest and pass the detector at later time. The separation between glycosylated and non-glycosylated proteins is largely due to the fact that glycans do not bind SDS; therefore, the expected charge-to-mass ratio is lower and its migration time is decreased. Hence, the glycoprotein always migrates slower compared to its non-glycosylated counterpart.

The first example is a monoclonal antibody (mAb) (Fig. 1) which has N-linked glycosylation sites(s) normally located on the heavy chain (HC) (~50 kDa). Figure 1a illustrates a generic mAb that has one N-linked glycosylation site with about 12% of non-glycosylated HC (NGHC). Figure 1b, c shows that two HC peaks are observed, the first HC1 is the expected HC as in Fig. 1a while the HC2 has an additional site of N-linked glycosylation (confirmed by Mass Spectrometry). Figure 1c has mostly HC2 species present containing two glycosylation sites. The presence of glycosylation can be further confirmed by using an enzyme (PNgase) to remove

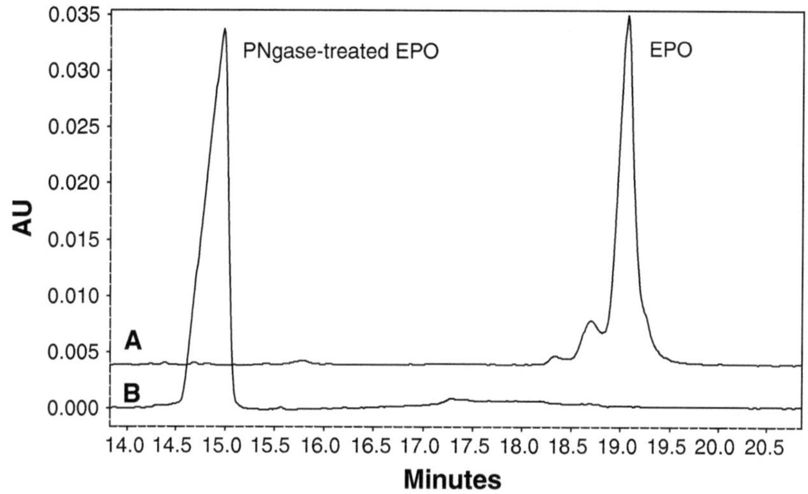

Fig. 2 CE-SDS electropherogram of yeast-derived EPO; (*A*) EPO sample contains at least three partially resolved *peaks* at 18–19.5 min indicating N-glycan heterogeneities; (*B*) EPO treated with PNgase for 18 h at 37°C shows that its *peak* at 14.7 min is more homogeneous indicating that no N-glycan is attached and it yields an expected MW of 18 kDa for non-glycosylated EPO. Note that PNGase-treatment is described in Subheading 3.3

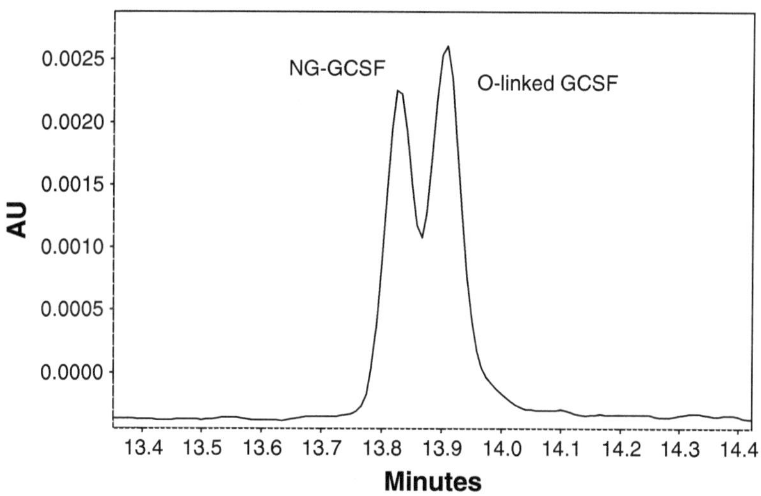

Fig. 3 Electropherogram of yeast-derived GCSF produced. The sample contains non-glycosylated GCSF (NG-GCSF) and O-linked GCSF. The assignment was done indirectly by HPLC and MS

the N-linked glycans. PNgase-treatment causes HC1 and HC2 to collapse into NGHC in Fig. 1d. The second example depicts N-linked glycans on EPO which is known to contain three N-linked sites. Figure 2 (trace A) illustrates that EPO has three partly resolved peaks which are likely due to glycan heterogeneities and after PNgase-treatment they collapse into one homogeneous almost symmetrical peak indicating that glycans are no longer attached (Fig. 2, trace B). The third example is O-linked glycan on granulocyte-colony stimulating factor (GCSF) which has at least one O-linked glycosylation site. Results for the GCSF protein are illustrated in Fig. 3 giving two partly resolved peaks. The first peak is the non-glycosylated form of GCSF and the second peak is the O-linked glycosylated form of GCSF. Since there is no specific enzyme for O-glycan cleavage, orthogonal methods such as HPLC and mass spectrometry were used to confirm the assignment (data not shown).

3.1.1 10× SDS Tris Sample Buffer Preparation

1. MilliQ water, 4 mL.
2. 10% SDS stock, 5 mL.
3. 1 M Tris buffer pH 8.0, 1 mL.
4. Vortex and store at 22–25°C for up to 1 year.

3.1.2 Reducing Condition CE-SDS Gel Sample Preparation

1. Place glycoprotein (0.5–1.5 mg/mL final) in 500 μL Eppendorf tube (see Note 3).
2. Add 10 μL of 10× SDS Tris sample buffer (see Note 4).
3. Add 5 μL of βME solution.
4. Add water necessary to obtain final volume of 100 μL (see Note 5).
5. Place in 70°C heating block for 10 min (see Note 6).
6. Equilibrate at room temperature for 10 min and centrifuge it to collect all liquid.
7. Transfer to CE PCR vial; avoid bubbles while transferring.

3.1.3 CE-SDS Gel Method

1. Install bare fused silica capillary according to Beckman user's manual (7) (see Note 7).
2. Precondition the capillary according to the Beckman IgG Purity Heterogeneity manual (8).
3. Rinse capillary with 0.1 N NaOH for 3 min at 70 psi in forward direction (see Note 8).
4. Rinse capillary with 0.1 N HCl for 1 min at 70 psi in forward direction (see Note 8).
5. Rinse capillary with water for 1 min at 70 psi in forward direction (see Note 8).
6. Rinse capillary with gel separation buffer for 10 min at 70 psi in reverse direction (see Note 8).
7. Place both ends of the capillary in water to remove residual viscous gel (see Note 8).
8. Inject the sample from the anode with reverse polarity at −5 kV for 25 s (see Note 5).

9. Place both ends of the capillary in water to remove residual sample (see Note 8).
10. Separate the sample at −15 kV with reverse polarity for 30 min and apply 20 psi pressure at both ends of capillary at the same time (see Note 9).

3.2 Imaged cIEF

This imaged cIEF method of separation is based on the total charge heterogeneities in glycoprotein. The method was developed to replace the labor-intensive traditional slab gel IEF. This technique provides a very fast, efficient, automatable, quantitative, robust, and highly reproducible method for the analysis of glycoproteins. This high-resolution technique is very useful in characterizing glycoproteins containing different forms of glycan with similar mass but different isoelectric points due to sialylation, phosphorylation, or sulfonation.

In imaged cIEF, a solution containing a complex mixture of ampholyte, p*I* markers and the protein mixture of interest is introduced to an FC-coated capillary. A high voltage is applied across the capillary and the ampholytes establish a pH gradient along the length of capillary. Under the influence of an electric field the p*I* markers and glycoprotein of interest will migrate along the capillary until they reach a pH value where their net charge is zero. Once focused, the iCE280 system uses whole-column imaging detection with a 280 nm CCD camera to monitor the IEF pattern of glycoprotein within the capillary column (9–11). A p*I* resolution of 0.05 pH unit can be achieved for glycoproteins.

Figure 4 (trace A) illustrates the isoform profile of a mAb containing multiple peaks with a p*I* ranging from 7.2 to 7.8. Sialidase treatment of the sample yields the electropherogram in Fig. 4 (trace B) in which some acidic peaks disappear confirming that several of the acidic variants are the result of a difference in sialylation of the glycoprotein. Figure 5 (trace A) shows the isoform distribution of yeast-derived EPO. EPO contains three potential N-glycosylation sites and different amounts of sialic acid capping of oligosaccharides attached to the different sites on the protein backbone lead to a mixture of isoforms of EPO. While the theoretical isoelectric point of the protein backbone is ~8.4, experimentally EPO ranges in p*I* from 4.5 to 5.1. After sialidase treatment (Fig. 5, trace B), one major and several minor species are visible at p*I* 8.5, indicating that the complexity of the EPO protein is mostly due to heterogeneity in sialic acids attached to oligosaccharide residues.

3.2.1 Solution Preparation

Methylcellulose Solution, 0.5%

1. 1% Methylcellulose solution, 10 mL.
2. MilliQ water, 10 mL.
3. Vortex and store at 2–8°C for up to 1 month.

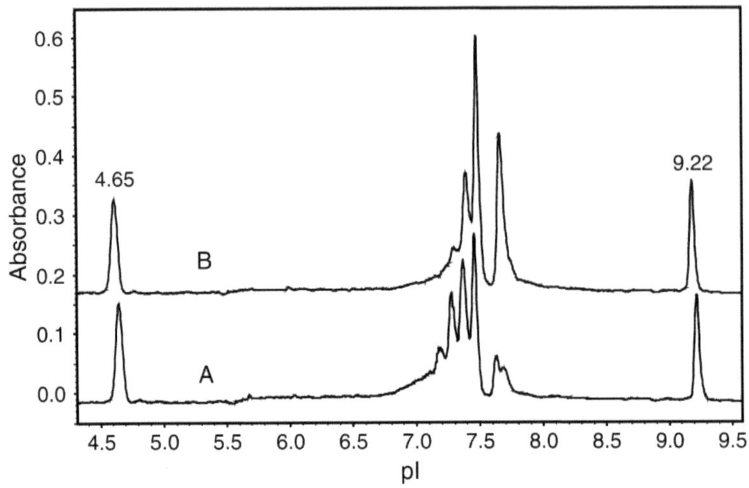

Fig. 4 Desialylation of a murine mAb. (*A*) Untreated mAb held at 4°C; (*B*) mAb treated with sialidase for 18 h at 37°C. Sample containing 20 μL mAb (20 mg/mL in PBS), 1 μL p*I* marker 4.65, 1 μL p*I* marker 9.22, 70 μL 1% methylcellulose, 100 μL MilliQ water, 10 μL pharmalyte 3–10 ampholytes; focusing time 8 min, 80 mM H_3PO_4 in anolyte, and 100 mM NaOH in catholyte

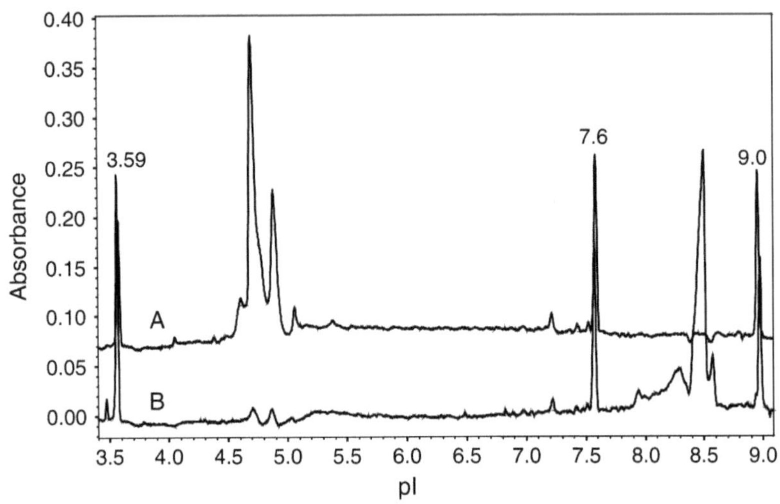

Fig. 5 Desialylation of yeast-expressed EPO. (*A*) Untreated EPO held at 4°C; (*B*) EPO treated with sialidase for 18 h at 37°C. Sample containing 70 μL EPO (2 mg/mL), 70 μL 1% methylcellulose, 100 μL MilliQ water, 10 μL pharmalyte ampholytes; focusing time 7 min. Untreated sample and sialidase-treated sample prepared using pH 3–10 pharmalytes; both samples contain 1 μL each of the following p*I* markers 3.59, 7.60, 9.0

Methylcellulose Solution, 0.1%

1. 1% Methylcellulose solution, 4 mL.
2. MilliQ water, 36 mL.
3. Vortex and store at 2–8°C for up to 1 month.

Phosphoric Acid (1 M)

1. 85% o-phosphoric acid, 6.7 mL.
2. MilliQ water, QS to 100 mL.
3. Vortex and store at 22–25°C for up to 1 year.

Anolyte Solution

1. 0.1% Methylcellulose solution, 2 mL.
2. 1 M phosphoric acid, 160 μL.
3. Vortex and prepare a fresh solution after 12 h of use.

Catholyte Solution

1. 0.1% Methylcellulose solution, 2 mL.
2. 1 M sodium hydroxide, 200 μL.
3. Vortex and prepare a fresh solution after 12 h of use.

3.2.2 Sample Preparation

1. Add 70 μL 1% methylcellulose to a microcentrifuge tube.
2. Add 10 μL pharmalyte ampholytes pH 3–10 to the methylcellulose solution (see Note 10).
3. Vortex to ensure that the ampholytes are dispersed homogeneously.
4. Dilute sample to 0.5 mg/mL in the methylcellulose/ampholyte solution (see Note 11).
5. Add water to bring the final volume in the vial to 200 μL.
6. Vortex to mix.
7. Add 1 μL of each p*I* marker needed to the above solution (see Note 12).
8. Vortex to mix.
9. Centrifuge at ambient temperature (25 ± 5°C) for 3 min at 6700 × *g*.
10. Remove 150 μL of sample supernatant and transfer to an appropriate vial (see Note 13).

3.2.3 Sialidase Digestion (See Note 14)

1. Add 100 μg of glycoprotein to a microcentrifuge tube.
2. Add MilliQ water to a total of 14 μL.
3. Add 4 μL 5× reaction buffer supplied from Glyko.
4. Add 2 μL sialidase A.
5. Incubate at 37°C overnight (16–20 h).

3.2.4 Imaged cIEF Method

1. Install column cartridge according to the iCE 280 Manual (12) (see Note 15).
2. Start up system according to iCE 280 Manual (12) (see Note 16).
3. Fill electrolyte tanks on the column cartridge with anolyte and catholyte solutions.

4. Set autosampler temperature at 8°C.
5. Set prefocus period 1 between 1 and 2 min (see Note 17).
6. Set focus period 2 between 5 and 20 min (see Note 17).
7. The sample and carrier ampholyte mixture is injected by the autosampler into the inlet capillary until the cartridge is overflowing.
8. Apply 3 kV DC voltage through electrodes.
9. Rinse capillary with 0.5% methylcellulose.

3.3 CZE for N-Glycan Analysis

Glycans are first released from the glycoprotein using the enzyme called PNgase, which cleaves specifically at the bond between the asparagine and the β-*N*-acetylglucosamine; the released glycans are then separated from protein, and are dried down. The dried glycans are then labeled with a fluorophore, APTS, which binds to the glycan reducing termini by reductive amination. The stoichiometry of the labeling reaction is such that only one APTS molecule is attached to each molecule of glycan. The labeled glycans are injected onto a PVA-coated capillary and separated by CZE. The mobility of the samples is compared to the maltose internal standard (13).

The CZE mode of separation is based on the charge-to-mass ratio of the glycans. Since all of glycans are labeled with APTS with 1-to-1 ratio, neutral glycans will contain the same charge. Therefore for neutral glycans, the separation is driven by the size and structure/shape, small glycans migrate past the detector first, and larger glycans migrate slower and are detected later. But for charged glycans, in particular sialylated glycans, the additional charge has an impact on the separation. For sialylated glycans, larger more sialyated glycans migrate faster than smaller or less sialylated glycans (14).

The glycosylation patterns are heterogeneous and dependent on the cell line the glycoprotein is expressed in and the fermentation conditions used during cell growth. CZE is a useful technique to profile the glycosylation pattern and can be used to monitor changes during cell line development and fermentation development. It can also be used to monitor lot-to-lot consistency of production material. A typical profile of a mAb produced in NS/0 cells is shown in Fig. 6a. The predominant glycan in this expression system is G1F. The identities of the glycoforms were confirmed using a combination of commercial glycan standards, specific endoglycosidases, HPLC, and MS (data not shown). As mentioned previously, the separation is based on the size and structure/shape of the glycan; the power of separation is clearly shown in distinguishing between G1F(1,6) versus G1F(1,3) and G2F+2αgal(1,6) versus G2F+2αgal(1,3) isomers which both have same size but different structures (see Fig. 7 for details of glycan structures). But in the case of sialylated glycans such as those present on EPO, the

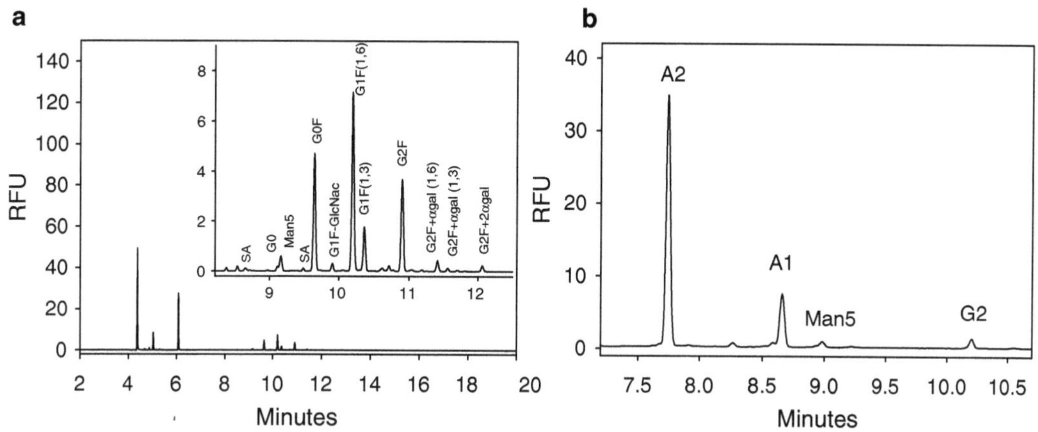

Fig. 6 (**a**) N-glycan electropherogram profile of mAb produced in NS/0. The whole trace shows that *peaks* at 4–6 min are excess dye and the glycan peaks of interest are located between 8 and 12 min. The *inset* figure is the expanded version and the assignments are based on the glycan standards, enzymatic digestion, and HPLC-MS. The glycan structures are illustrated in Fig. 7 and the sialic acid (SA) assignment is based on sialidase treatment. (**b**) N-glycan profile of yeast-derived EPO. The assignment is based on glycan standards and enzymatic digestion. Their glycan structures are also illustrated in Fig. 7

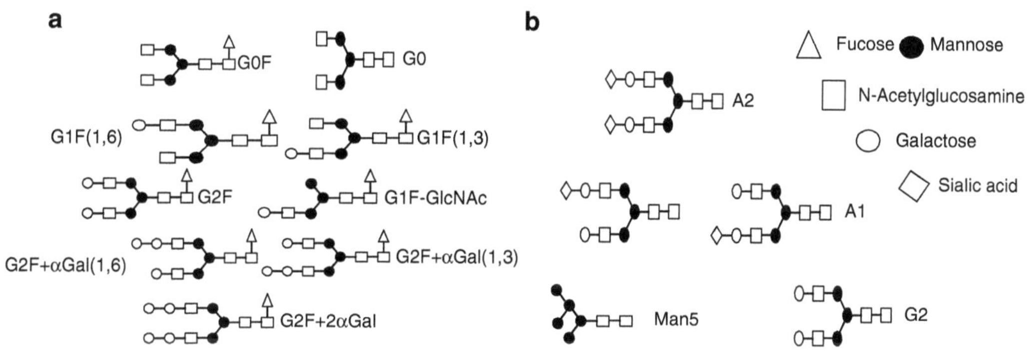

Fig. 7 Various biantennary glycan structures typically found in glycoproteins. Note there are two different A1 isomers

separation is also based on the additional charge from the sialic acid species. Figure 6b shows the typical profile of a yeast-derived sialylated EPO glycoprotein. The predominant glycan present on this molecule is A2 (containing two sialic acid moieties) which migrates first, then A1 (containing one sialic acid moiety), and finally the neutral glycans such as Man5 and G2 (see glycan structure in Fig. 7). The identities of the glycans were confirmed by using a combination of commercially available standards and endoglycosidases (data not shown).

3.3.1 PNgase Preparation (10,000 U/mL)

1. 10 μL of 500,000 U/mL PNgase stock.
2. 490 μL MilliQ water.
3. Vortex and store at 2–8°C for up to 3 months.

3.3.2 SDS Preparation (5%)

1. 10 mL MilliQ water.
2. 10 mL 10% (w/v) SDS.
3. Gently invert to mix and store at 22–25°C for up to 1 year.

3.3.3 APTS Labeling Dye Preparation (50 mg/mL)

1. 500 μL labeling dye solvent to 25 mg APTS.
2. Vortex to mix and store at ≤–70°C for up to 3 months.

3.3.4 βME (1:10) Solution

1. 1 mL βME.
2. 9 mL MilliQ water.
3. Vortex to mix and store at ≤–70°C for up to 2 years.

3.3.5 Maltose (1 nmol/μL) Internal Standard Preparation (Beckman Coulter)

1. Add 500 μL MilliQ water to the maltose (1 nmol/μL) quantitative control vial.
2. Vortex to mix and store at ≤–70°C for up to 2 years.

3.3.6 N-Glycan Sample Preparation

1. All samples will need to be buffer exchanged into either PBS or water (see Note 18).
2. Place 300 μg of protein into microcentrifuge tubes and dry completely using a centrifugal vacuum evaporator.
3. Add 45 μL PBS, 1.5 μL 5% SDS, and 1 μL βME (1:10) solution to the pellet.
4. Vortex to mix (see Note 19).
5. Incubate for 10 min at 37°C (see Note 20).
6. Equilibrate the tubes to 22–25°C.
7. Add 5 μL NP-40 and 10 μL PNgase.
8. Incubate for 14–20 h at 37°C (see Note 20).
9. Equilibrate the tubes to 22–25°C and add 3 volumes of cold ethanol.
10. Incubate the tubes on ice for 10–15 min.
11. Centrifuge at 19,000 × *g* for 7 min at 4°C (see Note 21).
12. Remove supernatant and add 5 μL maltose to each sample.
13. Dry completely in a centrifugal vacuum evaporator (see Note 22).
14. Add 2 μL sodium cyanoborohydride and 2 μL APTS to the pellet.
15. Incubate at 60°C for 2 h.
16. Stop the reaction with 46 μL water and vortex.
17. Dilute the samples 20-fold (5 μL sample + 95 μL water) in CE PCR vials.

3.3.7 N-Glycan CZE Method

1. Install a PVA-coated capillary according to the Beckman user's manual (7, 15) (see Note 7).
2. Rinse capillary with carbohydrate buffer for 2 min at 30.0 psi in the forward direction (see Note 23).
3. Inject sample under a pressure of 0.5 psi for 20.0 s (see Note 24).
4. Dip inlet side capillary in water and wait for 0.20 min to remove residual sample from the tip.
5. Perform the separation at −30 kV for 20 min using reverse polarity (see Note 25).

4 Notes

1. Use only βME as the reducing agent and work under a chemical hood since this has an unpleasant odor. Another commonly used alternative reducing agent dithiothreitol (DTT) does not work; it gives a background peak in the electropherogram.
2. The purity of the APTS is extremely important. To assess the purity of the APTS, look for mirror image "ghost peaks" (Fig. 8) that migrate at a later time. These peaks are likely due to the impurity of APTS that contains two sulfonate groups in addition to the three sulfonate groups.
3. If the salt concentration in the glycoprotein sample is >50 mM then desalting is necessary since this method is using electrokinetic injection. Protein desalting spin columns from Pierce can be used for desalting purposes.

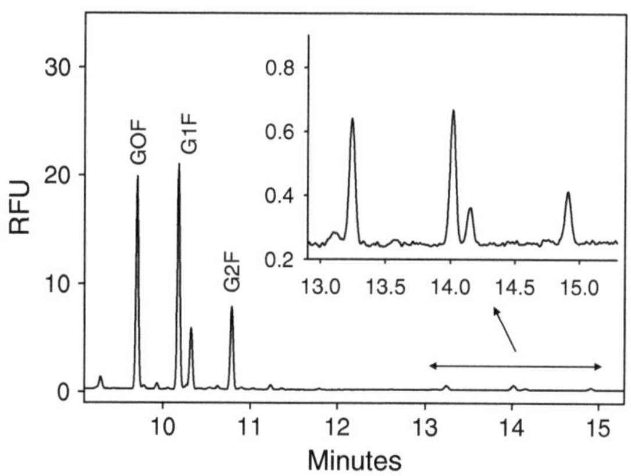

Fig. 8 Electropherogram illustrating APTS impurities in N-glycan analysis. The normal N-glycan *peaks* migrate between 8 and 12 min, but we observe a mirror image of the main glycoforms: G0F, G1F, and G2F between 13 and 15 min which is due to APTS containing two sulfonate groups (see expanded version, *inset*)

4. Sometimes for high glycoprotein concentrations such as in mAbs, precipitation occurs after adding the 10× SDS Tris buffer, but it disappears after adding the βME solution and heating for 10 min at 70°C. Final concentration of SDS currently is 0.5% but can be increased up to 1%.
5. Final volume of sample can be varied from 50 to 100 μL. Do not prepare sample solution in a volume greater than 100 μL. Injection times can be varied from 15 to 30 s; although injection times up to 90 s can be used, it is not recommended to use injection times greater than 30 s since resolution will be deteriorated.
6. Heating conditions for each glycoprotein normally need to be optimized. The current format of 10 min at 70°C is a good general starting point. For example, mAb is known to degrade at higher temperatures and longer heating times.
7. It is important when the capillary is installed that its length in inlet side is slightly shorter than the electrode.
8. Volume for NaOH and HCl in vial is 1.5 mL each, volume of water in vial is 1.6 mL, and gel separation buffer volume is 1.4 mL. These volume requirements help robustness of the run.
9. A typical current of 27 ± 3 μA should be observed during separation.
10. The method was not optimized to give the maximum resolution for each protein. In order to increase peak resolution optimal carrier ampholyte mixtures and focusing time are needed. It is known that different brands of ampholytes may provide different resolutions due to their differing compositions and mixtures of narrow range and wide-range ampholytes may be necessary to provide adequate resolution. For this reason, several brands of ampholytes should be evaluated during method development as well as different combinations of ampholytes mixed in various proportions. For mAb and EPO, wide-pH-range carrier ampholyte (pH 3.0–10.0) alone provided only partial resolution of the peaks. Of the various ampholytic conditions evaluated, optimal resolution was obtained by using a mixture of narrow-pH-range pharmalyte ampholytes such as pH 4–6.5, 5–8, and 8–10.5.
11. Sample at a concentration of 0.5 mg/mL was found to exhibit an optimal signal-to-noise ratio without a loss in resolution.
12. p*I* markers were chosen which bracket the sample of interest.
13. It is important to carefully add the sample solution to the vial so as to avoid the introduction of air bubbles into the solution. Air bubbles in the sample solution injected into the separation column will result in noise spikes in the electropherogram.

14. Sialidase A cleaves all nonreducing terminal sialic acid residues from complex carbohydrates and glycoproteins. The relative cleavage rates for different linkages are as follows: α(2–6) > α(2–3) > α(2–8), α(2–9). The enzyme will also cleave branched sialic acids, a unique property among sialidases. Higher concentrations of enzyme and prolonged incubation times may be required for cleaving branched residues (Glyko Sialidase A package insert). Therefore, it is necessary to optimize digestion conditions for each glycoprotein.
15. It is important to avoid contamination of the optical pass of the cartridge during the installation process. Hold the cartridge only by the two electrolyte tanks. Do not touch any other surfaces of the cartridge.
16. As part of the start-up procedure the iCE280 instrument should be checked to ensure optimal system performance. Several procedures are performed to ensure that the instrument is performing according to the manufacturer's specifications. The first is to perform a transfer time measurement. The transfer time is the time for the sample solution to travel from the sample transfer line to the separation column in the cartridge. This time is measured by tracing the current across the column under 500 V while a test sample is injected. The current-time plot should meet the acceptance listed in the iCE280 manual. The second procedure is to calibrate the light intensity. According to the manufacturer, the intensity of the maximum point in the background light intensity for a cartridge should be between 3,000 and 4,000 and the lowest point in the profile should be higher than 1/2 of the value at its maximum. The number of spikes should be under five (spikes in the profile are caused by defects in the column). The third procedure is to record an electropherogram of a protein standard and compare it with a reference electropherogram.
17. Focusing time should be optimized by observing the extent of focusing in the electropherograms when the sample is analyzed using several different focusing times. The sample should be prepared using the optimized ampholytic conditions. The focusing time chosen should be just long enough for complete focusing. When the focusing time is longer than necessary no improvement in focusing is observed with greater focusing times.
18. This can be easily done with desalting columns. PBS is the buffer of choice since not all proteins are stable in water. Most amine-containing buffers like Tris, Trizma, or histidine will interfere with the labeling reaction, so these buffers will need to be removed prior to labeling the released glycans. It is also recommended to measure the concentration of the samples

after the buffer exchange to ensure that a consistent amount of protein is dried each time.

19. In some cases the SDS will precipitate the protein; this is normal and will disappear upon the first incubation.
20. For some samples the incubation time will need to be optimized for both the SDS and PNgase treatment to ensure that all of the glycan is cleaved from the protein. Use the CE-SDS method in Subheading 3.1 of this chapter to perform this optimization.
21. Centrifuging the samples at 4°C helps keep the pellet visible, so it is easier to see when removing the supernatant.
22. It is important that the samples are completely dry before proceeding with the next step since any traces of water will interfere with the labeling reaction.
23. Volume of water and carbohydrate separation buffer is 1.6 mL and a waste vial should be filled with 800 μL water.
24. Injection times can be varied from 3 to 30 s and 1 psi pressure can be also used.
25. A current of between −15 and ±2 μA should be observed during the separation.

Acknowledgments

The authors would like to thank our colleagues in Bioprocess and GlycoFi for providing us with material and technical support. We gratefully thank Anna Mach for CE-SDS gel, Brian Peklansky for icIEF, and Catherine Lancaster for CZE N-glycan experimental support. Finally, we thank our management support Drs. Yang Wang and Michael W. Washabaugh.

References

1. Mechref Y, Novotny MV (2002) Structural investigations of glycoconjugates at high sensitivity. Chem Rev 102:321–369
2. Townsend RR, Hotchkiss AT Jr (1997) Techniques in glycobiology. Marcel Dekker, New York
3. Hanneman AJS, Rouse JC, Acworth I, Waraska J, Plante M (2009) Profiling and characterization of N- and O-linked glycans released from glycoproteins using PGC RP-HPLC with charged aerosol detection and mass spectrometry. In: The 57th American Society for Mass Spectrometry Poster Conference, Philadelphia, USA
4. Rustandi RR, Washabaugh MW, Wang Y (2008) Application of CE SDS gel in development of biopharmaceutical antibody-based products. Electrophoresis 29:3612–3620
5. Rustandi RR, Washabaugh MW, Sitrin RD, Wang Y (2005) Application of CE SDS gel technique in development of recombinant vaccines and therapeutic proteins. In: CE in the biotechnology & pharmaceutical industries 7th Symposium, Montreal, Canada, p. 18
6. Rustandi RR, Wang Y (2011) Use of CE-SDS gel for characterization of monoclonal antibody hinge region clipping due to copper and high pH stress. Electrophoresis 32:3078–3084
7. Beckman Coulter (2003) ProteomeLab™ PA800 User's Information Protein Characterization System, Maintenance procedures, pp. 23–38

8. Beckman Coulter (2004) IgG purity/heterogeneity assay SOP
9. Mao Q, Pawliszyn J (1999) Capillary isoelectric focusing with whole column imaging detection for analysis of proteins and peptides. J Biochem Biophys Methods 39: 93–110
10. Wu J, Wu XZ, Huang T, Pawliszyn J (2004) Analysis of proteins by CE, CIEF and microfluidic devices with whole-column-imaging detection. In: Strege MA, Lagu AL (eds) Methods in molecular biology: capillary electrophoresis of proteins and peptides, vol 276. Humana Press, New Jersey, pp 229–252
11. Anderson C, Wang Y, Rustandi RR (2012) Applications of imaged capillary isoelectric focusing technique in development of biopharmaceutical glycoprotein-based products. Electrophoresis 33:1538–1544
12. Convergent Bioscience (2008) iCE280 training manual
13. Evangelista RA, Liu M-S, Chen F-T (1995) Characterization of 9-aminopyrene-1,4,6-trisulfonate-derivatized sugars by capillary electrophoresis with laser-induced fluorescence detection. Anal Chem 67:2239–2245
14. Ma S, Nashabeh W (1999) Carbohydrate analysis of a chimeric recombinant monoclonal antibody by capillary electrophoresis with laser-induce fluorescence detection. Anal Chem 71:5185–5192
15. Beckman Coulter. Carbohydrate labeling and analysis guide, PA800 instruction manual

Chapter 12

Characterization of Glycoprotein Biopharmaceutical Products by Caliper LC90 CE-SDS Gel Technology

Grace Chen, Sha Ha, and Richard R. Rustandi

Abstract

Over the last decade, science has greatly improved in the area of protein sizing and characterization. Efficient high-throughput methods are now available to substitute for the traditional labor-intensive SDS-PAGE methods, which alternatively take days to analyze a very limited number of samples. Currently, PerkinElmer® (Caliper) has designed an automated chip-based fluorescence detection method capable of analyzing proteins in minutes with sensitivity similar to standard SDS-PAGE. Here, we describe the use and implementation of this technology to characterize and screen a large number of formulations of target glycoproteins in the 14–200 kDa molecular weight range.

Key words Caliper LC90, Caliper GX/GXII, CE-SDS gel, N-linked glycan occupancy, High-throughput methods, Molecular weight, Glycoprotein characterization, Monoclonal antibody, EPO, Fc-fusion protein

1 Introduction

Size-based protein separation analysis, such as SDS-PAGE gel, has been used over the last several decades to study glycoproteins (1–3). Typically, proteins bind to SDS with a stoichiometric ratio of 1.4 g SDS per gram protein or approximately one detergent molecule per two amino acid residues (4–6). This constant binding ratio provides identical size unit mobility in solution for all proteins; hence allows separation based on number of units and hydrodynamic size. Unique to the glycoprotein, additional separation of glycoprotein from protein can be obtained through SDS-PAGE based on the amount of SDS that binds to glycoprotein. This is because glycan moieties of glycoprotein are hydrophilic and do not bind SDS, whereas protein does bind. Therefore, the effective charge to mass ratio is lower for glycoprotein forms and the migration time becomes slower compared to non-glycosylated counterparts.

Alain Beck (ed.), *Glycosylation Engineering of Biopharmaceuticals: Methods and Protocols*, Methods in Molecular Biology, vol. 988, DOI 10.1007/978-1-62703-327-5_12, © Springer Science+Business Media New York 2013

Over the past 15 years, single capillary CE-SDS gel analysis has emerged as a replacement of traditional SDS-PAGE due to its advantages of automation, quantitation, and high resolution. The method has been used routinely in biopharmaceutical industries for release, stability, and characterization of drug intermediates and product (7–9). Even so, the single capillary CE-SDS method is considered slow if many samples need to be run. For example, the run time required for 20 samples in SDS-PAGE gel is almost equivalent to the run time of a single capillary CE-SDS gel (~20 h).

More recently in the last 5–10 years, CE-SDS gel, performed in microfluidic labchip technology developed by PerkinElmer/Caliper Life Science®, has been applied to analyze glycoproteins (10–12). This technology is capable of performing CE-SDS gel analysis of a full 96-well plate in about 2 h with comparable sensitivity and resolution to traditional single capillary CE-SDS gel. Thus, output by Caliper technology is much higher than traditional SDS-PAGE or single capillary CE-SDS gel methods.

In PerkinElmer/Caliper technology, all of the separation science occurs in a micro-sized capillary on a chip. Using a sipper that dips into a well on a 96-well plate containing prepared sample, denatured protein is loaded onto the chip. Sipped protein forms a complex with SDS in the sieving matrix and binds to the fluorescent dye noncovalently within the chip's capillaries. Size-based separation occurs in the polymer sieving matrix within the viewing window of the chip. At the end of the ~14 mm separation channel in the viewing window, SDS is rapidly diluted to below its Critical Micelle Concentration (CMC) before the detection occurs. This dilution process reduces the background signal due to fluorescent dye bound to SDS micelle; hence the fluorescent dye signal due to protein bound to SDS micelle increases significantly. Fluorescence signal is recorded for each sample. Here, the instrument and software control the movement of fluids via pressure and voltage. Automated data processing software generates a final report providing information comparable to that of traditional SDS-PAGE and single capillary CE-SDS gels.

Because of the unique separation properties of Caliper instrument, we have used this technology to support rapid development of our own therapeutic protein candidates, especially glycoproteins. The method described is presently being used in formulation, analytical, and process areas to support the development of glycoprotein candidates, such as monoclonal antibody (mAb), erythropoietin (EPO), and heavily glycosylated Fc-fusion protein.

2 Materials

2.1 Chemical

1. Protein Express Kit (Caliper) (see Note 1).
2. Dithiothreitol (DTT) (Thermo Scientific) (see Note 2).
3. Iodoacetamide (IAC) (Sigma) (see Note 2).

4. 70% Isopropyl Alcohol (IPA) (Decon).
5. Water (Hyclone).

2.2 Equipment

1. LabChip90 (LC90) (Caliper) (see Note 3).
2. Protein Chip (Caliper) (see Note 4).
3. Caliper LabChip GX software version 2.1.325.0 or greater (see Note 3).
4. Priming Station (Caliper).
5. 96-Well Plate (Fisher) (see Note 5).
6. Plate Cover (Nunc).
7. Centrifuge for 96-well plate (Sigma).
8. Microcentrifuge (Eppendorf).
9. Heating Block for 96-well plate and Ladder (Eppendorf).
10. Multichannel pipettes and Pipettes (Rainin).
11. NIST Thermometer (Fisher).

3 Methods

3.1 Preparation of In-Lab Reagents

3.1.1 Gel Dye

Add 18–20 μL Protein Express Dye to 520 μL Protein Express Gel Matrix. Vortex and transfer to spin filter. Centrifuge at 9,300 × *g* for 5 min.

3.1.2 De-stain Gel

Add 250 μL Protein Express Gel Matrix to a spin filter. Centrifuge at 9,300 × *g* for 5 min.

3.1.3 Reducing Sample Buffer

Add 50 μL water to the pre-weighed DTT solid. Mix until solid dissolves into water via pipette action and this gives 1 M DTT concentration. Protect from light and prepare fresh daily. Add 30 μL of 1 M DTT mix to 840 μL Protein Express Sample Buffer.

3.1.4 Non-reducing Sample Buffer

Add 600 μL of water to a vial of pre-weighed iodoacetamide solid (~56 mg). Vortex until solid dissolves into liquid. This should give 0.5 M IAC concentration. Protect from light and prepare fresh daily. Add 44 μL of 0.5 M IAC to 826 μL of Protein Express Sample Buffer.

3.1.5 Molecular Weight Ladder

Transfer 15 μL of Protein Express molecular weight (MW) Ladder to a microcentrifuge tube. Heat MW ladder at 100°C for 5 min. Immediately add 150 μL water to cool the sample. Load 150 μL water into well A of provided ladder strip. In the case of GX/GXII, use 12 μL ladder and 120 μL water and load 120 μL into the ladder well.

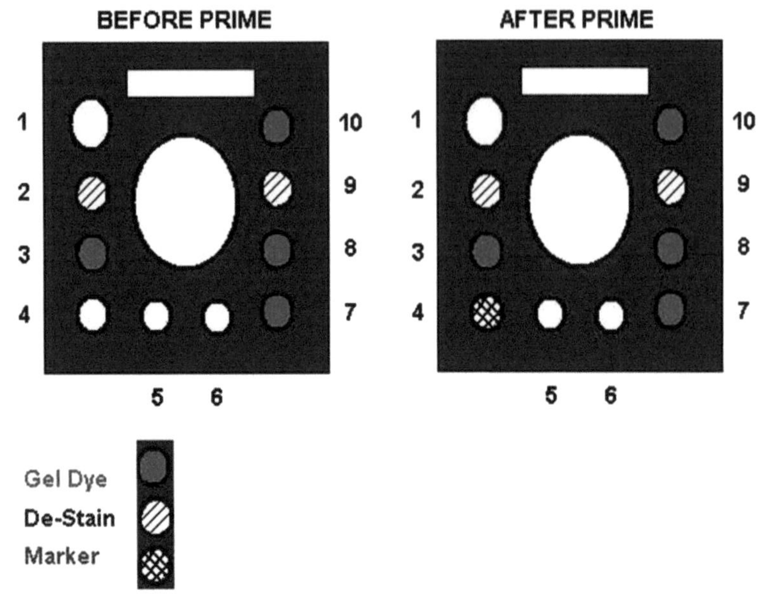

Fig. 1 Protein chip diagram before and after priming

3.1.6 Wash Buffer

Add 200 µL of Protein Express Wash buffer to each well of buffer strip. (In the case of the GX/GXII, use 750 µL wash buffer. Load into provided buffer well.)

3.2 Cleaning Chip Preparation

1. Fill cleaning chip with 1.2 mL water.
2. Insert chip into instrument and incubate for at least 2 min.
3. Remove chip and allow electrodes to dry for at least 5 min before inserting protein chip.

3.3 Preparation of Protein Chip (See Note 4)

1. Unwrap new protein chip. Remove para-film covering wells. View well numbers.
2. Rinse and aspirate each well at least twice with molecular biology grade water.
3. Add protein express reagents to wells before priming. Wells 2 and 9: Add 75 µL de-stain solution each. Wells 3, 7, 8: add 75 µL of gel dye. Well 10: add 120 µL of gel dye (see Fig. 1).
4. For LC90, prime chip (see Note 6) Use settings indicated on the chip identification label on chip box. Typically, settings are B7 for 10 min. (See chip label: Last two digits indicate required prime setting.) For GX/GXII, priming is internal to the instrument and not a separate step.
5. Aspirate wells 1 and 4. Well 4: Add 120 µL of protein marker (included in kit) for a maximum of three plates that can be run in a single preparation (40 µL per plate).
6. Clean both sides of the chip window with cloth (included in kit) and 70% IPA.

3.4 Preparation of Samples

1. Create plate map for the 96-well plate, using Microsoft Excel or similar program.
2. Add 14 μL of reducing sample buffer or non-reducing sample buffer (see Note 7) mix to the plate.
3. Add 4 μL of glycoprotein sample at recommended 1–3 mg/mL starting concentration (see Note 8).
4. Place plate in plate heating block at 70°C for 10 min.
5. Immediately add 70 μL of water. Pipette mix.
6. Cover and spin plate for 5 min at 1,800 × *g* or until there are no bubbles in plate.
7. Place the 96-wells plate in appropriate position and orientation on Caliper instrument. Have A1 well face appropriate demarcation on instrument.

3.5 CE-SDS Gel Run and Data Using Caliper LabChip90

After the chip and sample preparation steps are complete, load the chip and the 96-well plate onto the LC90 instrument. An appropriate selection of procedures has already been preprogrammed into the LC90 software script. The code takes the CE-SDS chip through steps, which include sample loading, staining, separation, de-staining, and detection. Detailed information about the LC90 instrument procedures has been described by Chow (13) and also in the manufacture instruction manuals (14) (see Note 9). Below, we provide several glycoprotein characterization data obtained using this LC90 technology.

Characterization of *N*-glycan occupancy is one of the critical parameters for glycoproteins. Currently, CE-SDS gel method is one of the fastest and easiest ways to quantitate *N*-glycan occupancy. To do this, PNGase F enzyme (see Note 10) is added to glycoprotein to remove the N-linked glycan, and thus deglycosylated protein peak migrates faster in CE-SDS gel than in glycosylated protein. It is known that mAb normally has one occupied N-linked glycan site at N297 position in heavy chain (HC). It is important that during mAb manufacturing the amount of non-glycosylated heavy chain (NGHC) is monitored closely, since it has been shown that NGHC impacts antibody stability and function (15, 16).

Figure 2a illustrates a mAb (trace 1) that has a relatively high NGHC (~15%) while a normal mAb usually contains <2% NGHC. Trace 2 shows that this particular chimeric mAb contains two N-linked glycan sites. The two heavy chain peaks disappear upon complete treatment with PNGase F (trace 3).

Figure 2b shows CE-SDS gel data from a different glycoprotein, EPO, which contains sialic acids and has three N-linked glycan sites. Upon complete treatment with PNGase F, the main broad EPO peak at 19 s originally seen in trace 1 migrates much earlier at 15 s and becomes much sharper as shown in trace 3. The incomplete PNGase F-treatment of EPO (Fig. 2b, trace 2) reveals

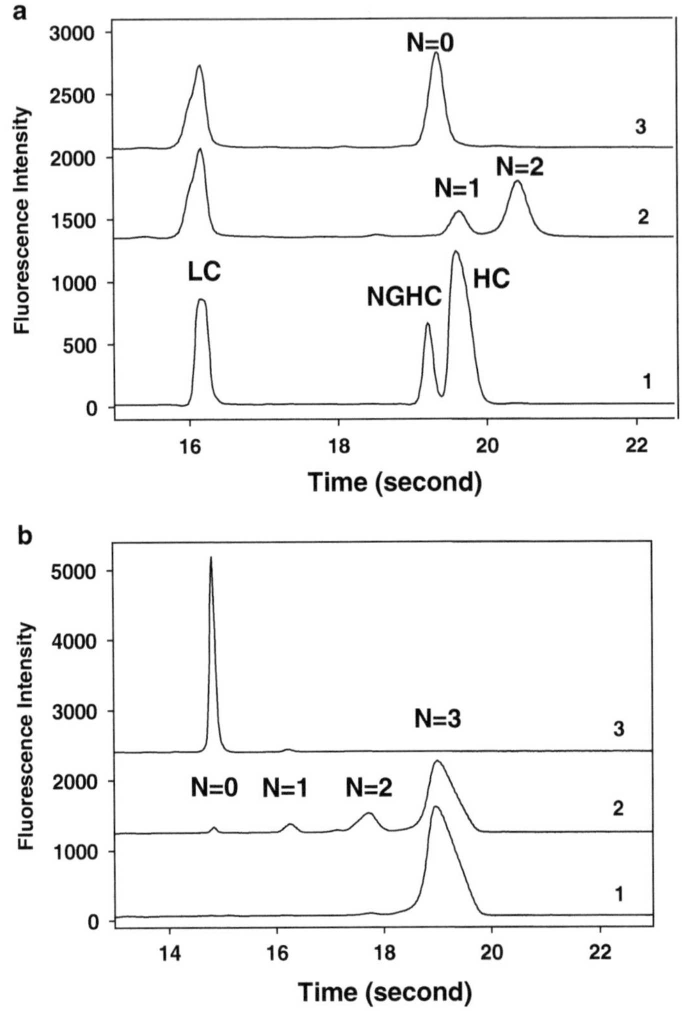

Fig. 2 **(a)** Caliper CE-SDS gel for mAbs, *trace 1* is mAb that contains LC, NGHC, and HC. *Trace 2* is mAb that has second glycosylation site ($N=2$) and *trace 3* is the PNGase F-treated of mAb in *trace 2*. (**b**) Caliper CE-SDS gel for EPO sample upon PNGase F-treatment, *trace 1* (before), *trace 2* (during), *trace 3* (after). N number is corresponding to the number of N-linked glycan sites

the presence of EPO molecule with 0, 1, 2, and 3 N-linked glycans, denoted by $N=0$, 1, 2, and 3. Thus, similar to single capillary CE-SDS gel, Caliper CE-SDS gel method is capable of separating EPO and mAb with multiple *N*-glycans. However, Caliper CE-SDS can perform the separation at a much higher throughput (9).

In Fig. 3, Caliper CE-SDS data of Fc-fusion protein, heavily glycosylated with N- and O-linked glycans, is shown. As depicted in Fig. 3a, this method can also be used to partially profile the main peak heterogeneities of this heavily glycosylated Fc-fusion protein. Three electropherograms shown in Fig. 3a (traces 1, 2, and 3) are

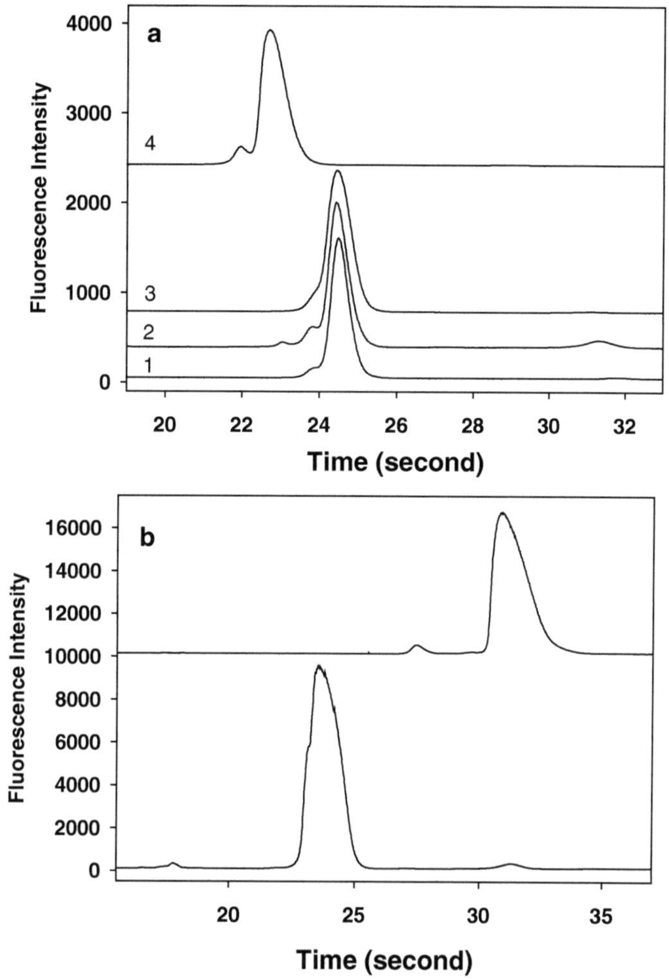

Fig. 3 Caliper CE-SDS gel of Fc-fusion protein that is heavily N- and O-linked glycosylated. (**a**) Lower concentration to illustrate the main peak heterogeneities due to glycan between two different clones (*traces 1* and *2*) and expression system (*trace 3*). Sample in *trace 3* was treated with PNGase F (*trace 4*). (**b**) Higher concentration of the same Fc-fusion protein to evaluate other impurities; reduced (*lower traces*) and non-reduced (*upper traces*) conditions were performed

the Fc-fusion protein obtained from different clones and expression systems. Trace 4 is the PNGase F-treated sample in trace 3 (see Note 11).

Figure 3b shows similar Fc-fusion protein samples as in Fig. 3a injected at higher concentration, under reducing (lower trace) and non-reducing (upper trace) conditions. Under reducing condition, non-product related impurity can be evaluated. This particular Fc-fusion glycoprotein samples has very low level of non-product related impurity (<1%). The CE-SDS gel Caliper was used to monitor these characteristics for clone selection and fermentation condition optimization with higher throughput.

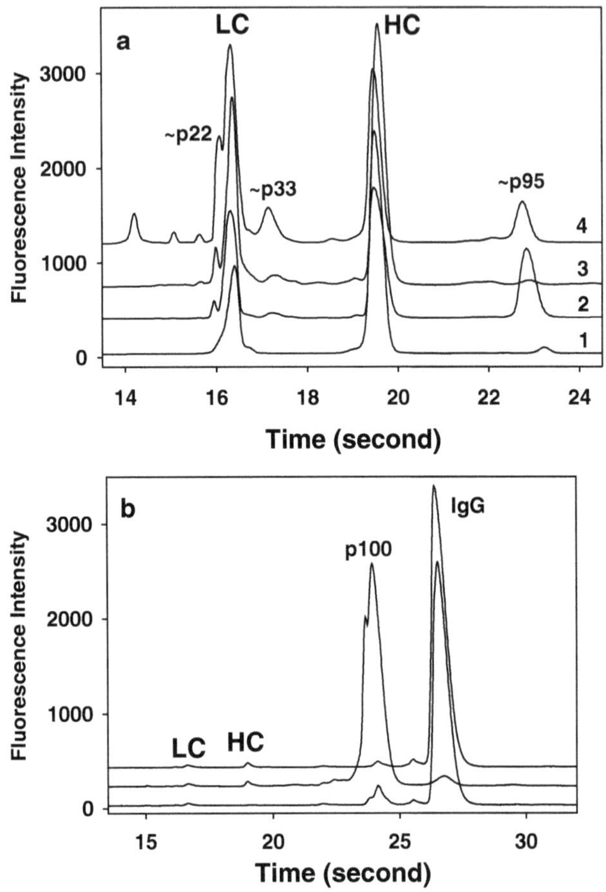

Fig. 4 **(a)** *Traces 1* and *2* are two mAbs that are in two different formulation buffers. *Traces 3* and *4* are the same mAb as in *trace 1* but it was subjected to heat stress. (**b**) Electropherograms of non-reducing mAb during clone selection (*bottom trace*), the enriched p100 peak (*middle trace*), and the enriched IgG peak (*top trace*) (see Note 13)

Figure 4a shows the usefulness of Caliper system to study the impact of different formulation and stress conditions on mAb stability. Traces 1 and 2 are mAbs in two different formulation buffers with different pH. Traces 3 and 4 are the heat-stressed mAb samples (see Note 12). Figure 4b shows electropherogram data of the same mAb, under non-reducing condition, for clone selection study. The bottom trace is the electropherogram of one mAb clone that contains slightly higher level of p100. In order to understand p100 peak, this peak is further enriched and shown in middle trace while the enriched IgG main peak is shown in top trace (see Note 13). Caliper LC90 method is used as a formulation and clone selection screening tool during product development.

4 Notes

1. Protein Express kit from Caliper consists of gel dye, gel matrix, sample buffer, spin filter, MW ladder, and marker.
2. Use the pre-weighed DTT and IAC.
3. The method and data described here are from the first generation instrument of Caliper LC90 and the second generation instrument is called GXII which has similar functionality but includes automated internal priming and CFR Part 11 Compliant software. In general, the described method here can also be applied similarly to Caliper GXII system. Caliper LC90 comes with different software than Caliper GX II; however, both equipment are compatible with the GXII data analysis software. The GX II software has enhanced features for sorting and labeling data. Thus, GX II version 2.1.325.0 or higher is recommended.
4. In our experience, a single protein chip can run three 96-well plates (although the manufacturer indicates 2,000 sample injections are acceptable). Prior to analysis, it is very important to examine the window of every new chip to make sure that the window is free of scratches, as scratches will distort data output. In microfluidic chip technology, it is very important to keep bubbles away from the channel by reverse pipette technique in reagent preparation. In addition, do not allow the metal sipper of the chip to dry out or else the sieving matrix may dry. Furthermore, it is important that the chip surface must be dry to avoid electrical interference. Finally, after first use do not store chip back in +4°C because now the chip contains SDS, which crystallizes at refrigerator temperatures. Store used chips at room temperature if you plan further use.
5. For Caliper LC90, it is preferred to use a round bottomed 96-well plate verses a flat bottomed 96-well plate. This is a standard practice, as the instrument is originally taught to read round-bottomed wells.
6. Priming Station electrode o-rings must be periodically wiped down with a damp cloth to avoid sticky residue build up. Residue can pull off o-rings over time. If an o-ring falls off, then the equipment will not have a good seal. Thus, the priming station will not be able to get good suction (vacuum pressure) to successfully prime the chip. This, in turn, can lead to air bubbles.
7. IAC is used to prevent disulfide shuffling upon SDS denaturation. This is particularly important when mAb is analyzed under non-denaturing condition, as omitting IAC will give fragmentation artifacts (17).

8. Since electro-kinetic injection is used in this LC90, the sensitivity is heavily influenced by sample buffer compositions. Low salt is recommended to increase signal. It is also recommended that samples are evaluated for possible matrix interference. Sample concentration used for injection is depended on the purpose of the experiment. Lower concentration is recommended if the resolution is critical, for instance, when evaluating product consistency (see Fig. 3).
9. The run is considered acceptable when the MW ladders have at least 5–7 sharp peaks and focusing data do not have negative peak. It is also highly recommended to place your own positive control and monitor its performance.
10. PNGase F is a specific enzyme that can cleave N-linked glycan from glycoproteins. The reaction needs to be optimized for each glycoprotein to get complete removal.
11. Profiling of main peak heterogeneities due to glycan can be used to analyze consistent product. This initial product screening provides significant time saving before performing detailed and lengthy N- and O-glycan analysis which takes about a week. This Fc-fusion glycoprotein contains many sites of N- and O-linked glycan therefore upon removing N-linked glycan the peak is still rather broad due to O-linked glycan but partially resolved (traces 3 and 4). The actual number of N- and O-linked glycan sites is not known for this glycoprotein.
12. The mAb in trace 2 contains higher pH formulation buffer compared to mAb in trace 1; small clipping at the HC hinge region producing p22 and p33 is observed and p95 is likely the thio-ether formation between LC and HC that has been observed previously (18, 19). When mAb with formulation buffer used in trace 1 is subjected to heat stress for a longer time (traces 3 and 4), the amount of p22 and p33 increases and some additional MW peaks appear likely due to more clipping in the HC hinge region.
13. The enrichment of p100 peak was achieved by collecting the later peak after the main monomer peak in size exclusion chromatography. The collected fractions were injected back to Caliper LC90 (middle and top traces). The p100 peak was confirmed to be HC dimer by mass spectrometry analysis.

Acknowledgments

The author would like to thank our colleagues in purification Bioprocess R&D for providing us with material and technical support. We also thank Yang Wang for management support.

References

1. Gibbons RA (1972) Physico-chemical methods for the determination of the purity, molecular size and shape of glycoproteins. Chapter 3, Section 1. In: Gottschalk A (ed) Glycoproteins their composition, structure and function, vol. 5, Elsevier Publishing Company, Amsterdam, pp. 31–128
2. Leach BS, Collawn JF, Fish WW (1980) Behavior of glycopolypeptides with empirical molecular weight estimation methods. Biochemistry 1980:5734–5741
3. Durchschlag H, Christl P, Jaenicke R (1991) Comparative determination of the particle weight of glycoproteins by SDS-PAGE and analytical ultracentrifugation. Prog Colloid Polym Sci 86:41–56
4. Pitt-Rivers R, Impiombato FSA (1968) The binding of sodium dodecyl sulphate to various proteins. Biochem J 109:825–829
5. Reynolds JA, Tanford C (1970) The gross conformation of protein-sodium dodecyl sulfate complexes. J Biol Chem 245:5161–5165
6. Guo XH, Chen SH (1990) The structure and thermodynamics of protein-SDS complexes in solution and the mechanism of their transports in gel electrophoresis process. Chem Phys 149:129–139
7. Hunt G, Nashabeh W (1999) Capillary electrophoresis sodium dodecyl sulfate nongel sieving analysis of a therapeutic recombinant monoclonal antibody: a biotechnology perspective. Anal Chem 71:2390–2397
8. Han M, Phan D, Nightlinger L, Taylor S (2006) Optimization of CE SDS method for antibody separation based on multi-users experimental practices. Chromatographia 64:335–342
9. Rustandi RR, Washabaugh MW, Wang Y (2008) Applications of CE SDS gel in development of biopharmaceutical antibody-based products. Electrophoresis 29:3612–3620
10. Bousse L, Mouradian S, Minalla A, Yee H, Williams K, Dubrow R (2001) Protein sizing on a microchip. Anal Chem 73:1207–1212
11. Chen X, Tang K, Lee M, Flynn GC (2008) Microchip assays for screening monoclonal antibody product quality. Electrophoresis 29:4993–5002
12. Chen X, Flynn GC (2009) A high throughput dimer screening assay for monoclonal antibodies using chemical cross-linking and microchip electrophoresis. J Chromatogr B Analyt Technol Biomed Life Sci 877: 3012–3018
13. Chow AW (2006) Protein separations. In: Henry CS (ed) Methods in molecular biology: microchip capillary electrophoresis methods and protocols, vol 339. Humana Press, Totowa, pp 145–158
14. Caliper Life Sciences (2008) LabChip® GX/GXII User Manual, P/N:127170 Rev. 00. http://www.perkinelmer.com/CMSResources/Images/44-133044MAN_LabChip%20GX.GX%20II%20User%20Manual.pdf
15. Mimura Y, Church S, Ghirlando R, Ashton PR, Dong S, Goodall M, Lund J, Jefferis R (2000) The influence of glycosylation on the thermal stability and effector function expression of human IgG1-Fc: properties of a series of truncated glycoforms. Mol Immunol 37:697–706
16. Ha S, Ou Y, Vlasak J, Li Y, Wang S, Vo K, Du Y, Mach A, Fang Y, Zhang N (2011) Isolation and characterization of IgG1 with asymmetrical Fc glycosylation. Glycobiology 21:1087–1096
17. Ma S, Nashabeh W (2001) Analysis of protein therapeutics by capillary electrophoresis. Chromatographia 53:S75–S89
18. Cohen SC, Price C, Vlasak J (2007) Beta-elimination and peptide bond hydrolysis: two distinct mechanisms of human IgG1 hinge fragmentation upon storage. J Am Chem Soc 129:6976–6977
19. Rustandi RR, Wang Y (2011) Use of CE-SDS gel for characterization of monoclonal antibody hinge region clipping due to copper and high pH stress. Electrophoresis 32: 3078–3084

Chapter 13

Hydrophobic Interaction Chromatography to Analyze Glycoproteins

Richard R. Rustandi

Abstract

Hydrophobic interaction chromatography (HIC) is one of many separation techniques that can be used to analyze proteins. The separation mechanism is based on the adsorption of the hydrophobic region of the protein to the hydrophobic ligands attached to the column in the presence of high salt. The proteins are then eluted by descending salt concentration. Here we describe the use of this HIC technique to evaluate the hydrophobicity of different monoclonal antibodies (mAbs) and to separate different heterogeneities that occur in mAb.

Key words Hydrophobic interaction chromatography, HPLC, mAb, CDR, N-linked glycan, Hydrophobicity, HIC

1 Introduction

Hydrophobic interaction chromatography (HIC) is a technique that is commonly performed during protein purification process, but it is still seldom used in analytical testing area as compared to other chromatography techniques such as size-exclusion, ion-exchange, and reverse phase. A significant advantage of HIC relative to its separation cousin, reverse phase, is that it is non-denaturing, so the native form is expected to be maintained, and the separated proteins can subsequently be collected for further bioassay activity. The separation concept of HIC, which is based on the salting out protein principle, was first described by Tiselius (1). Subsequently, Hjerten called this type of separation "hydrophobic interaction chromatography" (2). The retention mechanism theory of HIC is due to adsorption desorption equilibrium in the presence of salts and is described in detail in several publications (3–6).

In practice, proteins bind to the hydrophobic column in the presence of high salt of ammonium sulfate (see Note 1) and are then eluted in order of increasing hydrophobicity by decreasing the salt concentration. Historically, most of the separation was

Alain Beck (ed.), *Glycosylation Engineering of Biopharmaceuticals: Methods and Protocols*, Methods in Molecular Biology, vol. 988, DOI 10.1007/978-1-62703-327-5_13, © Springer Science+Business Media New York 2013

done using semi-preparative columns for fast protein liquid chromatography (FPLC), such as octyl or phenyl sepharose from General Electric (Pharmacia). Recently several polymer- and silica-based HPLC columns are commercially available and they provide high resolution that normally not observed for HIC (7, 8). HIC has been used to evaluate different types of protein (see ref. 8), but applications to mAbs are still limited. Several publications have described the use of HIC technique on mAbs specifically it has been applied to analyze Fc and Fab regions of mAb after papain enzyme digestion to simplify characterization (9–12). Mab is a glycoprotein and in general it has one N-linked glycan site in Fc region. Second N-linked glycan site in Fab region is sometime observed such as for commercial product recombinant chimeric Erbitux/cetuximab mAb (13, 14) and also for mouse mAbs the second site of N-linked glycan is usually in Fc region. This variation in number of N-linked glycan sites in full intact mAb can potentially be probed by HIC techniques (15) since it is expected that asparagine attached glycan is much more hydrophilic than either unmodified asparagine or aspartic acid (16).

In this chapter, we describe the use of HIC method to analyze the hydrophobicity of different types of mAb and the utility of this method to observe hydrophobic heterogeneities for different mAb clones or mAbs that have been heat and photo-stressed. The method illustrates the power of this new silica-based HIC HPLC column from Dionex company to separate full intact mAb with relatively high resolution.

2 Materials

2.1 Chemical

1. Ammonium sulfate, $(NH_4)_2SO_4$ (Sigma-Aldrich).
2. Sodium dibasic phosphate, Na_2HPO_4 (Sigma-Aldrich).
3. Sodium mono basic phosphate, NaH_2PO_4 (Sigma-Aldrich).
4. Sodium chloride, NaCl (Sigma-Aldrich).
5. MilliQ water.
6. 1 N NaOH (Lab Chem).
7. 37% HCl (Fisher).
8. Ammonium acetate, CH_3COONH_4 (Fisher).
9. Acetonitrile, CH_3CN (Acros).

2.2 Equipment

1. HPLC system (Thermo Separation Products) (see Note 2).
2. Hydrophobic interaction column, Propac HIC-10 (amide/ethyl chemistry); 4.6 mm ID × 100 mm length, 5 μm particle size, 300 Å pore size (Dionex) (see Note 3).
3. Pre-column filter, 0.5 μm (MacMod) (see Note 4).

4. Mobile Phase A (MPA): 50 mM NaPi, 75 mM NaCl, 1 M $(NH_4)_2SO_4$, pH 7.0 (see Note 5).
5. Mobile Phase B (MPB): 50 mM NaPi, 75 mM NaCl, pH 7.0 (see Note 5).
6. Wash Solution: 50 mM CH_3COONH_4, pH 5.4.
7. Storage Solution: 50 mM CH_3COONH_4 in 80/20 (v/v) CH_3CN/H_2O, pH 5.4.
8. Mobile phases filtration unit with 0.22 μm filter (see Note 6).

3 Methods

3.1 Solution Preparation

3.1.1 Mobile Phase A (1 L)

1. Weigh 132 g $(NH_4)_2SO_4$.
2. Add 152.5 mL 0.2 M Na_2HPO_4 (see Note 7).
3. Add 97.5 mL 0.2 M NaH_2PO_4 (see Note 7).
4. Weigh 4.39 g NaCl.
5. Adjust pH to 7.0 with 1 N NaOH.
6. Add MilliQ water necessary to obtain final volume of 1 L.
7. Filter through 0.22 μm filtration unit.

3.1.2 Mobile Phase B (1 L)

1. Add 152.5 mL 0.2 M Na_2HPO_4 (see Note 7).
2. Add 97.5 mL 0.2 M NaH_2PO_4 (see Note 7).
3. Weigh 4.39 g NaCl.
4. Adjust pH to 7.0 if necessary.
5. Add MilliQ water necessary to obtain final volume of 1 L.
6. Filter through 0.22 μm filtration unit.

3.1.3 Wash Solution (1 L)

1. Weigh 3.8 g CH_3COONH_4.
2. Add 800 mL MilliQ water.
3. Adjust pH to 5.4 with 37% HCl.
4. Add MilliQ water necessary to obtain final volume of 1 L.
5. Filter through 0.22 μm filtration unit.

3.1.4 Storage Solution (1 L)

1. Weigh 3.8 g CH_3COONH_4.
2. Add 180 mL MilliQ water.
3. Add 800 mL CH_3CN.
4. Adjust pH to 5.4 with 37% HCl.
5. Add MilliQ water necessary to obtain final volume of 1 L.
6. Filter through 0.22 μm filtration unit.

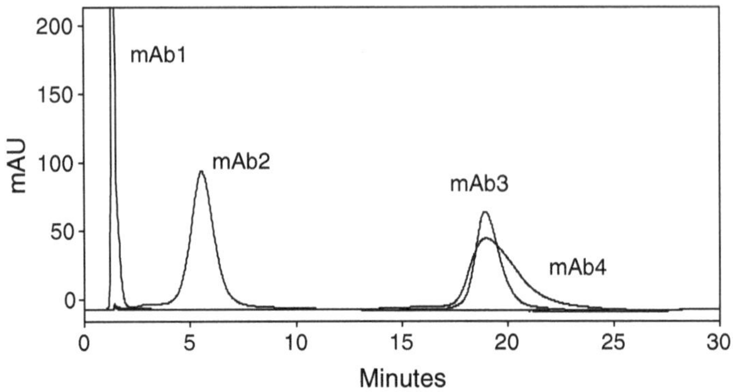

Fig. 1 Chromatograms of four different monoclonal antibodies were run under the same gradient condition. The chromatographic elution is as follow: 2 min 50% MPA and 50% MPB; 20 min linear gradient of MPB from 50–100%; 3 min at 100% MPB; flow rate 1 mL/min, 280 nm detection, and ambient temperature

3.2 Sample Preparation

All mAbs were produced in either CHO or NS/0 cell lines and purified according to standard mAb purification techniques with purity of 99% or above. All mAbs are IgG_1 type except mAb4 which is IgG_2m_4 (see Note 8). Samples were diluted to 1 mg/mL in Mobile Phase A for HIC HPLC. Heat-stressed mAb2 was prepared by incubation at 45°C for 1 week and photo-stressed mAb2 was generated by placing sample in photo chamber for 98 h following ICH guideline.

3.3 HIC HPLC

Assemble the HPLC system by connecting the pre-column filter (0.5 μm) and HIC column to the solvent delivery system. The system is equilibrated by the following condition:

Column Flushing: Wash Solution for 15 min at 0.5 mL/min.

Column Equilibration: Mobile Phase A for 30 min at 0.5 mL/min.

Detection: UV at 280 or 220 nm (see Note 9).

Temperature: Ambient (21–25°C).

After this step is completed (flat baseline), inject buffers blank and perform gradient run. It is highly recommended that this step is repeated twice to establish flat baseline using the intended gradient. Initial separation can be performed using linear gradient from 0 to 100% Mobile Phase B for 20 min and then continue at 100% Mobile Phase B for 5 min. At this condition, almost all mAbs will be eluted. The separation resolution can be adjusted by changing the gradient condition.

Four mAbs were tested for their relative hydrophobicity using chromatographic conditions described in Fig. 1. The first three

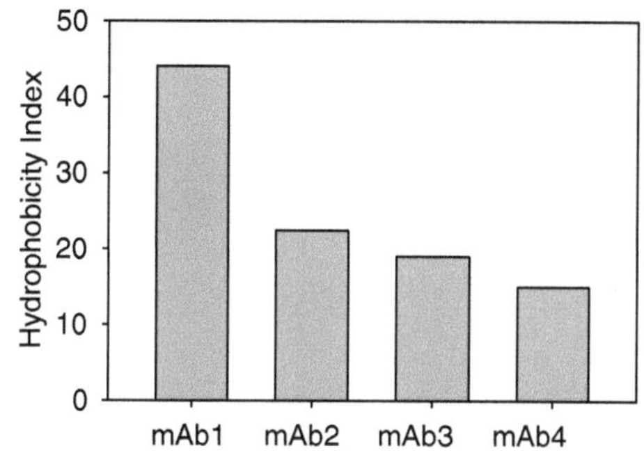

Fig. 2 Hydrophobicity index calculation for the four monoclonal antibodies using only their CDR domain amino acid residues. The lower the hydrophobic index number is the more hydrophobic the molecule is (see Note 10)

mAbs are derived from IgG_1 and therefore have the same backbone of amino acids except in CDR domains, while mAb4 is derived from IgG_2m_4. Figure 1 illustrates that mAb1 has the least hydrophobic patch followed by mAb2, then mAb3 and mAb4 have slightly similar hydrophobicity. In fact, mAb1 elutes at the void volume, indicating that this mAb is not retained in the column using this specific gradient condition. The observed order of mAb elution from HIC is in agreement with the predicted hydrophobic index calculated only using amino acids in CDR domains shown in Fig. 2 (see Note 10). This correlation clearly shows that the HIC mechanism in mAb can be described by the interaction of variable regions in CDR to the hydrophobic column. Hence, it is expected that the observed HIC peak heterogeneities are mostly due to the variation of amino acid residues and/or posttranslational modifications that occur in CDR domains.

Several types of mAb2 samples such as different clones, heat- and photo-stressed were evaluated using this HIC column with chromatographic condition described in Fig. 3 legend. The chromatograms illustrated in Fig. 3 indicate the power of this particular HIC column to separate different heterogeneities. Comparison of chromatographic traces between two different mAb2 clones (traces 1 and 2) shows that mAb2 clone 2 (trace 2) has two additional small peaks at about 16 and 17 min; in addition, clone 2 has a higher shoulder before the main peak. This is very useful information for providing clone selection or comparability studies. For heat-stressed mAb2 (trace 3), two prominent peaks appear before the main peak. The main peak is also slightly shifted, indicating some structural changes. Similarly, for photo-stressed mAb2 (trace 4), four significant peaks appear and the main peak is shifted even further compared to the heat-stressed sample. These observations demonstrate the utility of this HIC method

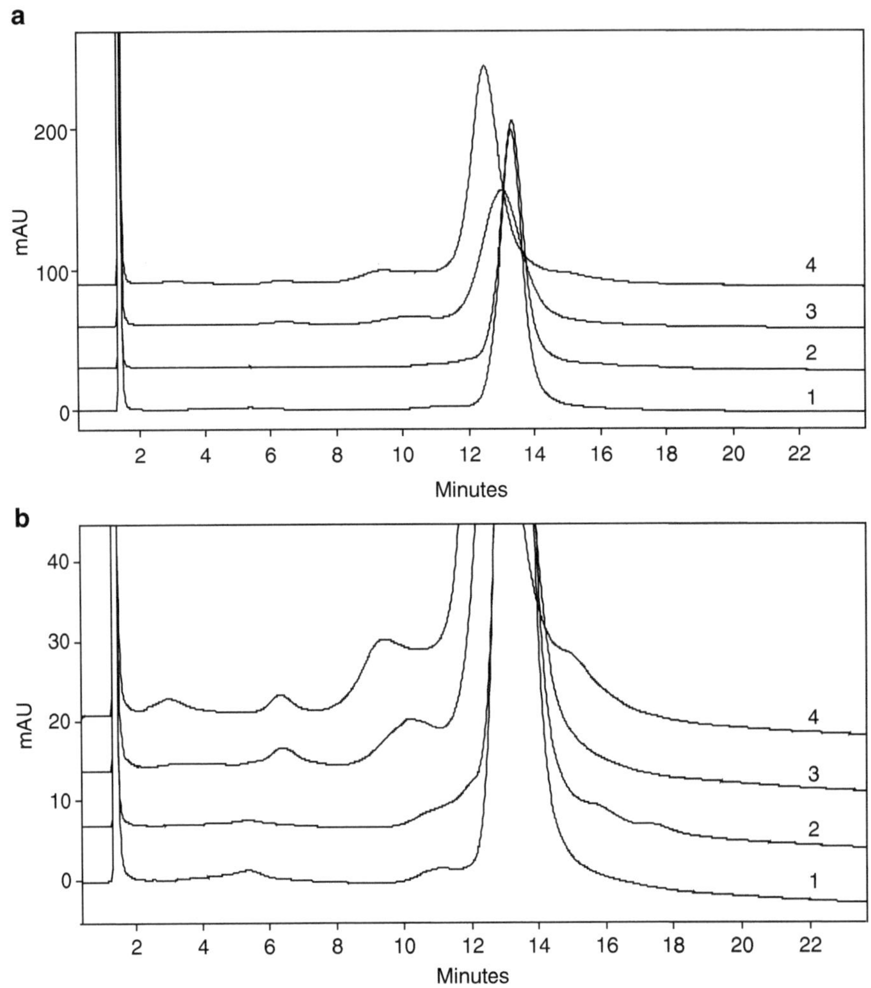

Fig. 3 (**a**) HIC separation of mAb2 samples that were produced from different clones (*traces 1* and *2*) and have been heat- (*trace 3*) and photo- (*trace 4*) stressed. (**b**) Expanded view of the same chromatograms as in (**a**). The chromatographic condition is as follow: 2 min 75% MPA and 25% MPB, 20 min linear gradient of MPB from 25 to 100%, 3 min at 100% MPB, flow rate at 1 mL/min, 220 nm UV detection, and ambient temperature. Peak at about 1.5 min is due to the histidine buffer

for monitoring stability of mAb. Figure 4 shows another example of hydrophobic heterogeneities from mAb3 drug product with two additional peaks that are well separated from the main peak. This particular mAb3 profile is quite different from those of mAb1 or mAb4, for which there are no additional peaks (data not shown). Although all of these additional peaks have not been characterized, the primary intent of this paper is to elucidate the power of this HIC technique with the availability of a new silica-based column (see Note 11).

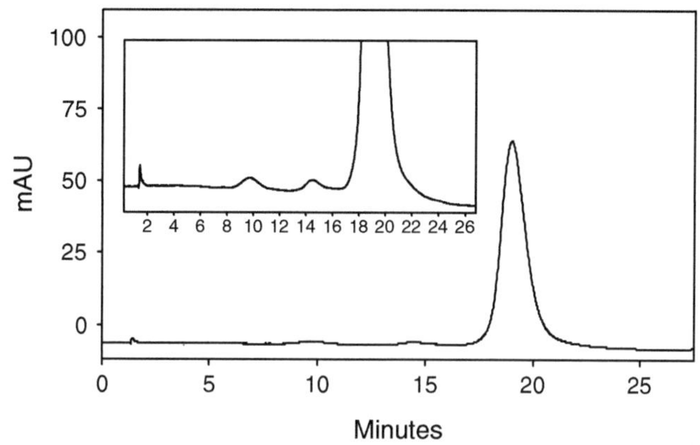

Fig. 4 HIC separation of mAb3 using the same chromatographic condition as in Fig. 1. The chromatogram illustrates that there are at least two additional heterogeneities for this mAb3 sample. Inset is the expanded view

4 Notes

1. Ammonium sulfate is the most common salt used in HIC. The cations or anions used influence the HIC in accordance with the well-known Hofmeister (lyotropic series). In general, the following salts strengthen hydrophobic interactions:

$$Na_2SO_4 > K_2SO_4 > (NH_4)_2SO_4 > NaCl > NH_4Cl > NaBr > NaSCN$$

2. Any HPLC system with a binary pump can be used, but a quaternary pump is preferred since wash and storage solutions can be pumped into the column directly without the need to change bottle solutions. It is very important that column is flushed with Wash Solution at the end of run, stored in storage solution, and detached from HPLC system. Subsequently, the HPLC system needs to be flushed thoroughly with Wash Solution and then water to remove high salt concentration from the MPA, MPB, wash, and storage solutions to prevent crystal formation inside HPLC.
3. Various HIC columns from different vendors have been evaluated and this particular silica-based HIC column provides the best resolution in terms of intact mAb. Further resolution improvement could potentially be achieved by connecting two columns in series or by employing a smaller diameter column (2.1 × 100 mm) with UPLC system.
4. Use of a pre-column filter is highly recommended since this column does not come with a guard column. This particular pre-column filter is hand tight and inexpensive.
5. The buffer composition for mobile phases needs to be evaluated and optimized for the separation purpose.

6. The filtration of mobile phases is important for running HIC since solid salts from sodium phosphate, sodium chloride, ammonium acetate, and especially solid ammonium sulfate may contain particles that could block the HPLC system lines, column, and detector.
7. These volume buffer components are based on the recipe described in (17) to achieve pH 7.0. Prepare 0.2 M Na_2HPO_4 and 0.2 M NaH_2PO_4 separately.
8. The IgG_2m_4 is created by using IgG_2 as the main frame amino acid sequence with several mutations at the region that binds C1q with residues from IgG_4.
9. Detection can be done either by UV at 280 and/or 220 nm or fluorescence mode ($\lambda_{ex} = 280$ nm, $\lambda_{em} = 350$ nm). Because of high salt concentration with gradient separation, there may be potential higher background noise especially at 220 nm; there is also baseline drift for gradient run. All of these issues can be overcome by performing fluorescence mode detection, which will also give better sensitivity. *Cautious*: UV detection at 280 nm or fluorescence may not give accurate % peak area composition for impurity and fragments due to different numbers of tryptophan and tyrosine residues.
10. The hydrophobicity index for each mAb was calculated based on the procedure described in literature (18). Hydrophobicity index is a measure of amino acid side-chain preference to be solvent exposed. Arginine (+3.95) and phenylalanine (−2.27) have the highest and lowest values, respectively, while the other amino acid residues have intermediate numbers. A negative index number represents hydrophobic amino acid residues, while a positive index number is for polar or charged amino acids.
11. Detailed characterization of mAb using this particular HIC column has been described in literature (19). The peaks that precede the main peaks are likely due to isoaspartate isomerization and/or fragmentation, while peaks after the main peaks could be due to succinimides and/or aggregations. However, conclusive studies need to be done for each individual mAb by collecting the fractions and performing further separation and/or molecular characterization using biophysical tools such as mass spectrometry.

Acknowledgments

The author would like to thank our colleagues in purification Bioprocess R&D for providing us with material and technical support. We gratefully thank Roxana Butoi for hydrophobicity index, Yang Wang for support of this work, and Lisa McCormick for critical reading of the manuscript.

References

1. Tiselius A (1948) Adsorption separation by salting out. Mineral Geol 26B:1
2. Hjerten S (1973) Some general aspects of hydrophobic interaction chromatography. J Chromatogr 87:325–331
3. Horvath C, Melander W, Molnar I (1976) Solvophobic interactions in liquid chromatography with nonpolar stationary phases. J Chromatogr 125:129–156
4. Melander W, Horvath C (1977) Salt effects on hydrophobic interactions in precipitation and chromatography of proteins: an interpretation of the lyotropic series. Arch Biochem Biophys 183:200–215
5. Melander WR, El Rassi Z, Horvath C (1989) Interplay of hydrophobic and electrostatic interactions in biopolymer chromatography. J Chromatogr 469:3–27
6. Queiroz JA, Tomaz CT, Cabral JMS (2001) Hydrophobic interaction chromatography of proteins. J Biotechnol 87:143–159
7. Kato Y, Kitamura T, Hashimoto T (1986) New resin-based hydrophilic support for high-performance hydrophobic interaction chromatography. J Chromatogr 360:260–265
8. Rao S, Bordunov A, Heckenberg A, Pohl C (2006) A hydrophobic interaction chromatography column with improved hydrolytic stability for protein separation and proteomics applications. Amer Biotech lab April: 8–10
9. Cacia J, Keck R, Presta LG, Frenz J (1996) Isomerization of an aspartic acid residue in the complementarity-determining regions of a recombinant antibody to human IgE: identification and effect on binding affinity. Biochemistry 35:1897–1903
10. Harris RJ, Kabakoff B, Macchi FD, Shen FJ, Kwong M, Andya JD et al (2001) Identification of multiple sources of charge heterogeneity in a recombinant antibody. J Chromatogr B Biomed Sci Appl 752:233–245
11. Beck A, Bussat M-C, Zorn N, Robillard V, Klinguer-Hamour C, Chenu S et al (2005) Characterization by liquid chromatography combined with mass spectrometry of monoclonal anti-IGF-1 receptor antibodies produced in CHO and NS0 cells. J Chromatogr B Analyt Technol Biomed Life Sci 819:203–218
12. Wakankar AA, Borchardt RT, Eigenbrot C, Shia S, Wang J, Shire SJ et al (2007) Aspartate isomerization in the complementarity-determining regions of two closely related monoclonal antibodies. Biochemistry 46:1534–1544
13. Beck A, Wagner-Rousset E, Bussat M-C, Lokteff M, Klinguer-Hamour C, Haeuw J-F, Goetsch L, Wurch T, Dorsselaer AV, Corvaïa N (2008) Trends in glycosylation, glycoanalysis and glycoengineering of therapeutic antibodies and Fc-fusion proteins. Curr Pharm Biotechnol 9:482–501
14. Qian J, Liu T, Yang L, Daus A, Crowley R, Zhou Q (2007) Structural characterization of N-linked oligosaccharide on monoclonal antibody cetuximab by the combination of orthogonal matrix-assisted laser desorption/ionization hybrid quadrupole–quadrupole time-of-flight tandem mass spectrometry and sequential enzymatic digestion. Anal Biochem 364:8–18
15. Grebenau RC, Goldenberg DM, Chang C-H, Koch GA, Gold DA, Kunz A, Hansen HJ (1992) Microheterogeneity of a purified IgG_1 due to asymmetric Fab glycosylation. Mol Immunol 29:751–758
16. Black SD, Mould DR (1991) Development of hydrophobicity parameters to analyze proteins which bear post or cotranslational modifications. Anal Biochem 193:72–82
17. Dawson RMC, Elliott DC, Elliott WH, Jones KM (1986) Data for biochemical research, 3rd edn. Oxford Science Publications, Oxford, p 432
18. Chiti F, Stefani M, Taddei N, Ramponi G, Dobson CM (2003) Rationalization of the effects of mutations on peptide and protein aggregation rates. Nature 424:805–808
19. Valliere-Douglass J, Wallace A, Balland A (2008) Separation of population of antibody variants by fine tuning of hydrophobic-interaction chromatography operating conditions. J Chromatogr A 1214:81–89

Chapter 14

Lectin Glycoprofiling of Recombinant Therapeutic Interleukin-7

Ludovic Landemarre and Eric Duverger

Abstract

Lectins array is a powerfull and complementary method of glycans analysis allowing fast identification of specific motifs on molecules or cells. This technology is of increased interest for the development of therapeutic recombinant glycoproteins and particularly relevant for a first study of lot-to-lot comparison, or detection of unwanted glycans. In this chapter, we describe a lectin array-type method specifically designed for the study of recombinant therapeutic interleukin-7 (rhIL-7). This specific method allows the analysis of the glycans motifs, the distribution of the glycoforms population, and the detection of potential immunogen glycans in rhIL-7 purified CHO-produced batches.

Key words Lectins, Glycosylation, Therapeutic glycoprotein, Interleukin-7, Glycoprofile, Glycans

1 Introduction

Glycosylation analysis of recombinant therapeutic glycoproteins require various strategies intended to decipher the glycans structures (monosaccharide content, linkage, etc.), the number of occupied sites, the quantification of each structure, and the resulting complexity of the glycoform population. Indeed, no one specific method is still able to determine all these parameters. Moreover, the more precise is the characterization of the glycosylation according to the expression system, the production, and the purification processes, the easier will be the control of the product and the filling of the regulation files. In the present method, we describe the relevance of an enzyme-linked-lectin sorbent assay used as a lectin microarray-type method, intended to identify specific glycans and analyze the global "glycosylation signature" of a therapeutic glycoprotein.

Lectins are carbohydrate-binding proteins which are widely used as tools for the study of glycoconjugates in solution or expressed at the cells surfaces (1–3). Lectins were initially used as ABO blood group typing reagent, then for the isolation of glycoproteins and separation of different glycoforms of glycoproteins. Thus, few years

Alain Beck (ed.), *Glycosylation Engineering of Biopharmaceuticals: Methods and Protocols*, Methods in Molecular Biology, vol. 988,
DOI 10.1007/978-1-62703-327-5_14, © Springer Science+Business Media New York 2013

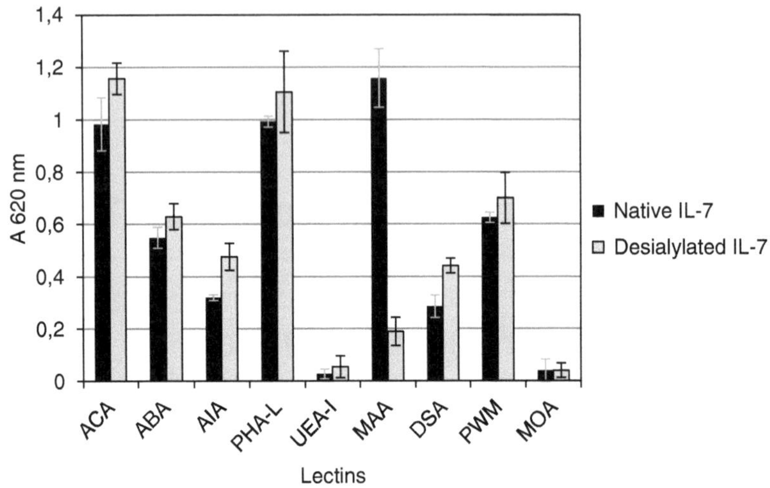

Fig. 1 Lectins microplate glycoprofiling of rhIL-7. The interaction of native and desialylated rhIL-7 with the nine lectins shows the specific "glycan signatures" resulting from the intensities of interaction of glycans motifs with each corresponding lectins. The profile obtained with the desialylated rhIL-7 shows a total decrease of the interaction with MAA lectin indicating the elimination of α2–3 neuraminic acid residues after neuraminidase digestion. The simultaneous slight increase of interaction observed for the other lectins correspond to a best accessibility to the glycans motifs after desialylation. Finally, no signal above the background was detected for the MOA lectin, indicating the lack of the corresponding recognized motif Galα1–3Galβ-, according to the limit of detection of the method (100 ng/ml)

ago, *Sambucus nigra* lectin was used to show that the inflammatory activity of IgG is the result of different degrees of sialylation (4). For the last 10 years, lectins were used as reagent for the development of microarray systems allowing a simultaneous and high throughput study of many discrete interactions leading to the establishment of specific glyco-profiles of biological samples (5–7). The chips usually consist of an array of different lectins immobilized by chemical coupling. After deposition and incubation of the glycoconjugate or the biological sample, the interactions with each lectin can be quantified by the means of enzymatic, colorimetric or fluorescence, thus resulting in the establishment of specific glycosylation signature, depending on the specificities of the lectins used, glycans content and accessibilities on the glycoconjugate studied. The lectin microarray method was proposed as a complementary and very useful method for various glycomics studies applied to glycoproteins (8, 9), cells (10–12), or tissues (13).

In this study, we present a lectin microplate method specifically dedicated to the study of the glycosylation profile of the purified CHO-produced rhIL-7 which is currently in phase II clinical trial. The IL-7 lectin microplate glycosylation analysis correspond to a customized assay developed according to the glycosylation characteristics (three potential N-glycosylation sites, one potential O-glycosylation site) of the glycoprotein and the addition of probes (lectins) intended to the detection of immunogen glycan like Galα1–3Galβ- (Galili) determinant (14) (Fig. 1).

2 Materials

2.1 Lectins Plates

96-wells microplates functionalized by lectins were obtained from GLYcoDiag. The lectprofile plates range of products is a customized service which allows the choice of lectins linked to the bottom of the raw of microwells. The selection and the number of different lectin are determined according to the glycoconjugate targeted and the test configuration needed. The products are supplied ready-to-use. For the purpose of the study, nine lectins described in Table 1 were selected.

2.2 Recombinant Human IL-7

Stock solution (4 mg/ml in 50 mM sodium acetate, pH 5.0) of purified rhIL-7 is provided by CYTHERIS.

2.3 Enzymes and Associated Buffers

Neuraminidase: Sialidase treatment is realized with the recombinant neuraminidase cloned from *Clostridium perfringens* (BioLabs, P0720). This neuraminidase catalyzes the hydrolysis of α2–3, α2–6 and α2–8 neuraminic acid residues. The enzyme is supplied at

Table 1
List of lectins used for the determination of the interaction profile of rhIL-7 glycans

Short name	Common name	Glycan specificity
ACA	*Amaranthus caudatus* agglutinin	Galβ3GalNAcα-O-R (T-antigen)
ABA	*Agaricus bisporus* agglutinin	Gal–GalNAcα-O-R, inhibited by O-linked glycans
AIA	*Artocarpus intergrifolia* agglutinin	Galα6 or Galβ3GalNAc (T-antigen)>>lactose
PHA-L	*Phaseolus vulgaris*	Galβ4GlcNAcβ6Man of branched structures of N-glycans, Galβ4GlcNAcβ2Man
UEA-I	*Ulex europeus* agglutinin I	Terminal fucose
MAA-I	*Maackia amurensis* agglutinin I	Neu5Ac/Gcα-3Galβ4GlcNAc/Glc-
DSA	*Datura stramonium* agglutinin	GlcNAcβ4GlcNAc oligomers, Galβ4GlcNAc
PWM	*Phytolacca americana*	GlcNAcβ4GlcNAc oligomers, Galβ4GlcNAc
MOA	*Marasmius oreades* agglutinin	Galα1–3Gal, Galα1–3Galβ1–4GlcNAc

50,000 U/ml with10× reaction buffer (1× G1 reaction buffer: 50 mM sodium citrate pH 6.0).

2.4 Buffers and Reagents

Dilution and incubation buffer: Except for neuraminidase treatment, complete PBS (PBS containing 1 mM $CaCl_2$ and 0.5 mM $MgCl_2$) supplemented or not with 0.5% BSA is used as dilution and incubation buffer.

Anti-IL-7 specific antibody: Biotinylated rat anti-human IL-7 monoclonal antibody is from BD PharMingen (554494). It is provided as an aqueous buffered solution containing 0.09% sodium azide. We used it diluted 1:500. The antibody solution is stored at 4°C, do not freeze.

ExtrAvidine-Peroxydase (Sigma, E2886) is provided as an aqueous buffered solution (PBS containing 0.01% thimerosal). We used it diluted 1:3,000. The antibody solution is stored at 4°C, do not freeze.

TMB (Sigma, T4444) 3,3′,5,5′-tetramethylbenzidine dihydrochloride liquid substrate is supplied as a ready-to-use one (see Note 1).

3 Methods

3.1 Neuraminidase Digestion

In standard conditions rhIL-7, 200 μg are treated with 500 U neuraminidase.

1. In an eppendorf microtube, add 50 μl stock solution of rhIL-7, 6.6 μl 10× G1 reaction buffer and 10 μl of neuraminidase.
2. Incubate the mixture at 37°C (oven) for 16 h.
3. Stop the reaction by chilling on ice and diluting with 434 μl cold complete PBS. At this time, the IL-7 concentration is 400 μg/ml (desialylated rhIL-7).

3.2 Lectin Glycoprofiling

Recombinant human IL-7 glycoprofile is established by using nine lectins on which native rhIL-7, desialylated rhIL-7 or PBS (negative control) is settled down. For each conditions, measurement is realized in quadriplicate ($n=4$).

3.2.1 Samples Preparation

rhIL-7, native or digested with neuramidase (desialylated rhIL-7) is used at 40 μg/ml in complete PBS.

1. Prepare 5 ml of native rhIL-7 *working solution* as follow: To 50 μl of stock solution add 4.95 ml of complete PBS.
2. Prepare 5 ml of desialylated rhIL-7 *working solution* as follow: Add 4.5 ml of complete PBS to chilled reaction mixture.

3.2.2 Samples Incubation

1. Add 50 μl of appropiate *working solution or PBS* in each well (according to your microplate map).

2. Incubate for 30 min at room temperature.
3. Wash three times with 200 µl of *washing solution* (see Note 2—provided in the GLYcoDiag Kit).

3.2.3 Revelation

1. Prepare 5 ml of *dilute antibody solution* by mixing 10 µl of Anti-IL-7 specific antibody with 4.99 ml of complete PBS containing 0.5% BSA.
2. Prepare 5 ml of *revealing solution* by mixing 6 µl of ExtrAvidine-Peroxydase with 4,994 ml of complete PBS.
3. After last washing (step 3, Subheading 3.2.2. *Samples incubation*) add 50 µl of *dilute antibody solution* in each well.
4. Incubate for 30 min at room temperature.
5. Wash three times with 200 µl of *washing solution.*
6. Add 50 µl of *revealing solution* in each well.
7. Incubate for 30 min at room temperature.
8. Wash three times with 200 µl of *washing solution.*
9. Add 50 µl of TMB solution in each well.
10. Incubate for 15 min incubation at room temperature in the dark and immediately read absorbance at 620 nm (see Subheading 4).

4 Notes

1. TMB: Following the reaction with peroxidase, a blue reaction product forms that may be read at 370 nm or between 620 and 655 nm. For end-point assays, the reaction can be stopped by the addition of a volume of 1 or 2 N hydrochloric acid or 1 N sulfuric acid equal to the volume of the substrate reaction in the well. The resulting yellow end product, which is stable for at least 1 h, can then be read at 450 nm.
2. The washing steps (steps 3, 5, and 8, Subheading 3.2.3 *revelation*) are performed manually (with eight multichannel pipet) with extreme precaution for the aspiration/dispensing. Until now, all microplate washer used for this analysis have a negative impact on the reliability of the method. Indeed, the affinity between glycoconjugates and lectins are very low and must be managed gently.

Acknowledgments

This work was carried out with the support of CYTHERIS S.A. The authors acknowledge Dr. Anne Grégoire and Dr. Yann Rancé for their very valuable help and advices.

References

1. Sharon N, Lis H (2004) History of lectins: from hemagglutinins to biological recognition molecules. Glycobiology 14:53R–62R
2. Sharon N (2007) Lectins: carbohydrate-specific reagents and biological recognition molecules. J Biol Chem 282:2753–2764
3. Wu AM, Lisowska E, Duk M, Yang Z (2009) Lectins as tools in glycoconjugate research. Glycoconj J 26:899–913
4. Kaneko Y, Nimmerjahn F, Ravetch JV (2006) Anti-inflammatory activity of immunoglobulin G resulting from Fc sialylation. Science 313: 670–673
5. Hirabayashi J (2008) Concept, strategy and realization of lectin-based glycan profiling. J Biochem 144:139–147
6. Gupta G, Surolia A, Sampathkumar SG (2010) Lectin microarrays for glycomic analysis. OMICS 14:419–436
7. Pilobello KT, Krishnamoorthy L, Slawek D, Mahal LK (2005) Development of a lectin microarray for the rapid analysis of protein glycopatterns. Chembiochem 6:985–989
8. Angeloni S, Ridet JL, Kusy N, Gao H, Crevoisier F, Guinchard S et al (2005) Glycoprofiling with micro-arrays of glycoconjugates and lectins. Glycobiology 15:31–41
9. Rosenfeld R, Bangio H, Gerwig GJ, Rosenberg R, Aloni R, Cohen Y et al (2007) A lectin array-based methodology for the analysis of protein glycosylation. J Biochem Biophys Methods 70:415–426
10. Pilobello KT, Slawek DE, Mahal LK (2007) A ratiometric lectin microarray approach to analysis of the dynamic mammalian glycome. Proc Natl Acad Sci U S A 104:11534–11539
11. Tateno H, Uchiyama N, Kuno A, Togayachi A, Sato T, Narimatsu H et al (2007) A novel strategy for mammalian cell surface glycome profiling using lectin microarray. Glycobiology 17:1138–1146
12. Tao SC, Li Y, Zhou J, Qian J, Schnaar RL, Zhang Y et al (2008) Lectin microarrays identify cell-specific and functionally significant cell surface glycan markers. Glycobiology 18: 761–769
13. Matsuda A, Kuno A, Ishida H, Kawamoto T, Shoda J, Hirabayashi J (2008) Development of an all-in-one technology for glycan profiling targeting formalin-embedded tissue sections. Biochem Biophys Res Commun 370: 259–263
14. Galili U, Shohet SB, Kobrin E, Stults CL, Macher BA (1988) Man, apes, and Old World monkeys differ from other mammals in the expression of alpha-galactosyl epitopes on nucleated cells. J Biol Chem 263:17755–17762

Chapter 15

Analysis of Monoclonal Antibodies by Sedimentation Velocity Analytical Ultracentrifugation

W. Blaine Stine Jr.

Abstract

Development of a thorough understanding of the solution polydispersity of therapeutic glycoproteins including monoclonal antibodies is an important and challenging undertaking. Degradation pathways involving fragmentation could result in loss of therapeutic efficacy. Protein aggregation on the other hand is frequently considered a critical quality attribute, and concerns exist that protein aggregates could result in undesirable immunological consequences (1). Sedimentation velocity analysis performed in the analytical ultracentrifuge (SV-AUC) provides a uniquely powerful first principal measure of the hydrodynamic size and shape of proteins under conditions that can come very close to the formulated drug product. This technique avoids the potential pitfalls associated with size exclusion chromatography (SEC) including on-column dilution, adsorption or disruption of species by a stationary phase, and the need to use high ionic strength mobile phases to screen unwanted electrostatic interactions (2, 3). Furthermore, not only does SV-AUC provide a quantitative size distribution analysis, but it also provides information about macromolecular conformation. For these reasons, use of SV-AUC for analysis of therapeutic monoclonal antibodies has become widespread throughout the biopharmaceutical industry and is one of the most common orthogonal techniques to SEC for measuring aggregate and fragment levels (4–9). The studies outlined in this chapter describe the basic principles of designing, collecting, and analyzing experimental data using SV-AUC with a focus on methods for therapeutic monoclonal antibodies and other similar biologics. Details are given that facilitate the acquisition of high quality data sets that in turn simplify data analysis resulting in robust and accurate measures of solution polydispersity.

Key words Analytical ultracentrifugation, Sedimentation velocity, Monoclonal antibody

1 Introduction

There are two general approaches to performing studies in the analytical ultracentrifuge—sedimentation equilibrium and sedimentation velocity. Sedimentation equilibrium studies are a powerful way to measure the solution molecular weight of a protein or other macromolecule, by analyzing the distribution of a sample in a centrifuge cell once the counteracting forces of sedimentation and diffusion have reached equilibrium. Typically experiments are

Alain Beck (ed.), *Glycosylation Engineering of Biopharmaceuticals: Methods and Protocols*, Methods in Molecular Biology, vol. 988,
DOI 10.1007/978-1-62703-327-5_15, © Springer Science+Business Media New York 2013

performed at multiple concentrations and at multiple rotor speeds in order to generate complete data sets. When factoring in the time needed to reach equilibrium under all of these conditions, the overall study can become a lengthy proposition. Furthermore, sedimentation equilibrium studies are not well suited for the discrimination and detection of low levels of protein fragments or aggregates. On the other hand, sedimentation velocity studies can be performed at a single rotor speed and at a single concentration (assuming a known well characterized noninteracting system). Since the data needed for quantitating the solution polydispersity in a sedimentation velocity experiment comes from measuring the evolution of a concentration gradient under relatively high centrifugal force, these experiments can be performed in several hours. If a very detailed understanding of a the nature of particular molecule is desired, it is recommended that even for sedimentation velocity studies, the experiment be performed at multiple protein concentrations, and replicated using different centrifuge cells inserted into different positions in the centrifuge rotor. If low molecular weight or high molecular weight aggregates are detected, the experiment can be repeated at higher or lower rotor speeds to optimize the sedimentation of that particular species, thereby providing a more comprehensive and accurate picture of the size and shape of the macromolecules that make up a particular sample.

2 Materials

2.1 Analytical Ultracentrifuge (Beckman Coulter Inc.)

1. Recommended Model: ProteomeLab XL-I (equipped with both Rayleigh interference and scanning UV/visible absorbance optics) (see Note 1).
2. Optional Model: XL-A (equipped with scanning UV/visible absorbance optics only).
3. Recommended Rotor: An-60 Ti four-place titanium analytical rotor rated for a maximum speed of 60,000 rpm which translates to 262,000 × g at the center of the cell. This rotor will accommodate three sample cells and one counterbalance cell (necessary for absorbance scanning) (see Note 2).
4. Optional Rotor: An-50 Ti eight-place titanium analytical rotor rated for 50,000 rpm. This rotor will accommodate seven sample cells and one counterbalance cell (necessary for absorbance scanning) (see Note 3).
5. Recommended Sample Cells: Each analytical cell consists of the following parts:
6. Cell Housing—Aluminum double-sector.
7. Window holder—black anodized aluminum.
8. Window gasket—Vinylite.

9. Window liner—Bakelite.
10. Sapphire windows (see Note 4).
11. 12 mm Double-Sector Flow-Through Charcoal-filled Epon Centerpieces (see Note 5).
12. Screw-ring washer—Bakelite.
13. Screw ring—black anodized aluminum (see Note 6).
14. Housing plug screw—brass (see Note 7).
15. Plug gasket—red vinyl dots.

2.2 Additional Tools and Equipment

1. Cell Alignment Tool (Beckman Coulter).
2. Screwdriver (size 3.5 slotted) or torque driver (Wiha TorqueVario®-S 15–80 in-oz set to ~30 in-oz with a 0.6×3.5 slotted tip) (see Note 8).
3. Curved sharp tipped tweezers (Techni-Tool P/N: 758TW564).
4. Soft picks (wrapped foam swabs, Coventry P/N:20080).
5. Small disposable plastic weigh boats for cleaning cells.
6. Kimwipes™.
7. Hellmanex III (Hellma) or PCC-54 Detergent (Pierce/Thermo Scientific) (see Note 9).
8. Optical Tissue (Opticwipes™ P/N:C920 Chemtronics).
9. Black Sharpie™ marker (see Note 6).
10. Spinkote™ (Beckman Coulter).
11. Torque wrench (0–150 in-pounds model MD1501 Sturtevant Richmont) (see Note 10).
12. Analytical Torque Stand (Beckman Coulter) (see Note 10).
13. 200 μL pipette.
14. Gel loading pipette tips (see Note 11).
15. Nitrogen gun with filter (for use with house nitrogen supply—Innotech TA-N2-2000FT) (see Note 12).
16. Illuminated magnifier (Prolite 886IN522 with a 3/10 diopter lens) (see Note 13).

(*Note that many of these specialty items can be ordered through Techni-Tool Inc.*)

3 Methods

3.1 Experimental Design

Bulk drug substance can be formulated under a wide variety of conditions. Most typical formulations are compatible with analysis by sedimentation velocity; however, there are some considerations that should be taken into account. Formulation buffers with very low ionic strength (less than 10 mM salt for example) may not sufficiently screen ionic interactions between protein molecules leading to solution non-ideality. Since nonideal solutions are not well accounted for using existing sedimentation velocity data analysis tools, solutions containing at least 10 mM salts are recommended. If diafiltration and ultrafiltration are used to exchange the drug substance into formulation buffer, then the diafiltration buffer used during this process provides an ideal diluent and reference solution for the experiment.

Formulation excipients that either sediment or float, forming dynamic gradients during the centrifuge run, can greatly complicate data analysis. It has been shown recently that the commonly used excipient sorbitol can affect aggregate detection in formulations (10). If possible, initial studies should be carried out in buffer systems that do not form dynamic gradients. Often these excipients are necessary for long-term stability and can be excluded from the formulation for sedimentation velocity analysis. However, if such excipients are necessary, then data analysis tools do exist that model the sedimentation of macromolecules in inhomogeneous solvents (11).

Sample concentration should be calculated to yield a final absorbance of 1–1.2 OD at 280 nm in the 1.2 cm path-length ultracentrifuge cells. Typically for antibodies with absorbance coefficients of ~1.4 this results in concentrations in the 0.6–0.7 mg/mL range.

If antibody aggregates or dimers are thought to be in reversible equilibrium with monomer, the following experiments should be considered. One approach would be to compare results obtained acquiring scan data as rapidly as possible after sample dilution to results obtained after an extended equilibration (overnight or longer) after sample dilution. If there is a slow reversible equilibrium between species the distributions will differ. Another approach would be to perform sedimentation velocity runs at a series of increasing concentrations. This approach will detect species in rapid equilibrium. If a concentration dependent change in sedimentation coefficient is measured, there is an increased likelihood of an interacting system.

3.2 Cell Assembly

Assemble the cells as described by the manufacturer (see *An-50 and An-60 Ti Analytical Rotor, Cells, and Counterbalance* publication LXL/A-TB-003 G, Beckman Coulter Inc.) taking extra precaution with respect to the following areas:

1. Keep all of the components for each cell together as an individual set. This will allow cell-dependent variability across experiments to be tracked.
2. Use compressed filtered nitrogen (or air) to blow off any dust or fibers immediately before assembling the cells.

3. Assemble the window assemblies first noting the scribe mark on the windows indicating the optical axis of the sapphire crystal. Align the scribe mark with the keyway slot in the window holder to ensure the generation of optimal fringe patterns when collecting interference data and run-to-run reproducibility (and aid with troubleshooting damaged windows) when using absorbance optics.
4. Use as little Spinkote as possible to create a thin film on the screw ring threads and the screw ring washer. Some Spinkote is necessary on these parts to ensure that these surfaces do not bind during the torque adjustment of the cell.
5. When torquing the cells, use one slow smooth motion building up torque until reaching the desired value. Although Beckman recommends a value of 120 in.-pounds, we routinely torque the cells to 130 in.-pounds and almost never encounter leaks. It is important that the final applied torque building from 100 to 130 in.-pounds be applied slowly, trying to maximize the rotation of the screw ring in one continuous motion until 130 in.-pounds is reached.
6. Store the cells assembled and torqued. If the cells have been stored, re-check cell torque before filling with sample and reference.
7. Marking the sample and reference filling holes on the aluminum housing (with an "S" for sample and "R" for reference, for example) with a black Sharpie marker can help avoid mix-ups.
8. Use a P-200 Pipette to fill the cells with sample and reference. The gel loading tips in combination with the twin port external loading flow-through centerpieces greatly simplify this process. For typical sedimentation velocity experiments load 435 μL (3 × 145 μL) of sample and 445 μL (2 × 145 μL and 1 × 155 μL) of reference solution. Take care not to get any liquid between the centerpiece and the housing otherwise it will rapidly wick by capillary action throughout the cell. If this happens, disassemble, clean, and reassemble the cell. Also, each time liquid is added to the cell use a clean dry pipette tip and add liquid to the same filling hole position for each sector so that one of the two filling holes on each sector remains dry allowing air to freely escape during loading. Otherwise liquid will get pushed back out the filling hole and become wicked away between the housing and centerpiece.
9. Once the sample and reference sectors have been filled, carefully place one red vinyl plug gasket over each filling hole making sure it is centered and not in contact with the threads. Gently tighten the brass housing plug with a slotted (size 3.5) screwdriver. The torque for each screw must be consistently applied for each cell and during each run. If the brass housing

plugs are not tight enough the cells may leak. If they are torqued too much the aluminum housing will become deformed and not fit smoothly in the centrifuge rotor. If multiple people will be assembling cells on an infrequent basis, a specialized screwdriver can be ordered with adjustable torque settings that will ensure that the brass housing plugs are torqued the same amount each time. (See 2.2 Additional Tools and Equipment).

3.3 Loading Cells in the Rotor

1. The red anodized counterbalance should be loaded in cell position #4 (for an An-60Ti rotor). Load cells 1–3 in their respective positions making sure that they are in the correct orientation. The cells should easily glide into the rotor holes and should not be forced. If the cells do not easily slide into the rotor holes then this is likely the result of over tightening the brass plugs causing the aluminum housing to deform. Since the housing is constructed from a soft aluminum, they may eventually come back to round after several runs in the centrifuge.
2. All cells must weigh within 0.5 g of each other. If there is any doubt, check the weights on a balance and adjust the size of the weight in the counterbalance if necessary. If you wish to achieve better than 0.5 g accuracy between cells, additional counterbalance screws can purchased and cut down in size to achieve intermediate weights that can provide a better match.
3. Alignment of the cells is an important aspect of generating high quality data sets. Cell misalignment can result in convection during the run that can disturb the sedimentation boundary. This will significantly complicate data analysis. A mirror is provided with the instrument to assist with aligning the scribe marks on the sample cell and the rotor. Optionally, the rotor can be handheld and inspected with an illuminated magnifier that allows very precise alignment of the scribe marks. It should be noted, however, that the cell alignment is only as good as the original positioning of the scribe marks, and if these are not manufactured to within very precise tolerances the cell may still be slightly out of alignment. Once the cells are carefully aligned, be sure to avoid disturbing the cells when placing the rotor back in the centrifuge.

3.4 Equilibration

1. Sedimentation velocity analysis of monoclonal antibodies can be successfully performed at 20°C. It is very important that the centrifuge and rotor assembly fully equilibrate before beginning the run.
2. Once the rotor and optical arm have been returned to the centrifuge, start the thermal equilibration by setting the run time to several hours and the speed to 0 rpm using the controls on the front of the centrifuge. By using a speed of 0 rpm and

beginning a run, the centrifuge will pump down the chamber first using the rough pump, and then the diffusion pump, to reach a final value of "0 microns" on the display.

3. Before the vacuum in the centrifuge reaches ~50 μm, the reported temperature is a combination of the air temperature in the centrifuge and the rotor surface temperature. Once the vacuum reaches a reported value of 0 μm, an accurate surface temperature of the rotor can be measured.
4. To ensure that the centrifuge is fully equilibrated from the surface of the rotor to the solutions within the cells, it is strongly recommended that the system be allowed to equilibrate one to two additional hours after the reported temperature reaches 20.0 ± 0.2°C at 0 μm.
5. Complete equilibration prior to beginning a run will help avoid thermal changes that can cause convection. As with improper cell alignment, convection caused by incomplete thermal equilibration will greatly complicate data analysis, as the theoretical models used to describe sedimentation velocity do not take into account changes caused by convection.

3.5 Run Conditions

The run conditions for sedimentation velocity analysis of monoclonal antibodies should be determined based on the goals of the study. Typically, speeds between 36,000 and 42,000 rpm are well suited for sedimentation of 150 kDa proteins. Advantages of using speeds closer to 36,000 rpm are that more scan data can be collected before the sedimentation boundary becomes overly influenced by back diffusion of protein collecting at the bottom of the cell. These slower speeds are also better suited for analysis of high molecular weight aggregates. Advantages of using speeds closer to 42,000 rpm are that lower molecular weight fragments are more easily resolved from intact antibodies and boundary spreading due to diffusion is less apparent. Either way, 100–300 scans per cell should be acquired over a time interval where the sedimentation boundary is not disturbed by back diffusion. Typical experiments are carried out at 20°C with a radial increment of 0.003 cm, one replicate, and continuous scanning settings enabled.

Although some researchers recommend performing a slow speed leak check before starting the run, we have found that with proper attention to cell assembly and maintenance, leaks can be avoided. If there are still concerns that the cells might leak, the vacuum level can be closely watched as the rotor builds up speed at the start of a run. If a cell does begin to leak, then the vacuum level will rapidly increase and the run should be aborted. Furthermore, the meniscus position can be monitored during the first few scans to ensure that the meniscus does not migrate towards the bottom of the cell. If it does, abort the run.

3.6 Cell Cleaning and Maintenance

1. After completion of a run the cells should be carefully disassembled and thoroughly cleaned. Proper attention to detail during this process will ensure long life for the cells and reproducibility between experiments. Protein deposits or scratches to either the windows or the centerpiece channels can result poor data quality.
2. There should be two approaches to cleaning the windows and centerpieces. First is to use a suitable detergent such as PCC54 (Pierce), or HELMANX (Hellma) which are highly effective for cleaning the windows. Second is a gentile non-abrasive method to mechanically clean the channels in the centerpieces. We have used the cotton tipped swabs (Chemitronix) or Kimwipes soaked with diluted detergent with good success. The use of swabs may not be necessary after every run depending on the protein being analyzed and as long as protein doesn't precipitate during the experiment.
3. To clean the windows and centerpieces, place one folded Kimwipe in the bottom of a disposable weigh boat filled with diluted detergent solution. Each cell should have its own weigh boat of cleaning solution to avoid mixing up the parts of the cells. The purpose of the Kimwipe to provide a soft surface for the centerpieces and windows to rest on.
4. Allow the parts to soak for at least 1 h or longer depending on the amount of cleaning needed. After soaking is complete, rinse the parts in distilled water several times then rinse with a 70% ethanol solution.
5. Blow-dry the windows and centerpieces using compressed filtered nitrogen. There should be no signs of haze or precipitate when examined under magnification.
6. Reassemble the cells before storage applying a very thin film of Spinkote on the threads and base of the screw ring. If there are any marks or scratches in the black anodized coating on the screw ring that could scatter light, these can be touched up with a black permanent marker.
7. Storing the cells assembled will help keep dust out of the centerpiece channels and facilitate the next experiment.
8. Finally, the interior of the centrifuge should also be cleaned and kept free of oil from the diffusion pump. We routinely wipe down the interior with a large Kimwipe sprayed with a 70% ethanol solution and allow to dry after every run.

3.7 Data Analysis (Figs. 1, 2, and 3)

There are many excellent programs available for analyzing sedimentation velocity data each having its own strengths and weaknesses. A complete analysis of all of the methods and software available is beyond the scope of this chapter; therefore, we focus on $c(s)$ analysis as implemented in SEDFIT (12). This approach to deriving information from sedimentation velocity data uses direct fitting of the Lamm Equation while modeling the influence of diffusion on the data to enhance resolution.

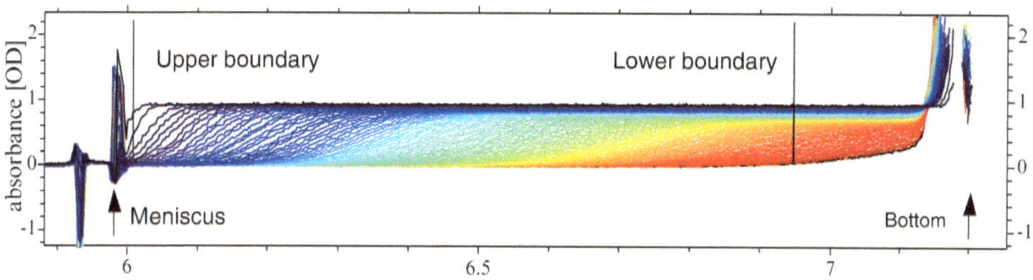

Fig. 1 Fitted absorbance data plot using SEDFIT from a typical sedimentation velocity run showing the approximate positions of the meniscus, *bottom* of the cell, and *upper* and *lower* fitting boundaries. The fitting parameters for *c*(*s*) analysis will differ depending on experimental design. However, for general antibody analysis using *c*(*s*), some starting values are as follows. The resolution parameter can be set to a lower value such as 50 for rapid fitting of data and a quick look at the results. Once other settings have been optimized the resolution can be increased to between 100 and 300. The greater this value, the longer the time needed to fit the data. The min and max values for sedimentation coefficient should bracket the expected size range of the species present in your sample. A range of s min = 1 and s max = 20 will cover antibody monomer, fragments, and soluble aggregates. If very low molecular weight species or large aggregates are expected, or if the data are not well fit using this range, then it should be expanded. The frictional ratio (f/f_0) should be a fitted parameter (*box checked*) with a starting value of around 1.4. For globular proteins a starting value of 1.2 would be more appropriate; however, with antibodies having a more asymmetric conformation a starting value of 1.4 is closer to the expected fitted result. Baseline, RI Noise, Time Independent Noise and Meniscus should all be fitted (*boxes checked*). The distance to the bottom of the cell should be fixed at 7.2. The confidence level (F-ratio) should be kept at two standard deviations (0.95) to avoid fitting artifacts; however, in some cases species can be resolved better using one standard deviation (0.68). Once these values are entered in the parameters dialog box, set and confirm the upper and lower fitting boundaries and the meniscus. If the cells are filled according to the procedures outlined in this chapter the starting point for the meniscus should be at the apex of the positive peak (caused by optical skewing) at around 5.975 cm. The upper fitting boundary should be set far enough into the solution column to avoid artifacts near the meniscus and the lower fitting boundary should be set far enough above the bottom of the cell to avoid any region that contains signal originating for sample back diffusion. Typically these values are around 6.1 cm for the upper boundary and 7 cm for the lower boundary. In general the more data that can be collected from the longest possible solution columns will provide the most accurate results

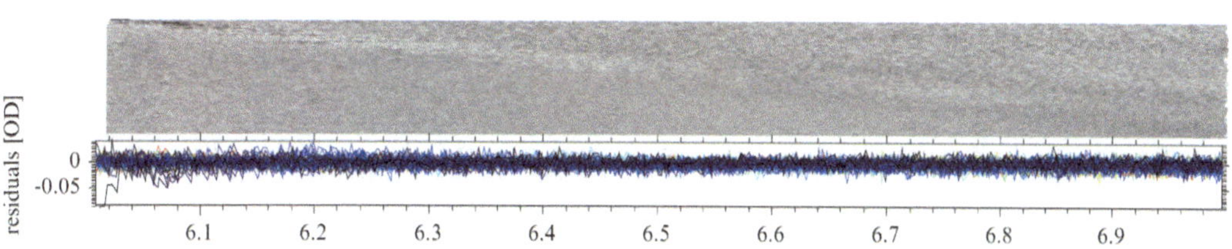

Fig. 2 Bitmap and plot of residuals after fitting sedimentation velocity data using the *c*(*s*) method demonstrating minimal bias and a random distribution pattern. The extent to which the data fit the model using the *c*(*s*) method can be determined using the reported rmsd of the fit (typically less than 0.01) and importantly any bias observed in the residuals. Both the residual plot and residual bitmap should have a random distribution with regard to time and radial position (Fig. 2). If a prominent horizontal line is present in the residual bitmap image, then the model is not describing the data well and further work needs to be done optimizing fitting parameters or generating better quality raw data

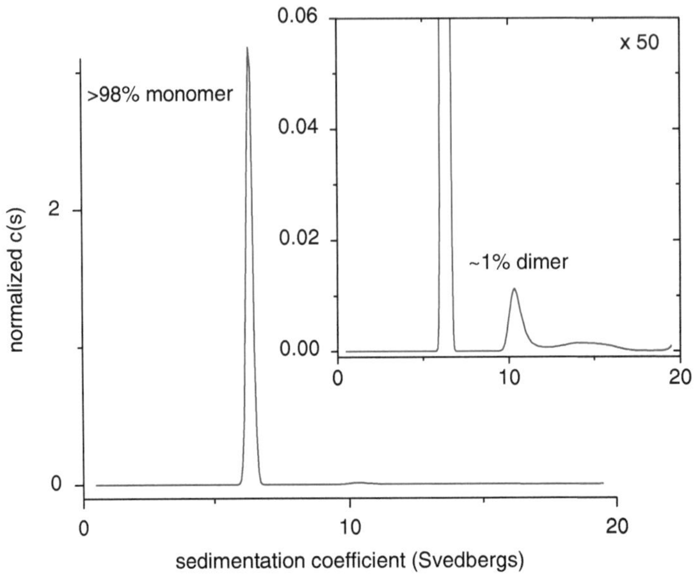

Fig. 3 The calculated *c*(*s*) distribution from a typical monoclonal antibody sedimentation velocity run. The final *c*(*s*) distribution (Fig. 3) can be generated by copying the distribution data out of SEDFIT and plotting in using ORIGIN. Utilities are built into SEDFIT for integrating peak area (used to calculate the relative percent), calculating hydrodynamic shape (or frictional coefficient), and molar mass. Substantially more information regarding the theory and application of the *c*(*s*) method can be found at http://analyticalultracentrifugation.com and users are strongly recommended to spend time with this resource. As pointed out earlier in this chapter, sedimentation velocity analysis is a technique of accuracy more so than precision. Therefore, analyzing multiple replicates of the same sample is strongly recommended. Intra-assay replicates are beneficial in measuring any potential cell-dependent differences. For example, if the % aggregate values are consistently higher for a sample replicate in position three, then the experiment should be repeated with a new cell in that position and the centerpiece and windows should be checked for scratches or damage. At a minimum, triplicate analysis should be performed for each sample. Species such as small aggregates or fragments that do not replicate in all three cells should be interpreted with caution. The parameters outlined in this chapter should only be considered as starting points for sedimentation velocity studies of monoclonal antibodies. Development of a comprehensive understanding of experimental design and data analysis techniques is imperative to generate high quality sedimentation velocity results. Fortunately many other resources exist such as hands on workshops, online tutorials, discussion groups, symposia, and a wealth of published literature

$$\frac{dc}{dt} = \frac{1}{r}\frac{d}{dr}\left[rD\frac{dc}{dr} - s\omega^2 r^2 c\right]. \tag{1}$$

The Lamm Equation 1 describes the changes to a concentration distribution over time $c(r, t)$ of an object in a sector shaped solution column placed in a centrifugal field $[\omega^2 r]$. The object or macromolecule has a specific a sedimentation coefficient (s) and diffusion coefficient (D). This partial differential equation describes the influence of both diffusion $[rD \times dc/dr]$ and sedimentation $[s\omega^2 r^2 c]$ on the macromolecule over time. Once the sedimentation coefficient (s) and diffusion coefficient (D) are measured, the molar mass can be calculated using the Svedberg equation:

$$s = \frac{MD(1 - \bar{\upsilon}\rho)}{RT} \quad (2)$$

where M is the molar mass of the protein, $\bar{\upsilon}$ is its partial specific volume, ρ is the solvent density, R is the universal gas constant and T is absolute temperature. This calculation can provide a fairly accurate measure of molar mass when the $c(s)$ distribution is dominated by single species with well determined sedimentation and diffusion coefficients. However, for minor species that may have much lower signal/noise ratios or different frictional coefficients, the conversion to molar mass is much less accurate.

In order to calculate a $c(s)$ distribution there are three protein solution parameters that have to be determined including partial specific volume, buffer density, and buffer viscosity. These experimental parameters can be calculated using utilities such as SEDNTERP (public domain software available at http://www.rasmb.bbri.org/) or directly measured. When using SEDNTERP to calculate $\bar{\upsilon}$, be sure to include post-translational modifications in the calculations as they can impact the final value especially for highly glycosylated antibodies. We routinely measure buffer density and viscosity prior to running sedimentation velocity analysis. It is important to perform these measurements at the same temperature as the centrifuge run; typically 20°C. Density measurements can be performed using an oscillating tube densitometer (Anton Paar model DMA 4500) and viscosity measurements can be performed using a falling ball capillary viscometer (Anton Paar model AMVn) with a 0.3–10 cP capillary. Both measurements are carried out according to the procedures supplied by the manufacturer.

3.8 Additional Support

Several resources are available to assist researchers with performing and analyzing data from the analytical ultracentrifuge. The Reversible Associations in Structural and Molecular Biology (RASMB) group provides an active discussion forum and a comprehensive archive and is currently hosted by the Boston Biomedical Research Institute (BBRI). The archives and subscription information can be found at:

http://www.bbri.org/rasmb/rasmb.html

This institute has also been the host of several sessions of the Advanced Analytical Ultracentrifugation Workshop and details for these sessions can be found at their website.

Additional information about SEDFIT and SEDPHAT along with comprehensive tutorials, explanations of sedimentation velocity theory, and a SEDFIT listserv group can be found at:

http://analyticalultracentrifugation.com/default.htm

This site also contains information about a workshop on hydrodynamic and thermodynamic analysis of macromolecules run by Peter Shuck and his group at the National Institutes of Health in Bethesda, MD.

In addition to these sites there are several other excellent resources available to researchers interested in the theory and practice of performing experiments with the analytical ultracentrifuge and readers are encouraged to seek them out.

4 Notes

1. Sedimentation velocity data collected from UV absorbance is preferred for routine analysis and requires minimal sample preparation (e.g., dilution) and the buffer used for dilution can be used in the reference cell. For collecting data using interference optics, ideally the sample should be dialyzed at the target concentration to equilibrium against excess buffer and the dialysate should be used in the reference cell. Dialysis ensures that the sample and reference cells contain the identical buffer matrix; however, the additional time necessary to perform these steps may not be preferred for routine analysis. For more in-depth analysis, analyzing data acquired using both interference and absorbance optics can provide a more robust understanding of the hydrodynamic properties of the sample.
2. The three cells available in the An-60Ti rotor make it easy to acquire triplicate data sets for each sample being analyzed and minimize the time in between scans for each sample cell.
3. One downside of the larger eight hole rotor is that there can be significant sample sedimentation occurring between the time a scan is completed for a given cell, and the time necessary to scan the remaining six cells before the next scan can be initiated for that given cell.
4. Quartz windows are also available, and if necessary offer better optical performance than sapphire windows below 240 nm. However, sapphire windows are less prone to scratching and produce better interference fringe patterns than quartz and therefore should be used for all runs where interference data is acquired. The enhanced durability, lower levels of light refraction producing better fringe patterns, and good transmission at 280 nm are the reasons why sapphire windows are recommended for these studies.
5. A variety of other Epon and aluminum centerpieces are available; however, the charcoal-filled Epon material offers a more inert surface than aluminum and the flow-through design greatly simplifies sample loading and removal from the assembled cell.
6. A black permanent marker can be used to touch up any scratches in the black anodized coating to minimize possible light scattering interference.

7. The brass housing plugs are made from a soft metal and should be tightened carefully with the proper size screwdriver.
8. The slotted torque driver (screwdriver) will ensure that the brass plugs do not become over-tightened which deforms the aluminum housing and can make it difficult to insert or remove from the rotor.
9. Proper cleaning of the windows and centerpieces is essential for accurate data collection in the analytical ultracentrifuge. It is very important to remove all residual surface contaminants or protein without causing any damage to the windows or delicate carbon-epon centerpieces—especially the polished trapezoidal inner surfaces of the centerpieces.
10. Proper use of the torque wrench, stand, and a thin film of Spinkote on the Bakelite window liners ensure consistent and accurate cell assembly which will minimize leaks and extend the usable life of your centrifuge cells.
11. The gel loading tips make it easy to fill the centerpieces; however, make sure that one of the two small holes in the centerpieces remains free from liquid so that air can escape when filling the cell.
12. The compressed nitrogen gun greatly facilitates cell cleaning and maintenance. Dust and debris within the cell can severely compromise sedimentation velocity data.
13. The illuminated magnifier also helps with inspection of the centerpieces and windows to ensure that there is no residual contamination on the critical surfaces. There should also be no small scratches on these surfaces which can distort the sedimentation boundaries.

References

1. Rosenberg AS (2006) Effects of protein aggregates: an immunologic perspective. AAPS J 8:E501–E507
2. Chirino AJ, Mire-Sluis A (2004) Characterizing biological products and assessing comparability following manufacturing changes. Nat Biotechnol 22:1383–1391
3. Philo JS (2006) Is any measurement method optimal for all aggregate sizes and types? AAPS J 8:E564–E571
4. Berkowitz SA (2006) Role of analytical ultracentrifugation in assessing the aggregation of protein biopharmaceuticals. AAPS J 8:E590–E605
5. Gabrielson JP, Randolph TW, Kendrick BS, Stoner MR (2007) Sedimentation velocity analytical ultracentrifugation and SEDFIT/c(s): limits of quantitation for a monoclonal antibody system. Anal Biochem 361:24–30
6. Arthur KK, Gabrielson JP, Kendrick BS, Stoner MR (2009) Detection of protein aggregates by sedimentation velocity analytical ultracentrifugation (SV-AUC): sources of variability and their relative importance. J Pharm Sci 98(10):3522–3539
7. Gabrielson JP, Brader ML, Pekar AH, Mathis KB, Winter G, Carpenter JF, Randolph TW (2007) Quantitation of aggregate levels in a recombinant humanized monoclonal antibody formulation by size-exclusion chromatography, asymmetrical flow field flow fractionation, and sedimentation velocity. J Pharm Sci 96: 268–279

8. Pekar A, Sukumar M (2007) Quantitation of aggregates in therapeutic proteins using sedimentation velocity analytical ultracentrifugation: practical considerations that affect precision and accuracy. Anal Biochem 367: 225–237
9. Liu J, Shire SJ (1999) Analytical ultracentrifugation in the pharmaceutical industry. J Pharm Sci 88:1237–1241
10. Gabrielson JP, Arthur KK, Kendrick BS, Randolph TW, Stoner MR (2009) Common excipients impair detection of protein aggregates during sedimentation velocity analytical ultracentrifugation. J Pharm Sci 98:50–62
11. Schuck P (2004) A model for sedimentation in inhomogeneous media. I. Dynamic density gradients from sedimenting co-solutes. Biophys Chem 108:187–200
12. Schuck P (2000) Size-distribution analysis of macromolecules by sedimentation velocity ultracentrifugation and lamm equation modeling. Biophys J 78:1606–1619

Part III

Glycoprotein Complexes Characterization

Chapter 16

Noncovalent Mass Spectrometry for the Characterization of Antibody/Antigen Complexes

Cédric Atmanene, Elsa Wagner-Rousset, Nathalie Corvaïa, Alain Van Dorsselaer, Alain Beck, and Sarah Sanglier-Cianférani

Abstract

Monoclonal antibodies (mAbs) have taken on an increasing importance for the treatment of various diseases including cancers, immunological disorders, and other pathologies. These large biomolecules display specific structural features, which affect their efficiency and need therefore to be extensively characterized using sensitive and orthogonal analytical techniques. Among them, mass spectrometry (MS) has become the method of choice to study mAb amino acid sequences as well as their posttranslational modifications with the aim of reducing their chemistry, manufacturing, and control liabilities.

This chapter will provide the reader with a description of the general approach allowing antibody/antigen systems to be characterized by noncovalent MS. In the present chapter, we describe how recent noncovalent MS technologies are used to characterize immune complexes involving both murine and humanized mAb 6F4 directed against human JAM-A, a newly identified antigenic protein (Ag) overexpressed in tumor cells. We will detail experimental conditions (sample preparation, optimization of instrumental parameters, etc.) required for the detection of noncovalent antibody/antigen complexes by MS. We will then focus on the type and the reliability of the information that we get from noncovalent MS data, with emphasis on the determination of the stoichiometry of antibody/antigen systems. Noncovalent MS appears as an additional supporting technique for therapeutic mAbs lead characterization and development.

Key words Monoclonal antibody, Antigen, Noncovalent mass spectrometry, Antibody/antigen complexes, Binding stoichiometries

1 Introduction

Sensitive and orthogonal analytical methods are required both for mAbs and recombinant target antigen structural assessment and quality control. Among them, mass spectrometry has mainly been applied to characterize mAbs at the amino acid level including their glycosylation patterns, their disulfide linkages and other posttranslational modifications, which may have an impact on their pharmacology, pharmacokinetic, and immunogenicity (1). Mass spectrometry

Alain Beck (ed.), *Glycosylation Engineering of Biopharmaceuticals: Methods and Protocols*, Methods in Molecular Biology, vol. 988, DOI 10.1007/978-1-62703-327-5_16, © Springer Science+Business Media New York 2013

has also been used to identify antigen epitopic and antibody paratopic regions (2–4). However, although soft ionization techniques have been extensively used to investigate protein–protein assemblies (5–8), only a limited number of publications have reported so far the detection of intact immune complexes using either matrix-assisted laser desorption/ionization mass spectrometry (MALDI-MS) or electrospray ionization mass spectrometry (ESI-MS). While MALDI-MS requires preliminary chemical cross-linking of the interacting partners as well as the use of a high-mass detector (4, 9, 10), ESI-MS presents the advantage to allow the direct observation of noncovalent immune complexes without any chemical modification (11–16).

In the present chapter noncovalent MS technologies including automated chip-based nanoESI-MS were used to study the formation of immune complexes involving mumAb 6F4 and its humanized form rose against recombinant human JAM-A extracellular domain (ECD). This state-of-the-art mass spectrometry technology first allowed to assess the quality of different recombinant antigen batches and to determine its oligomerization state. As a proof of concept, formation of immune complexes between JAM-A ECD and each mAb was then investigated in order (1) to determine mAb:Ag binding stoichiometry, (2) to assess mAbs recognition selectivity of the target antigen, (3) to compare mumAb and hzmAb relative binding affinity towards JAM-A ECD, and (4) to support Ag/Fab co-crystallization studies.

2 Materials

2.1 Buffers

- Milli Q water.
- Buffers: Ammonium acetate ≥99.0% puriss. p.a. for mass spectroscopy (Fluka), ammonium bicarbonate or ammonium carbonate.
- Acetonitrile (Carlo Erba).
- Formic acid.
- Cesium iodide puriss. p.a (Fluka) for calibration of the MS instrument.

2.2 Desalting Procedure

- Micro-concentration on centrifugal filter units: Centricon or Microcon (Millpore), Vivaspin (Sartorius).
- Gel filtration: NAP-5, NAP-10, PD-10 gel filtration columns (GE Healthcare), Zeba (Thermo Scientific).
- Equilibrium dialysis: Slide-A-Lyzer (Thermo Scientific).

2.3 Mass Spectrometry

- Any ESI-TOF or ESI-Q-TOF instrument.
- Analysis under classical "denaturing conditions."
  - Calibration of the mass spectrometer in the positive ion mode on the mass range *m/z*500–5,000 using $Cs_{(n+1)}I_n$ clusters produced by spraying a 3 mg/ml solution of cesium iodide in 2-propanol/water (1/1).
  - Dilute the sample to 1–5 μM in a H_2O/CH_3CN—1/1—solution + 1% HCOOH.
  - Injection into the mass spectrometer.
  - Record mass spectra on the mass range *m/z*500–5,000.
- Analysis under "non-denaturing conditions."
  - Calibration of the mass spectrometer in the positive ion mode on the mass range *m/z*1,000–10,000 using a 3 mg/ml solution of cesium iodide in 2-propanol/water (1/1).
  - Dilute the sample to 1–20 μM in ammonium buffer.
  - Injection into the mass spectrometer.
  - Adjust the accelerating voltages and the pressure in the interface to get optimal transmission and desolvation without complex disruption (see Note 9).
  - Record mass spectra on the mass range *m/z*1,000–10,000.

2.4 Materials for Antigen Production

The His-tagged ECD of the human junctional adhesion molecule A (His6-JAM-A ECD) was expressed in *Escherichia coli* and purified from either inclusion bodies or the soluble fraction as detailed elsewhere (11). Briefly, a construct including the 242 amino acid ECD of human JAM-A receptor was generated using conventional PCR methods. Purification of the soluble fractions of JAM-A ECD was achieved on an IMAC column (NiSepharose FF, GE Healthcare, Piscataway, USA) followed by tangential flow filtration. Purification of JAM-A ECD from inclusion bodies was performed by affinity and size exclusion chromatographies; refolding was achieved using 1% linear gradient from 6 to 0 M urea.

2.5 Materials for the mAb Production

Humanized antibody hzmAb (IgG4, anti-hJAM-A) and a negative control (IgG4, anti-hIGF-1R) were purified from CHO and NS0 cell supernatants, respectively (17), whereas mumAb (anti-hJAM-A) was purified from a murine hybridoma. The Fab fragment was obtained after papaïn treatment and further purified on a SP Sepharose HP column followed by a Mono Q column.

2.6 mAb Deglycosylation Protocol

Glycans were released from mAbs by incubation with 3 μl of PNGase F (5×10^5 U/ml, New England Biolabs, Hitchin, UK) per mg of glycoproteins in their initial buffer, overnight at 37 °C. For desalting and purification, CarbographTM cartridges (Alltech, Goussainville, France) were used. N-glycans were eluted from the cartridges within 1 ml of a solution containing acetonitrile/water (70v/30v) + 0.02% TFA.

3 Methods

3.1 mAb and Antigen Sample Preparations for Noncovalent MS Analysis

3.1.1 Why Is Desalting an Inescapable Step to Run Noncovalent MS Analyses?

Usually, buffers used to purify or extract proteins or noncovalent complexes (phosphate buffers, Tris, HEPES, etc.) are nonvolatile salts which are not compatible with ESI-MS analysis, even at low concentration. Therefore, a prerequisite to perform noncovalent complex analysis by MS is to exchange the purification buffer, a procedure also called "desalting step." The new buffer must fulfill two conditions: (1) being compatible with the ESI ionization process, i.e., volatile buffers are required and (2) integrity of noncovalent assemblies in solution must be preserved. Ammonium buffers best fulfill these requirements. Indeed, classical buffers used for noncovalent MS analysis include ammonium acetate, ammonium carbonate. Those buffers allow the pH of the solution to be ranging from 5.0 to 8.5. Further pH adjustments towards more acidic or basic pHs can be achieved by adding small volumes of formic acid or ammonia, respectively. The ionic strength of the buffer can also range from 10 to 500 mM depending on the stability of the complex (18). In most studies, solutions between 10 and 200 mM ammonium buffers are used, ensuring optimal ESI mass spectra quality (see Note 1).

Classical methods used for small volume sample desalting include size exclusion chromatography either by gravity (NAP-5™, NAP-10™, PD-10™ gel filtration columns, GE Healthcare) or with spin columns (Zeba® Spin Desalting Columns, Thermo Scientific), micro-concentration on centrifugal filter units (Centricon®, Microcon®, from Millipore; Vivaspin from Sartorius) and equilibrium dialysis (Slide-A-Lyzer, Thermo Scientific). These devices are all used according to supplier recommendations (see Note 2).

3.1.2 Antigens and Antibodies Desalting Procedures

To illustrate the relevance of the desalting step, Fig. 1 presents ESI mass spectra obtained for hzmAb with four different desalting techniques (ultrafiltration, dialysis, and gel filtration on either spin columns or by gravity) and different buffers (ammonium acetate or bicarbonate buffers) in order to determine which condition gives the best compromise between satisfactory desalting efficiency, preservation of protein ternary/quaternary structure, and protein recovery.

Effect of the Desalting Technique on the Overall Desalting Efficiency

First, four different desalting techniques were tested using the ammonium bicarbonate buffer (150 mM ammonium bicarbonate buffer at pH 7.2). ESI mass spectra depicted on Fig. 1a–d can be compared in terms of peaks width, i.e., resolution of the different glycoforms, to assess the desalting efficiency of each technique.

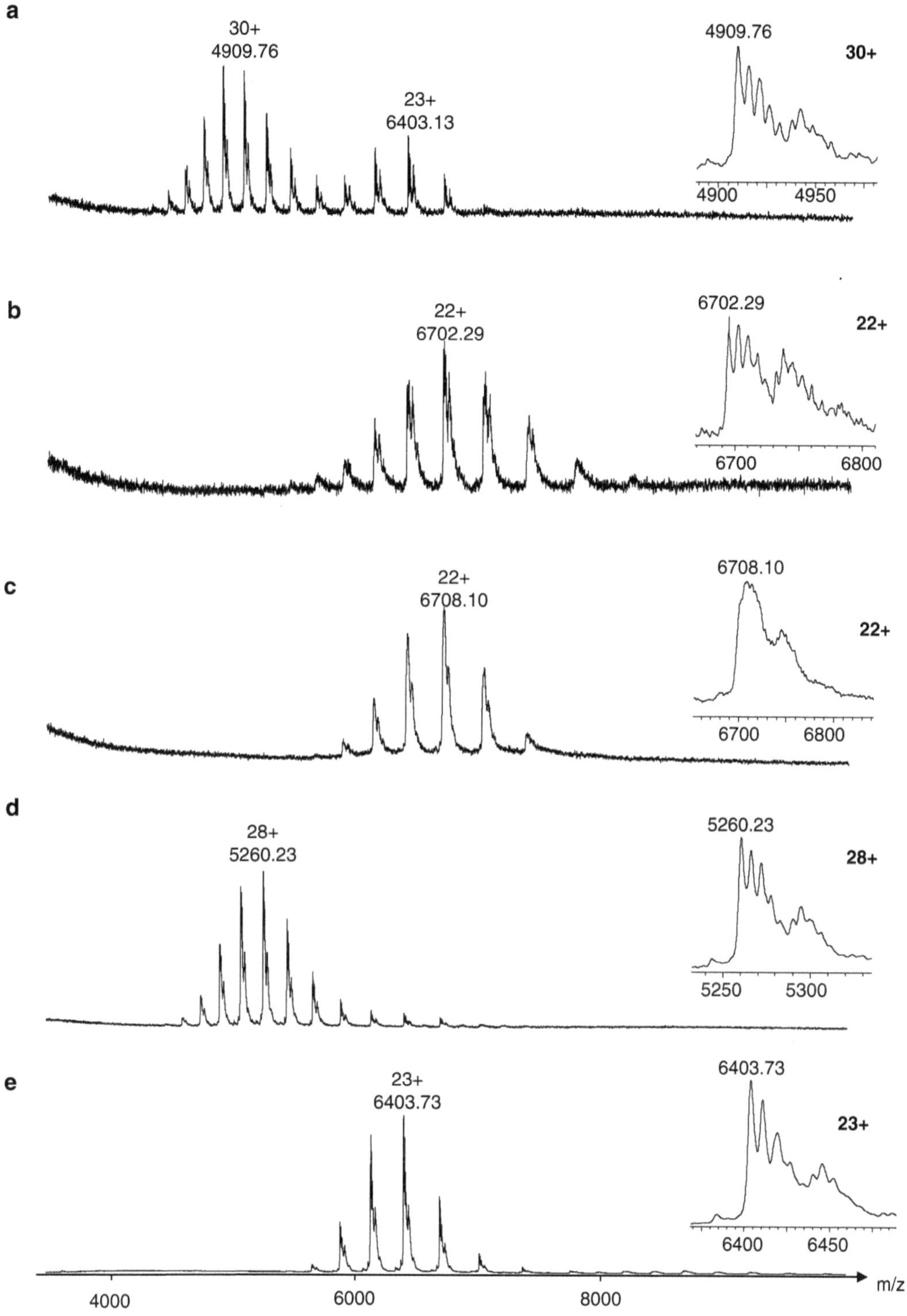

Fig. 1 Optimization of the desalting conditions for hzmAb. Humanized mAb buffer was exchanged against a 150 mM ammonium bicarbonate buffer at pH 7.2 by (**a**) ultrafiltration, (**b**) dialysis, (**c**) gel filtration using a Zeba column and (**d**) gel filtration using a NAP-5 column. (**e**) Humanized mAb after desalting against a 150 mM ammonium acetate buffer at pH 7.2 using NAP-5 column. After buffer exchange the protein was diluted to 10 μM in the ammonium buffer used for the desalting procedure and subsequently analyzed under optimized instrumental conditions (Vc = 200 V, Pi = 6 mbar, V_{ESI} = 1.8 kV, P_{ESI} = 1 psi)

- Figure 1c which was obtained using a Zeba column clearly shows that the different glycoforms are not resolved indicating thus the presence of residual nonvolatile salts after desalting which induce peak broadening and subsequent poor mass resolution.
- Concerning dialysis (Fig. 1b), although the mass spectrum is quite noisy, hzmAb heterogeneity is much more obvious but not as well as in the case of ultrafiltration (Fig. 1a) and NAP-5 gel filtration (Fig. 1d). Indeed mass spectra obtained after ultrafiltration (Fig. 1a) and after NAP-5 gel filtration (Fig. 1d) display well resolved glycoforms indicating thus that, in the present case, these techniques are equivalent in terms of desalting efficiency (see Note 3). However, in the case of NAP-5, the protein recovery (estimated from Bradford assays) is twice as high as for ultrafiltration (see Table 1) which makes of this former technique (gel filtration) the most interesting one in our case.

Buffer Effect on the Overall Desalting Efficiency

Two volatile buffers were also tested: ammonium bicarbonate (NH_4HCO_3) and ammonium acetate (NH_4Ac). In both cases, buffer concentration was set to 150 mM and the pH was adjusted to 7.2 either with ammonia or with formic acid. These conditions were chosen since they are close to physiological conditions (150 mM pH 7.4) and because the pH is not too close from protein's isoelectric points.

Comparison of the charge state distributions displayed in Fig. 1 indicates that hzmAb probably starts unfolding (see Note 4) after buffer exchange against ammonium bicarbonate buffer using NAP-5 column. Indeed, Fig. 1a shows a bimodal charge state distribution: one centered on the 30+ charge state and the other one on the 23+. This charge increase can be related to protein unfolding since less compact (more unfolded) structures make more protonation sites available during the ionization/desorption process. So desalting performed in 150 mM NH_4HCO_3 using ultrafiltration leads to conformational heterogeneity of the protein: one folded conformation centered on the 23+ charge state, and one less-folded structure represented by the charge state distribution centered on the 30+ charge state. Figure 1d also shows a distribution centered on the 28+ charge state. On the contrary, Fig. 1b, c display a narrow charge state distribution centered on the 22+ charge state of the protein, in agreement with a more-folded conformation.

Since partial unfolding occurs only for the two most efficient desalting techniques (Table 1), it could be hypothesized that hzmAb is not stable in ammonium bicarbonate buffer. Therefore, the most efficient desalting technique, i.e., NAP-5 gel filtration, was tested with ammonium acetate. In this case, Fig. 1e displays both a narrow charge distribution centered on the +23 charge state and well-resolved glycoforms.

Table 1
Summary of the results obtained for the optimization of hzmAb desalting procedure

Desalting technique	Buffer	Protein recovery (%)	Desalting efficiency	Preservation of the protein quaternary/ternary structure
Ultrafiltration	NH_4HCO_3	35	++	–
Dialysis	NH_4HCO_3	63	+	+
Gel filtration (Zeba)	NH_4HCO_3	96	–	+
Gel filtration (NAP-5)	NH_4HCO_3	76	++	–
Gel filtration (NAP-5)	NH_4Ac	74	++	+

Conclusions

According to these results, the following two-step desalting procedure was considered as the best suited for mAbs and antigens buffer exchange:

1. Buffer exchange against a 150 mM NH_4Ac buffer at pH 7.2 using NAP-5 gel filtration columns (NAP-5, GE Healthcare, Little Chalfont, UK). A single desalting step was sufficient to efficiently remove nonvolatile salts.
2. Samples were subsequently concentrated using centrifugal concentrators (Vivaspin, 3 kDa (JAM-A) and 30 kDa (mAbs) molecular-weight cutoff membranes, Sartorius, Göttingen, Germany). Protein concentration was determined from the Bradford assay using a 2 g/l BSA solution as standard.

To conclude, buffer exchange is an essential step in sample preparation (see Notes 5 and 6). It provides the protein sample free of nonvolatile salts, allowing acquisition of high quality ESI mass spectra and accurate mass measurements.

3.2 Characterization of Individual Antigens and Antibodies by Native MS

Prior to interaction assays involving antigen/antibody complexes, each individual partner was analyzed by MS in order to check purity and homogeneity of the samples:

- Mass spectrometry experiments were performed on an electrospray time-of-flight mass spectrometer (LCT, Waters, Manchester, UK) via an automated chip-based nanoelectrospray device (Triversa Nanomate, Advion Biosciences, Ithaca, USA) (see Subheading 2.3).
- External calibration was performed using the singly charged ions produced by a 3 mg/ml solution of cesium iodide in 2-propanol/water (1/1).

3.2.1 Antigen Analysis

In the context of the present study, recombinant human JAM-A ECD was expressed in *E. coli* and purified either from the soluble fraction (JAM-A SF) or from inclusion bodies (JAM-A IB).

Mass Measurements Under Denaturing Conditions

Desalted antigen samples were diluted to 2 μM in water/acetonitrile/formic acid 50/50/1. In these conditions, noncovalent interactions are disrupted in solution, which allows the measurement of the molecular weight of the monomeric subunits with a good precision (≥0.01%).

- Samples were infused into the mass spectrometer through the chip-based microfluidic system at a flow rate of ~100 nl/min.
- Mass spectra were recorded in the positive ion mode on the *m/z* range 500–4,000. Accelerating voltage was set to 30 V and the pressure Pi in the interface region of the mass spectrometer was 2.5 mbar.
- Data analysis was performed with MassLynx 4.0 (Waters, Manchester, UK). These nanoESI-MS experiments revealed highly pure and homogeneous protein preparations for both JAM-A SF and IB, as only one major ion series was detected. JAM-A SF and JAM-A IB measured molecular weights are 24,540.3 ± 0.5 Da and 24,542.5 ± 0.6 Da, respectively, which are both in agreement with the theoretical value calculated from the amino acid sequence (24,539.4 Da—Table 2). Interestingly, JAM-A IB measured mass is 2 Da higher than JAM-A SF. Compared to JAM-A SF, JAM-A IB also displays a charge state distribution shifted towards lower *m/z* (Fig. 2a, b). Both observations may be correlated with the absence of a disulfide bond in the case of JAM-A IB (for more details see ref. 11) (see Notes 7 and 8).

Mass Measurements Under Non-denaturing Conditions

In some cases, small modifications of pharmaceutical proteins may have dramatic consequences on protein properties and activities (19), especially when protein conformations are altered (20). Given the differences observed in denaturing conditions between JAM-A SF and JAM-A IB, it was of interest to check if the disulfide bridge pairing heterogeneity is likely to induce differences in the ternary/quaternary organization of the protein.

- ESI-TOF instrument was calibrated over the extended mass range (*m/z* 1,000–8,000) through a separate injection of a solution of a 3 mg/ml solution of cesium iodide in 2-propanol/water (1/1).
- Desalted antigen samples were diluted to 10 μM in 150 mM ammonium acetate buffer at pH 7.2 (Vc = 120 V; Pi = 6 mbar). In these conditions, noncovalent interactions are maintained, which allows the determination of antigen oligomerization state (see Note 9).

Table 2
Antigen molecular weights measured in denaturing and non-denaturing conditions

Denaturing conditions	Glycoforms	Calculated MW (Da)	Measured MW (Da)	ΔMW (Da)	ΔMW (ppm)
JAM-A SF	/	24,539.5[a]	24,540.3 ± 0.5	0.8	33
JAM-A IB	/	24,539.5[a]	24,542.5 ± 0.6	3.0	122
Control Ag	/	12,767.2	12,767.5 ± 0.1	0.3	23
Non-denaturing conditions	**Binding stoichiometry**	**Calculated MW (Da)**	**Measured MW (Da)**	**ΔMW (Da)**	**ΔMW (%)**
JAM-A SF	Monomer	24,539.5[a]	24,541 ± 1	1.5	0.01
	Dimer	49,079.0[a]	49,083 ± 1	4	0.01
JAM-A IB	Monomer	24,539.5[a]	24,543 ± 1	3.5	0.01
	Dimer	49,079.0[a]	49,088 ± 1	9	0.02

[a]Molecular weights are calculated assuming the presence of two disulfide bridges/no glycosylated protein detected

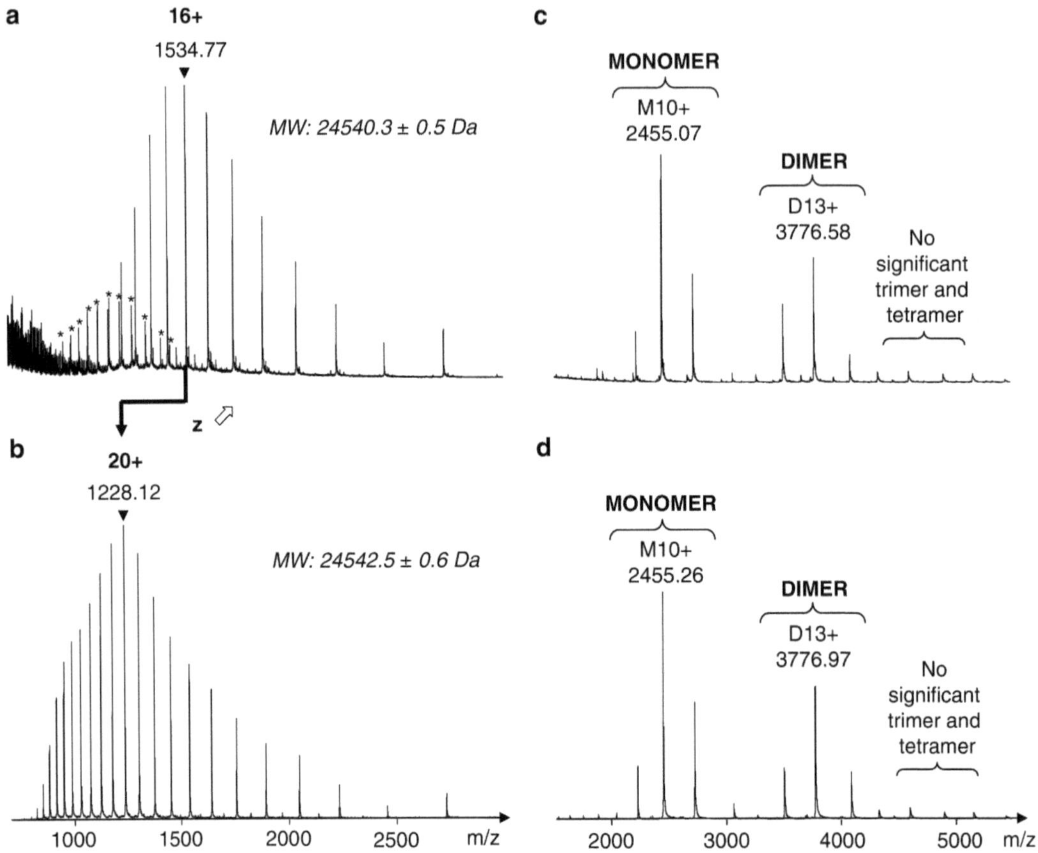

Fig. 2 ESI-MS analysis of JAM-A antigen in denaturing versus non-denaturing conditions. JAM-A SF (**a**, **c**) and JAM IB (**b**, **d**) were diluted either to 2 μM (monomer concentration) in 50/50/1 ACN/H_2O/HCOOH mixture for ESI-MS analysis in denaturing conditions for (**a**) and (**b**) (Vc = 30 V; Pi = 2.5 mbar) or to 10 μM (monomer concentration) in the ammonium acetate buffer for ESI-MS analysis in non-denaturing conditions for (**c**) and (**d**) (Vc = 120 V; Pi = 6 mbar). Asterisks correspond to a co-purified protein

- Samples were infused into the mass spectrometer through the chip-based microfluidic system at a flow rate of ~100 nl/min.
- Mass spectra were recorded in the positive ion mode on the *m/z* range 1,000–8,000. Accelerating voltage was set to 120 V and the pressure Pi in the interface region of the mass spectrometer was 6.0 mbar (see Note 9).
- Data analysis was performed with MassLynx 4.0 (Waters, Manchester, UK). nanoESI-MS analyses of both JAM-A SF and JAM-A IB (Fig. 2c, d respectively) lead to similar mass spectra with the detection of two main ion distributions in the *m/z* ranges 2,000–3,000 and 3,500–4,500. These species correspond respectively to JAM-A monomer (24,541 ± 1 Da for JAM-A SF and 24,543 ± 1 Da for JAM-A IB) and JAM-A dimer (49,083 ± 1 Da for JAM-A SF and 49,088 ± 1 Da for JAM-A IB) (see Note 10).

Data Interpretation and Conclusions: Recombinant JAM-A ECD Displays a Heterogeneous Disulfide Bridge Pairing

Several factors like the expression system, the cell culture or fermentation conditions, and the purification process are known to have a crucial impact on protein ternary structure and especially on disulfide bridge pairing. Therefore, correct S–S bonds formation between cysteine residues needs to be in-depth investigated (21, 22). In the context of the present study, recombinant human JAM-A ECD was expressed in *E. coli* and purified either from the soluble fraction (JAM-A SF) or from inclusion bodies (JAM-A IB). Both antigen preparations were first analyzed for quality control purposes under denaturing conditions by nanoESI-MS in order to check their purity and homogeneity.

Recombinant JAM-A SF and IB batches hence contain different cysteine oxidation states, revealing thus an inter-batch heterogeneity (see Note 11). Taken together these results highlight: (1) a crucial impact of the extraction process (i.e., from inclusion bodies or soluble fraction) on cysteine oxidation state of the recombinant antigen; (2) JAM-A is mostly detected as a dimer (see Note 12); and (3) heterogeneities linked to cysteine oxidation state and disulfide bridge pairing do not seem to affect the global antigen conformation as indicated by the detection of very similar, narrow charge state distributions.

3.2.2 Antibody Analysis

Similarly mAbs (glycosylated and PNGaseF-deglycosylated proteins) integrity and homogeneity were first checked by denaturing mass spectrometry analyses. Identical procedures as described in Subheadings 3.2.1 were employed except for injected concentrations (see Fig. 3, legend).

Under denaturing conditions, each mAb displays a broad charge state distribution in the *m/z* range 2,500–5,000 (Fig. 3a, b) giving rise to typical mAbs glycosylation profiles on deconvoluted mass spectra (inset Fig. 3a) (23). Mass measurements were performed with a mean relative error below 100 ppm for the different glycoforms (Table 3).

In mass spectrometry analyses conducted under non-denaturing conditions, mAbs are detected in the *m/z* range 5,500–7,000 with narrow charge states distributions centered on the 22+ and 23+ charge states (Fig. 3c, d). This observation suggests that a folded native-like structure of mAbs is retained in solution (24). In addition, efficient desalting and ion desolvation are achieved as indicated by the detection of resolved glycoforms (insets Fig. 3c). For each mAb, mass measurements of the most abundant glycoforms fall within +0.05% of the corresponding theoretical molecular weights (Table 3).

3.3 Determination of the Binding Stoichiometries of Antibody/Antigen Complexes Using Noncovalent MS

The knowledge of mAb:Ag binding stoichiometry is of primary importance to understand recognition mechanisms involved in immune complexes and to gain insights in possible mechanism of action. For several FDA and EMEA-approved mAbs, size-exclusion chromatography (SEC) was successfully used to investigate noncovalent mAb:Ag immune complexes (see Note 12). We also

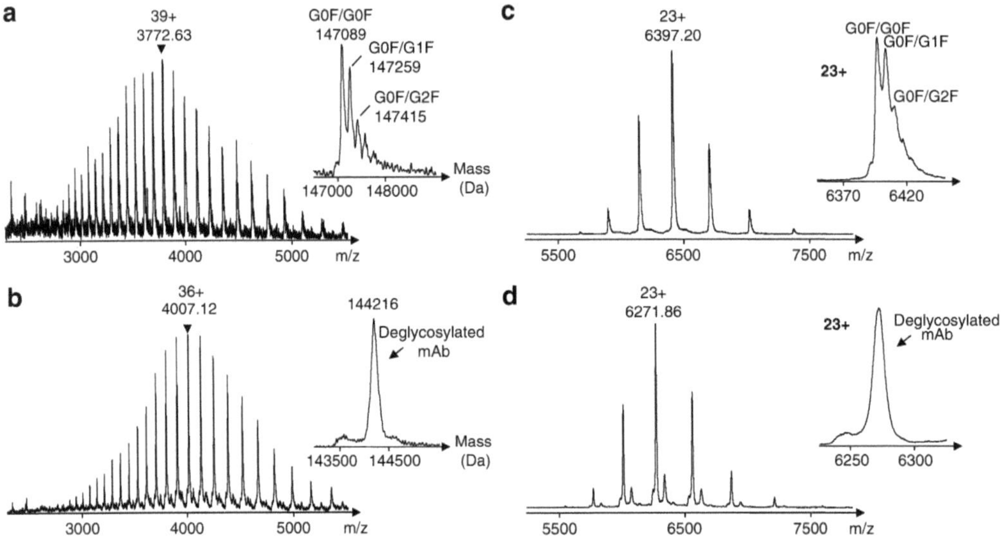

Fig. 3 ESI-MS analysis of glycosylated versus deglycosylated murine mAb. Glycosylated (**a**, **c**) and PNGase-deglycosylated mumAb (**b**, **d**) were diluted either to 5 μM (**a**, **b**) in 50/50/1 H_2O/ACN/HCOOH mixture (Vc = 70 V, Pi = 4 mbar) or to 5 μM (**c**, **d**) in a 150 mM ammonium acetate buffer at pH 7.2 (Vc = 200 V, Pi = 6 mbar). Insets in (**a**) and (**b**) correspond to deconvoluted mass spectra (MaxEnt 1 algorithm from Masslynx 4.0, Waters, Manchester, UK). Insets in (**c**) and (**d**) correspond to enlarged mass spectra of the +23 charge state

used SEC to characterize mumAb and hzmAb:JAM-ECD complexes, which yielded complicated profiles difficult to interpret from a ratio point of view (data not shown). Noncovalent nano-ESI-MS titration experiments were then carried out in order to address the stoichiometry question.

3.3.1 Sample Preparation and Instrumental Optimizations

- For each titration point, a fixed amount of desalted mAb (5 μM) was incubated with increasing amounts of desalted JAM-A IB (from 2.5 up to 40 μM).
- Mass spectrometry experiments were performed as described in Subheading 3.2.1.2 using an electrospray time-of-flight mass spectrometer (LCT, Waters, Manchester, UK) fitted with an automated chip-based nanoelectrospray device (Triversa Nanomate, Advion Biosciences, Ithaca, USA).
- Optimal accelerating voltages applied on the sample cone (Vc) and a well-adapted pressure in the interface (Pi) were used to allow fragile noncovalent assemblies to be transferred intact from solution to the gas phase while achieving efficient ion desolvation and ion transmission through the mass spectrometer (see Note 9). For antibody/antigen complex detection, optimal settings were found to be Vc = 120 V and Pi = 6 mbar.

Table 3
Molecular weights measured in denaturing and non-denaturing conditions for antibodies

Denaturing conditions	Glycoforms	Calculated MW (Da)	Measured MW (Da)	ΔMW (Da)	ΔMW (ppm)
Humanized Fab	/	47,117.6	47,117.6 ±0.6	+0.3	6
Murin mAb	Deglycosylated	144,185	144,216 ±3.0	+31	215
	G0F/G0F	147,073	147,099 ±5	+26	177
	G0F/G1F	147,236	147,256 ±8	+20	136
	G0F/G2F	147,398	147,414 ±8	+16	109
Humanized mAb	Deglycosylated	144,352	144,371 ±3	+19	132
	G0F/G0F	147,240	147,248 ±3	+8	54
	G0F/G1F	147,402	147,411 ±6	+9	61
	G0F/G2F	147,565	147,570 ±3	+5	34
	G1F/G2F	147,727	147,728 ±7	+1	7
	G2F/G2F	147,889	147,888 ±6	–1	7
Negative control mAb	G0F/G0F	148,576	148,583 ±3	+7	47
	G0F/G1F	148,738	148,748 ±4	+10	67
	G0F/G2F	148,901	148,912 ±4	+11	74
	G1F/G2F	149,063	149,074 ±3	+11	74
	G2F/G2F	149,225	149,232 ±2	+7	47
	G2F/G2F + Gal	149,387	149,388 ±6	+1	7
	G2F/G2F + 2Gal	149,549	149,542 ±10	–7	47

Note: no glycosylated protein detected

3.3.2 Data Interpretation

Antigen/Full-Size mAb Complexes

- 5 μM mumAb + 5 μM JAM-A IB: As shown in Fig. 4b, when mumAb and JAM-A IB are present in equimolar concentrations, three species are detected: the intact mAb (MW = 147,099 ± 5 Da) and two additional complexes corresponding to the formation of 1:1 and 1:2 mAb:Ag assemblies (MW = 171,783 ± 15 Da and 196,368 ± 8 Da, respectively, see Table 4).
- 5 μM mumAb + 10 μM JAM-A IB: Increasing JAM-A IB concentration up to two molar equivalents induces an almost quantitative shift of the equilibrium towards the formation of the 1:2 mumAb:Ag complex (Fig. 4c).
- 5 μM mumAb + 40 μM JAM-A IB: Finally, addition of an eightfold molar excess of JAM-A IB leads mainly to the formation of a 1:4 mumAb:Ag complex (MW = 245,485 ± 15 Da, see Table 4) with a small proportion of 1:3 complex (MW = 220,878 ± 15 Da) (Fig. 4d). Interestingly, in spite of this high antigen molar excess, no higher order complexes are detected which indicates that mumAb is able to bind up to four antigen molecules. The observed binding stoichiometries may result from mumAb association with JAM-A either as a monomer or as a dimer.
- The same kind of experiments was also performed with hzmAb (measured masses are summarized in Table 4). Similarly to mumAb, results depicted in Fig. 4e–h demonstrate that hzmAb is able to bind up to four JAM-A IB molecules.
- Same experiments were carried out with JAM-A SF (data not shown) and led to similar results, which ascertains that the previously described antigen heterogeneity (see Subheading 3.2.1) does not affect 6F4 mAbs recognition (see Note 14).

Antigen/Full-Size Deglycosylated mAb Complexes

- Similar titration experiments were performed in presence of deglycosylated hz- and mu- mAb with JAM-A SF (see Fig. 5) and led to similar results, which enabled to conclude that as expected, mAb glycosylation (located in the Fc portion) has no effect on antigen recognition and binding affinity linked to the Fab moiety. This would probably be different for mAbs like cetuximab having a second N-glycosylation site in the variable domain (25).

Antigen/mAb Fab Fragment Complexes

To gain further insights into the formation of these immune complexes, titration experiments were also performed with the humanized papaïn-generated Fab fragment (hzFab). Since the corresponding hzmAb is bivalent, experiments used twice as much hzFab concentration (10 μM) such that hzFab and hzmAb titrations could be compared.

- 10 μM hzFab + 5 μM JAM-A IB: The main ion distribution is related to the 11+ to 13+ charge states of unbound hzFab (MW = 47,122 ± 2 Da) (Fig. 6b). Two additional low-intense species are detected with molecular weights of 71,687 ± 2 Da

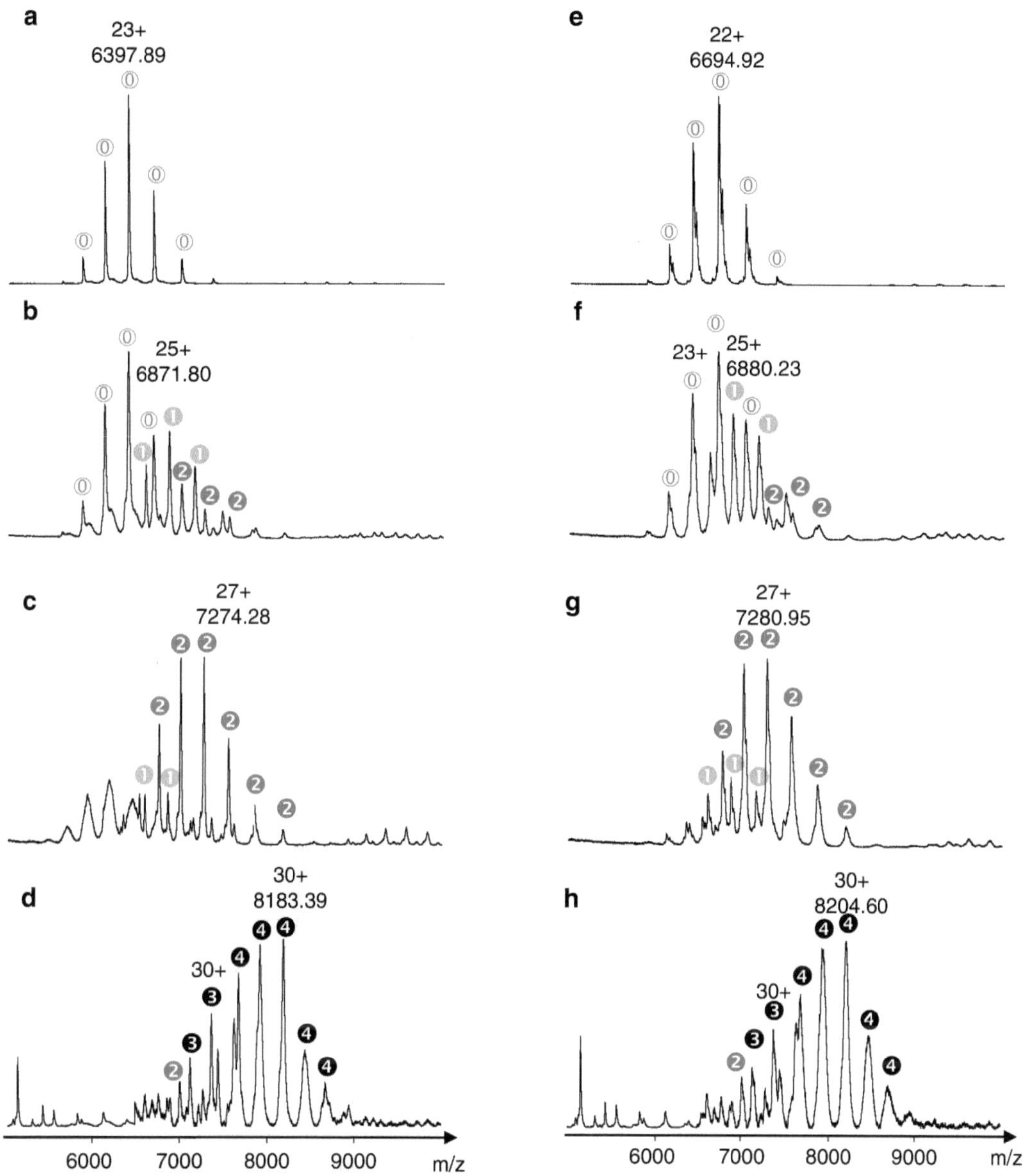

Fig. 4 Detection and determination of binding stoichiometries of intact full-size mAb/Ag inmmune complexes. (**a–d**) mumAb and (**e–h**) hzmAb were diluted to 5 μM in 150 mM ammonium acetate buffer (7.2) either (**a**, **e**) alone or in presence of (**b**, **f**) 5 μM, (**c**, **g**) 10 μM and (**g**, **h**) 40 μM JAM-A IB. Vc = 120 V, Pi = 6 mbar. Circled numbers refer to the number of Ag molecules bound per mAb

and 143,410 ± 4 Da which correspond to 1:1 and 2:2 hzFab:JAM-A complexes, respectively.

– 10 μM hzFab + 10 μM JAM-A IB: The relative intensities of 1:1 and 2:2 complexes increase significantly when the concentration of JAM-A IB is raised. Since the hzFab fragment appears quasi-exclusively monomeric in absence of JAM-A (Fig. 6c),

Table 4
Molecular weights measured in non-denaturing conditions for mAb/JAM complexes

	mAb:Ag binding stoichiometry	Calculated MW (Da)	Measured MW (Da)		Relative error (%)
Murin mAb:JAM-A IB	1:1	171,663	171,783	±15	0.1
	1:2	196,205	196,368	±8	0.1
	1:3	220,748	220,878	±15	0.1
	1:4	245,290	245,485	±15	0.1
Humanized mAb: JAM-A IB	1:1	171,814	172,000	±16	0.1
	1:2	196,356	196,522	±26	0.1
	1:3	220,899	221,187	±22	0.1
	1:4	245,441	246,067	±31	0.3
Humanized Fab: JAM-A IB	1:0	47,118	47,122	±2	0.01
	1:1	71,665	71,687	±3	0.03
	2:2	143,329	143,410	±4	0.06
	1:2	96,207	96,214	±4	0.01

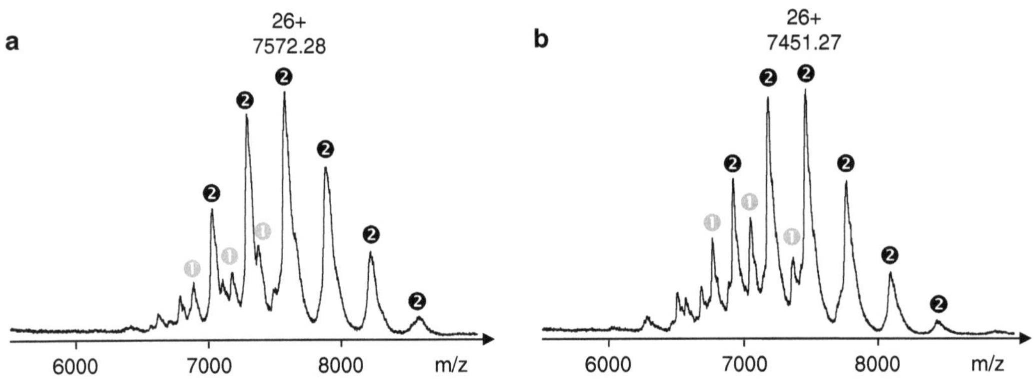

Fig. 5 Comparison of antigen binding ability of glycosylated versus deglycosylated hzmAb. (**a**) hzmAb and (**b**) PNGaseF-treated hzmAb were diluted to 5 μM in 150 mM ammonium acetate buffer (pH 7.2) in presence of 10 μM JAM-A IB. Vc = 120 V, Pi = 6 mbar. Circled numbers refer to the number of Ag molecules bound per mAb

the detection of 2:2 complex is most likely due to the binding of two Fab fragments per JAM-A homodimer.

- 10 μM hzFab + 40 μM JAM-A IB: Increasing JAM-A IB concentration to 40 μM induces a quasi-quantitative shift of the equilibrium towards the formation of the 1:2 hzFab:JAM-A. Almost no higher order assemblies are detected even in the presence of such a large antigen excess.

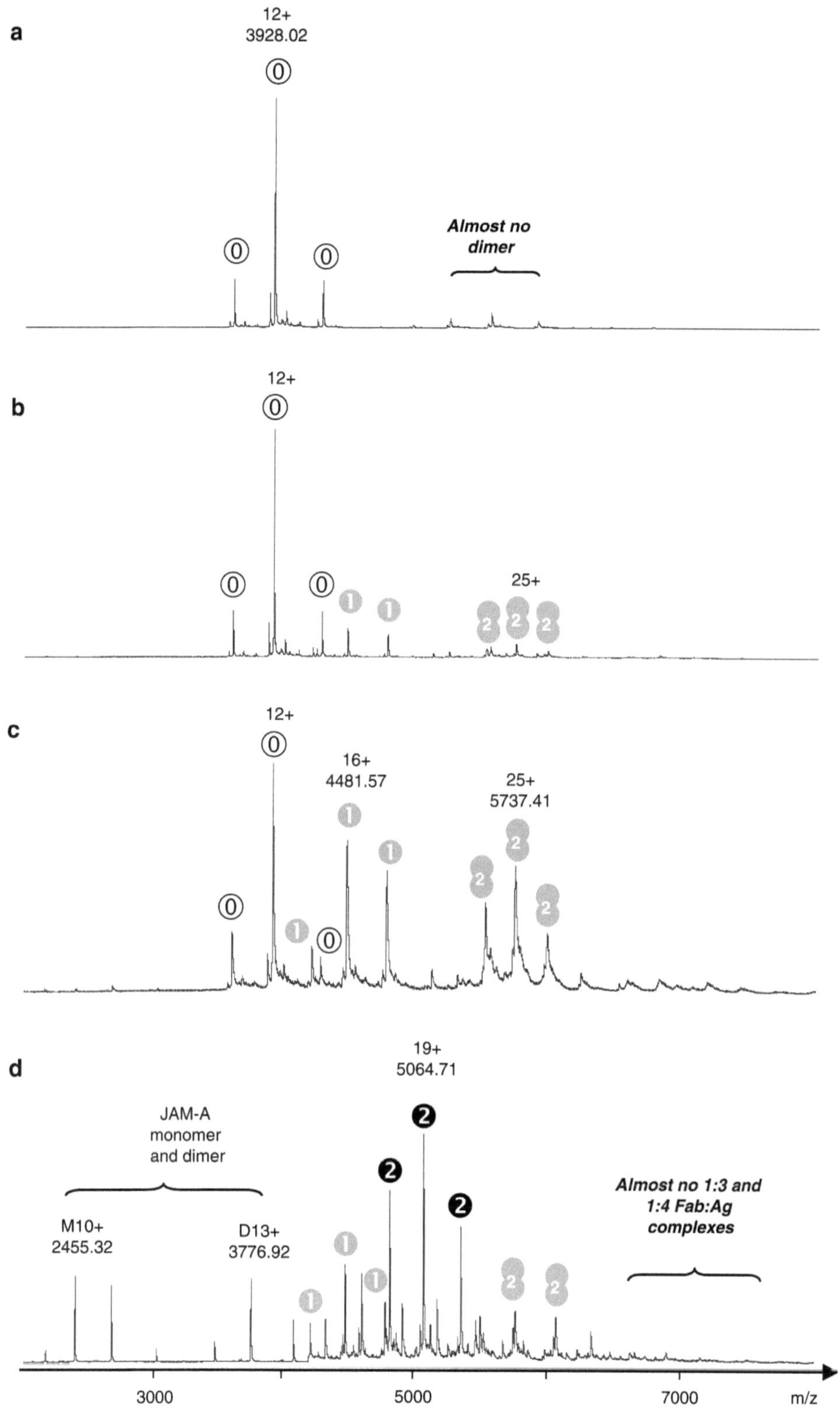

Fig. 6 Detection and determination of binding stoichiometries of hzFab/Ag inmmune complexes. hzFab was diluted to 10 μM in 150 mM ammonium acetate (pH 7.2) either (**a**) alone or in presence of (**b**) 5 μM, (**c**) 10 μM and (**d**) 40 μM JAM-A IB. Vc = 120 V, Pi = 6 mbar. Circled numbers refer to the number of Ag molecules bound per hzFab

Conclusion: Noncovalent MS Reveals the Formation of 1:4 mAb:JAM-A Complexes

- MS-based structural insight evidenced that recombinant JAM-A heterogeneous disulfide bridge pairings alter neither its native structure nor mAbs 6F4 recognition properties.
- Investigations focused on mAb:Ag complexes revealed that similarly to mumAb, hzmAb 6F4 binds up to four antigen molecules with a similar affinity, confirming in this way the reliability of the humanization process. According to these results, the formation of 1:4 mAb:Ag complexes must result from the recognition of two JAM-A homodimers by the antibody, each one interacting with one of the Fab arms (see Note 15).

3.4 Determination of Binding Specificities and Selectivities

3.4.1 Are Interactions Detected by Noncovalent MS Representative of the Ones That Preexist in Solution?

The essential prerequisite to the use of ESI-MS for the determination of binding stoichiometries of noncovalent complexes is that the peaks observed on mass spectra *in vacuo* are related to species effectively present in solution. Great care in data acquisition as well as in data interpretation must be taken, because it is established that the solution-phase image might be distorted during the ESI-MS analysis due to several factors, in particular during the evaporation of the ions in the gas phase, or during the transfer from the ion source to the analyzer through the interface region of the mass spectrometer (see Note 16).

Control experiments (involving different interacting partners, or different experimental and instrumental conditions) should always be performed in order to avoid any misinterpretations.

3.4.2 Results and Conclusions: 6F4 mAbs Selectively Recognize JAM-A

The great interest aroused by mAbs in the therapeutic field is largely due to their high selectivity towards their target antigen. In such a context, control experiments were designed in order to assess the selectivity of previously detected mAb:Ag complexes.

- For that purpose, *E. coli* produced 12.8 kDa beta2-microglobulin antigen and humanized IgG4, which specifically targets another transmembrane receptor (same subclass as our studied mAb) were selected as negative controls (see Tables 2 and 3).
- Each molecular entity was first analyzed under denaturing and native conditions alone and in combination with either JAM-A or mAbs (mAb/Ag, 5 μM/10 μM).
- Analyses were performed under strictly identical experimental conditions as applied to detect previously described noncovalent immune complexes.
- Negative control mAb + JAM-A: No complex is detected between JAM-A IB and the negative control mAb (Fig. 7a, b).
- mAb + negative control antigen: As shown in Figs 7c, d, neither mumAb nor hzmAb bind the negative control antigen.
- Conclusions: These results unambiguously confirm that both mumAb:JAM-A and hzmAb:JAM-A complexes arise from selective recognition occurring in solution and not from any artifact of the MS technique.

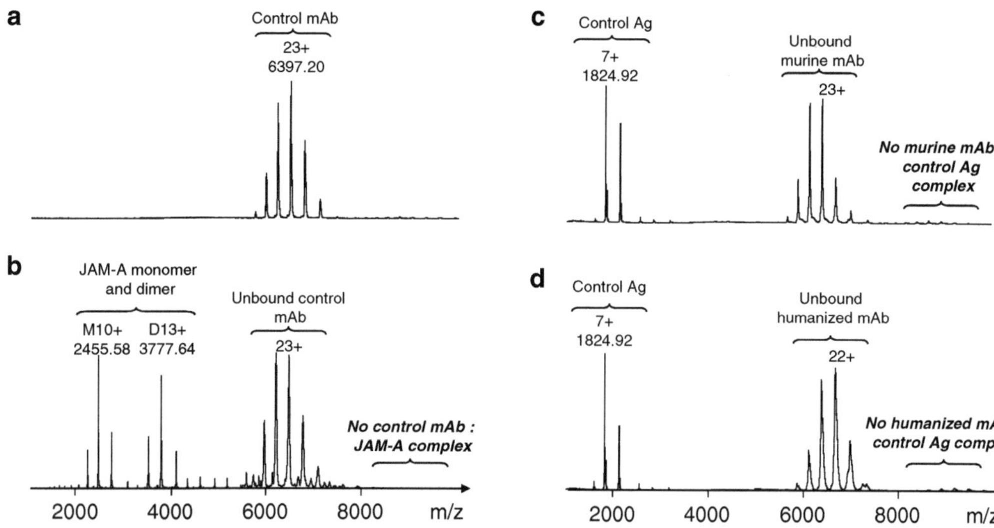

Fig. 7 Assessment of mAb:Ag binding selectivity. (**a**, **b**) Negative control mAb was diluted at 5 μM either (**a**) alone or (**b**) in presence of 10 μM JAM-IB. (**c**) mumAb and (**d**) hzmAb were incubated at 5 μM in presence of 10 μM negative control Ag

3.5 Comparison of Relative JAM-A Binding Affinity for hzmAb and mumAb

Along with specificity, affinity of mAb towards its target is crucial for its therapeutic efficiency (19, 26, 27). Many different biophysical methods are commonly used to assess the stability of mAb:Ag interactions and more generally of noncovalent complexes involving proteins (see Note 17) (28, 29). Immune complexes usually display dissociation constants in the picomolar and nanomolar ranges while noncovalent ESI-MS is much more suited to complexes with high nanomolar up to high micromolar dissociation constants (30, 31).

- Moreover, considering ESI response factors which include ionization efficiency, ion transmission through the mass spectrometer and detection, discrepancies are likely to occur between free mAbs and mAbs:Ag complexes due to their significant mass differences (32–34).
- Nevertheless, because mumAb and hzmAb have similar molecular weights, it seems reasonable to compare the relative intensity of their complexes for a given binding stoichiometry (34).
- Titration experiments in the present study (Fig. 4) showed that mumAb and hzmAb generated similar complex relative intensities for a given antigen concentration suggesting similar affinity of both mAbs for JAM-A.
- This comparison aims at roughly estimating relative affinities and would not be reliable to detect subtle affinity changes. This approach is, however, valuable in the present case to ensure that the humanization process of mAb 6F4 is meaningful and does not lead to a dramatic loss of affinity for JAM-A.

4 Notes

1. The choice of the desalting buffer is sample-dependent. In our laboratory, standard desalting procedure consists of ammonium acetate, 50 mM, pH 6.8. Its ionic strength is increased to 100 or 200 mM when assemblies are stable only at high salt concentrations. pH can be adjusted using ammonia (no NaOH to avoid contamination with Na^+ ions) or acetic/formic acid.
2. The choice of the desalting technique is sample-dependent and cannot be predicted. In our laboratory, we first try desalting with spin gel filtration columns, which are less time-consuming than micro-concentration (often 4–10 dilution/concentration steps are required) or overnight dialysis using micro-dialysis units (precipitation of the protein may occur overnight). All the desalting devices are used according to supplier recommendations.
3. Performing buffer exchange using micro-concentrators is often a tedious work. Proteins can stick to the ultrafiltration membrane, which requires the device to be changed regularly (it is common to use at least two devices per desalting). However, this type of desalting affords very high-quality ESI-MS spectra.
4. Noncovalent ESI-MS has been widely used to study protein folding/unfolding (35–44). Changes in the protein conformation can be evidenced by alteration of the ESI charge state distribution. An unfolded protein in solution leads to the formation of higher charge states than the same protein in a tightly folded conformation (37–41).
5. A relevant "trick" to perform MS analysis is to use fresh biological material. Freezing in ammonium acetate or even in the purification buffer should be avoided to preserve proteins or complexes stability. In our laboratory, samples are usually analyzed by mass spectrometry the day after purification and immediately after desalting.
6. After buffer exchange, it is highly advised to check the activity of the protein/complex in the ammonium buffer in order to ensure that conditions used for mass spectrometry analysis do not affect its biological activity.
7. Absence of disulfide linkages, resulting in wide open protein structure under denaturing conditions, was reported to increase the number of protonation sites available during the ESI ionization/desorption process (45).
8. The presence of disulfide bridges in JAM-A was evidenced in the protein crystal structure which showed two S–S bonds involving four out of the five cysteines present in the protein sequence (46, 47).

9. Optimization of interface parameters of the mass spectrometer for the detection of noncovalent complexes: A crucial point to maintain noncovalent interactions during the ionization/desorption process is the optimization of the instrumental settings controlling the energy communicated to the ions in the first pumping stage of the mass spectrometer. This is a key step to ensure that the integrity of noncovalent complexes is preserved between the ion source of the instrument at atmospheric pressure and the high-vacuum region of the analyzer. This region of intermediary pressure is called "interface" and corresponds physically to the zone of the first hexapole (for schematic representation of the interface of our instruments, see refs. 48, 49). Two parameters are of utmost importance and need to be optimized for each new system to obtain optimum sensitivity and high-quality ESI mass spectra while preventing complex disruption: (1) the pressure in the interface (Pi) region which affects collision efficiency; (2) the accelerating voltage (Vc) which controls the kinetic energy communicated to the ions in the source of the instrument. Several groups have reported that transmission of high *m/z* ions requires elevated pressures in the first vacuum stages of mass spectrometers (49–54). On our LCT instrument (Waters, Manchester, UK), the first vacuum stage of the instrument is located between the sample cone and the extraction cone (for schematic representation see refs. 48, 49). The pressure in this region (Pi) is regulated with a speedivalve that throttles pumping by the rotary pump and allows the Pi value to be adjusted between 1 and 8 mbar. Pi is directly linked to the internal energy communicated to the ions via collisions with residual gaseous molecules present in this part of the mass spectrometer. As the distance between two consecutive collisions (mean free path) with ambient gaseous molecules is inversely proportional to Pi, lower pressures in the interface (1–3 mbar) imply longer distances between two successive collisions. Consequently, gas phase ions have enough time to be "warmed up" and to accumulate energy which further results in "destructive" collisions. Although ion desolvation is improved, such energetic collisions may lead to the dissociation of labile noncovalent subassemblies. Inversely, increasing the pressure in the interface yields to more frequent but lower energy and less "destructive" collisions after which the "thermalized" ions (corresponding to large macromolecules) are transferred without any damage to the analyzer. Elevating the pressure is also associated to less efficient ion desolvation, which is observed on ESI mass spectra by significant peak broadening. Higher pressures also improve substantially high *m/z* ion transmission. In summary, increased Pi values provides improved collisional cooling and focusing of large ions in the multipole guides and, therefore, better transmission

through the interface. Varying the accelerating voltage (Vc) induces changes in the kinetic energy communicated to the ions in the electrospray source. At high accelerating voltages, ions have higher kinetic energies that cause strong energetic collisions and possibly dissociation of weak interactions. Decreasing the Vc leads to a considerable loss in sensitivity due to: (1) nonoptimal transmission of high m/z ions and (2) much less efficient desolvation, resulting altogether in dramatically reduced mass accuracy. Better desolvation and focalization of high m/z ions at high accelerating voltages makes interpretation of the recorded mass spectra much easier. Accordingly, the peak broadening effect previously mentioned for high interface pressures can be reduced by increasing the Vc. Fine-tuned instrumental settings providing the best compromise between sufficient desolvation, optimal transmission of intact high m/z ions and nondestructive gas-phase collisions is required to detect specific noncovalent edifices of high molecular weights.

Vc and Pi are not independent parameters and should be optimized together to get the best compromise between sufficient ion desolvation, good transmission of high m/z ions without destruction of the noncovalent framework. For each noncovalent assembly, careful optimization of the interface pressure (Pi) and the accelerating voltage (Vc) different for each noncovalent assembly, is necessary to get the best results. In practice, for each studied complex, several ESI mass spectra are recorded for different (Pi, Vc) couples.

10. Increasing antigen concentration from 5 to 40 μM favors dimer formation which relative intensity increases from 35 to 61% (data not shown). In addition, it is noteworthy that the small proportion of protein trimer and tetramer which appears at 5 μM does not significantly increase at 40 μM (<10%) suggesting that these higher oligomerization states arise from a nonspecific aggregation occurring in solution and/or during the ESI process.
11. In order to check the presence of intra-batch heterogeneities, nanoESI-TWIMS (ion mobility coupled to mass spectrometry) and LC-MS experiments were carried out for both antigen batches. These experiments are not detailed in the present chapter. For more details, see ref. 11.
12. Detection of dimeric JAM-A is consistent with the X-ray crystallography structure showing two antigen molecules interacting together through extensive ionic and hydrophobic interactions (47).
13. In the case of mAb:TNFalpha complexes, different stoichiometries were found for adalimumab (human IgG1, 3:3 ratio),

infliximab (chimeric IgG1, 1:1 ratio), and etanercept (Fc-fusion protein, 6:3 ratio) and were correlated to different neutralizing mechanisms and pharmacokinetic profiles (55, 56). In the case of mAb:EGFR ECD complexes both panitumumab (human IgG2) and cetuximab (chimeric IgG1) were shown by SEC to have a 1:2 molecular ratio (19).

14. Note also that absence of heterogeneous glycosylation patterns for recombinant JAM-A ECD corresponds to a favorable case and that the use of noncovalent nanoESI-MS to monitor the formation of immune complex is limited to low-heterogeneity samples.

14. This unusual mAb/antigen stoichiometry with the previously described TNF and EGFR mAbs binding ratios or with bevacizumab Fab/VEGF-A complexes (2 Fabs for 1 VEGF-A dimer, as assessed by co-crystallography) (57). Interestingly, the fact that 6F4 mAbs are able to bind four JAM-A ECD chains suggest that these mAbs may bind JAM-A tetramers in a cellular context.

16. Possible reasons for discrepancies between MS data and solution data: As described in the literature (58–60), the stability of noncovalent complexes upon the ESI-MS process strongly depends on the type of interaction (electrostatic contacts, hydrogen bonds, Van der Waals interactions) involved in the formation of the complex. During ion transfer from the solution to the gas phase both electrostatic interactions and hydrogen bonds are strengthened. In contrast, complexes that are stabilized in solution by hydrophobic effects appear to be weakened (61, 62). To understand those effects, one should remember that water molecules evaporate when passing into the gas phase of the mass spectrometer. Without water molecules around the complex, it is reasonable to assume that hydrophobic interactions do not contribute significantly to any complex stabilization in the gas phase of the mass spectrometer. This assumption has been verified by several groups comparing X-ray crystallography and ESI-MS results (7, 8, 63–65).

17. As for example surface plasmon resonance (SPR) (66), enzyme-linked immunosorbent assays (ELISA) (67) or SEC (19, 56).

Acknowledgments

This work was funded by the CNRS, the University of Strasbourg UdS, and the Region Alsace. C.A. was supported by a studentship from the French Ministère de la Recherche.

References

1. Beck A, Wurch T, Bailly C, Corvaia N (2010) Strategies and challenges for the next generation of therapeutic antibodies. Nat Rev Immunol 10:345–352
2. Hager-Braun C, Katinger H, Tomer KB (2006) The HIV-neutralizing monoclonal antibody 4E10 recognizes N-terminal sequences on the native antigen. J Immunol 176:7471–7481
3. Kiselar JG, Downard KM (1999) Direct identification of protein epitopes by mass spectrometry without immobilization of antibody and isolation of antibody-peptide complexes. Anal Chem 71:1792–1801
4. Pimenova T, Nazabal A, Roschitzki B, Seebacher J, Rinner O, Zenobi R (2008) Epitope mapping on bovine prion protein using chemical cross-linking and mass spectrometry. J Mass Spectrom 43:185–195
5. Heck AJ (2008) Native mass spectrometry: a bridge between interactomics and structural biology. Nat Methods 5:927–933
6. Heck AJ, van den Heuvel RH (2004) Investigation of intact protein complexes by mass spectrometry. Mass Spectrom Rev 23:368–389
7. Loo JA (1997) Studying noncovalent protein complexes by electrospray ionization mass spectrometry. Mass Spectrom Rev 16:1–23
8. Loo JA (2000) Electrospray ionization mass spectrometry: a technology for studying noncovalent macromolecular complexes. Int J Mass Spectrom 200:175–186
9. Bich C, Scott M, Panagiotidis A, Wenzel RJ, Nazabal A, Zenobi R (2008) Characterization of antibody-antigen interactions: comparison between surface plasmon resonance measurements and high-mass matrix-assisted laser desorption/ionization mass spectrometry. Anal Biochem 375:35–45
10. Nazabal A, Wenzel RJ, Zenobi R (2006) Immunoassays with direct mass spectrometric detection. Anal Chem 78:3562–3570
11. Atmanene C, Wagner-Rousset E, Malissard M, Chol B, Robert A, Corvaia N et al (2009) Extending mass spectrometry contribution to therapeutic monoclonal antibody lead optimization: characterization of immune complexes using noncovalent ESI-MS. Anal Chem 81:6364–6373
12. Lu X, DeFelippis MR, Huang L (2009) Linear epitope mapping by native mass spectrometry. Anal Biochem 395:100–107
13. Oda M, Uchiyama S, Noda M, Nishi Y, Koga M, Mayanagi K et al (2009) Effects of antibody affinity and antigen valence on molecular forms of immune complexes. Mol Immunol 47: 357–364
14. Oda M, Uchiyama S, Robinson CV, Fukui K, Kobayashi Y, Azuma T (2006) Regional and segmental flexibility of antibodies in interaction with antigens of different size. FEBS J 273:1476–1487
15. Tito MA, Miller J, Walker N, Griffin KF, Williamson ED, Despeyroux-Hill D et al (2001) Probing molecular interactions in intact antibody: antigen complexes, an electrospray time-of-flight mass spectrometry approach. Biophys J 81:3503–3509
16. Zhang Z, Pan H, Chen X (2009) Mass spectrometry for structural characterization of therapeutic antibodies. Mass Spectrom Rev 28:147–176
17. Beck A, Bussat MC, Zorn N, Robillard V, Klinguer-Hamour C, Chenu S et al (2005) Characterization by liquid chromatography combined with mass spectrometry of monoclonal anti-IGF-1 receptor antibodies produced in CHO and NS0 cells. J Chromatogr B Analyt Technol Biomed Life Sci 819:203–218
18. Vis H, Dobson CM, Robinson CV (1999) Selective association of protein molecules followed by mass spectrometry. Protein Sci 8:1368–1370
19. Rehder DS, Chelius D, McAuley A, Dillon TM, Xiao G, Crouse-Zeineddini J et al (2008) Isomerization of a single aspartyl residue of anti-epidermal growth factor receptor immunoglobulin gamma2 antibody highlights the role avidity plays in antibody activity. Biochemistry 47:2518–2530
20. Bobst CE, Abzalimov RR, Houde D, Kloczewiak M, Mhatre R, Berkowitz SA et al (2008) Detection and characterization of altered conformations of protein pharmaceuticals using complementary mass spectrometry-based approaches. Anal Chem 80:7473–7481
21. Beck A, Klinguer-Hamour C, Bussat MC, Champion T, Haeuw JF, Goetsch L et al (2007) Peptides as tools and drugs for immunotherapies. J Pept Sci 13:588–602
22. Haeuw JF, Beck A (2004) Proteomics for development of immunotherapies. In: Hondermack H (ed) Proteomics: biomedical and pharmaceutical applications. Lille/Kluwer Academic, France/The Netherlands, pp 243–278
23. Beck A, Wagner-Rousset E, Bussat MC, Lokteff M, Klinguer-Hamour C, Haeuw JF et al (2008) Trends in glycosylation, glycoanalysis and glycoengineering of therapeutic antibodies and Fc-fusion proteins. Curr Pharm Biotechnol 9:482–501
24. Chowdhury SK, Katta V, Chait BT (1990) Probing conformational changes in proteins by

mass spectrometry. J Am Chem Soc 112: 9012–9013

25. Qian J, Liu T, Yang L, Daus A, Crowley R, Zhou Q (2007) Structural characterization of N-linked oligosaccharides on monoclonal antibody cetuximab by the combination of orthogonal matrix-assisted laser desorption/ionization hybrid quadrupole-quadrupole time-of-flight tandem mass spectrometry and sequential enzymatic digestion. Anal Biochem 364:8–18
26. Bachmann MF, Kalinke U, Althage A, Freer G, Burkhart C, Roost H et al (1997) The role of antibody concentration and avidity in antiviral protection. Science 276:2024–2027
27. Usinger WR, Lucas AH (1999) Avidity as a determinant of the protective efficacy of human antibodies to pneumococcal capsular polysaccharides. Infect Immun 67:2366–2370
28. Hensley P (1996) Defining the structure and stability of macromolecular assemblies in solution: the re-emergence of analytical ultracentrifugation as a practical tool. Structure 4:367–373
29. Neri D, Montigiani S, Kirkham PM (1996) Biophysical methods for the determination of antibody-antigen affinities. Trends Biotechnol 14:465–470
30. Griffey RH, Sannes-Lowery KA, Drader JJ, Ve M, Swayze EE, Hofstadler SA (2000) Characterization of low-affinity complexes between RNA and small molecules using electrospray ionization mass spectrometry. J Am Chem Soc 122:9933–9938
31. Sannes-Lowery KA, Griffey RH, Hofstadler SA (2000) Measuring dissociation constants of RNA and aminoglycoside antibiotics by electrospray ionization mass spectrometry. Anal Biochem 280:264–271
32. Ayed A, Krutchinsky AN, Ens W, Standing KG, Duckworth HW (1998) Quantitative evaluation of protein-protein and ligand-protein equilibria of a large allosteric enzyme by electrospray ionization time-of-flight mass spectrometry. Rapid Commun Mass Spectrom 12:339–344
33. Gabelica V, Galic N, Rosu F, Houssier C, De Pauw E (2003) Influence of response factors on determining equilibrium association constants of non-covalent complexes by electrospray ionization mass spectrometry. J Mass Spectrom 38:491–501
34. Peschke M, Verkerk UH, Kebarle P (2004) Features of the ESI mechanism that affect the observation of multiply charged noncovalent protein complexes and the determination of the association constant by the titration method. J Am Soc Mass Spectrom 15:1424–1434
35. Babu KR, Moradian A, Douglas DJ (2001) The methanol-induced conformational transitions of beta-lactoglobulin, cytochrome c, and ubiquitin at low pH: a study by electrospray ionization mass spectrometry. J Am Soc Mass Spectrom 12:317–328
36. Dobo A, Kaltashov IA (2001) Detection of multiple protein conformational ensembles in solution via deconvolution of charge-state distributions in ESI MS. Anal Chem 73:4763–4773
37. Grandori R, Matecko I, Mayr P, Müller N (2001) Probing protein stabilization by glycerol using electrospray mass spectrometry. J Mass Spectrom 36:918–922
38. Konermann L, Douglas DJ (1997) Acid-induced unfolding of cytochrome c at different methanol concentrations: electrospray ionization mass spectrometry specifically monitors changes in the tertiary structure. Biochemistry 36:12296–12302
39. Konermann L, Douglas DJ (1998) Unfolding of proteins monitored by electrospray ionization mass spectrometry: a comparison of positive and negative ion modes. J Am Soc Mass Spectrom 9:1248–1254
40. Konermann L, Douglas DJ (1998) Equilibrium unfolding of proteins monitored by electrospray ionization mass spectrometry: distinguishing two-state from multi-state transitions. Rapid Commun Mass Spectrom 12:435–442
41. Sogbein OO, Simmons DA, Konermann L (2000) Effects of pH on the kinetic reaction mechanism of myoglobin unfolding studied by time-resolved electrospray ionization mass spectrometry. J Am Soc Mass Spectrom 11:312–319
42. Konermann L, Silva EA, Sogbein OF (2001) Electrochemically induced pH changes resulting in protein unfolding in the ion source of an electrospray mass spectrometer. Anal Chem 73:4836–4844
43. Liang Y, Du F, Sanglier S, Zhou BR, Xia Y, Van Dorsselaer A et al (2003) Unfolding of rabbit muscle creatine kinase induced by acid. A study using electrospray ionization mass spectrometry, isothermal titration calorimetry, and fluorescence spectroscopy. J Biol Chem 278:30098–30105
44. Heam MT, Quirino JP, Whisstock J, Terabe S (2002) Thermal unfolding of proteins studied by coupled reversed-phase HPLC-electrospray ionization mass spectrometry techniques based on isotope exchange effects. Anal Chem 74:1467–1475
45. Loo JA, Edmonds CG, Udseth HR, Smith RD (1990) Effect of reducing disulfide-containing proteins on electrospray ionization mass spectra. Anal Chem 62:693–698

46. Kostrewa D, Brockhaus M, D'Arcy A, Dale GE, Nelboeck P, Schmid G et al (2001) X-ray structure of junctional adhesion molecule: structural basis for homophilic adhesion via a novel dimerization motif. EMBO J 20: 4391–4398
47. Prota AE, Campbell JA, Schelling P, Forrest JC, Watson MJ, Peters TR et al (2003) Crystal structure of human junctional adhesion molecule 1: implications for reovirus binding. Proc Natl Acad Sci U S A 100:5366–5371
48. Sanglier S, Atmanene C, Chevreux G, Dorsselaer AV (2008) Nondenaturing mass spectrometry to study noncovalent protein/protein and protein/ligand complexes: technical aspects and application to the determination of binding stoichiometries. Methods Mol Biol 484:217–243
49. Sanglier S, Ramstrom H, Haiech J, Leize E, Van Dorsselaer A (2002) Electrospray ionization mass spectrometry analysis revealed a ~310 kDa noncovalent hexamer of HPr kinase/phosphatase from Bacillus subtilis. Int J Mass Spectrom 219:681–696
50. Chernushevich IV, Thomson BA (2004) Collisional cooling of large ions in electrospray mass spectrometry. Anal Chem 76: 1754–1760
51. Krutchinsky AN, Chernushevich IV, Spicer VL, Ens W, Standing KG (1998) Collisional damping interface for an electrospray ionization time-of-flight mass spectrometer. J Am Soc Mass Spectrom 9:569–579
52. Schmidt A, Bahr U, Karas M (2001) Influence of pressure in the first pumping stage on analyte desolvation and fragmentation in nano-ESI MS. Anal Chem 73:6040–6046
53. Sobott F, Hernandez H, McCammon MG, Tito MA, Robinson CV (2002) A tandem mass spectrometer for improved transmission and analysis of large macromolecular assemblies. Anal Chem 74:1402–1407
54. Tahallah N, Pinkse M, Maier CS, Heck AJ (2001) The effect of the source pressure on the abundance of ions of noncovalent protein assemblies in an electrospray ionization orthogonal time-of-flight instrument. Rapid Commun Mass Spectrom 15:596–601
55. Kim MS, Lee SH, Song MY, Yoo TH, Lee BK, Kim YS (2007) Comparative analyses of complex formation and binding sites between human tumor necrosis factor-alpha and its three antagonists elucidate their different neutralizing mechanisms. J Mol Biol 374: 1374–1388
56. Santora LC, Kaymakcalan Z, Sakorafas P, Krull IS, Grant K (2001) Characterization of noncovalent complexes of recombinant human monoclonal antibody and antigen using cation exchange, size exclusion chromatography, and BIAcore. Anal Biochem 299:119–129
57. Fuh G, Wu P, Liang WC, Ultsch M, Lee CV, Moffat B et al (2006) Structure-function studies of two synthetic anti-vascular endothelial growth factor Fabs and comparison with the AVASTIN(TM) Fab. J Biol Chem 281: 6625–6631
58. Hernandez H, Robinson CV (2001) Dynamic protein complexes: insights from mass spectrometry. J Biol Chem 276:46685–46688
59. Robinson CV, Chung EW, Kragelund BB, Knudsen J, Aplin RT, Poulsen FM et al (1996) Probing the nature of noncovalent interactions by mass spectrometry. A study of protein–CoA ligand binding and assembly. J Am Chem Soc 118:8646–8653
60. Smith RD, Light-Wahl KJ (1993) The observation of non-covalent interactions in solution by electrospray ionization mass spectrometry: Promise, pitfalls and prognosis. Biol Mass Spectrom 22:493–501
61. Li YT, Hsieh YL, Henion JD, Senko MW, McLafferty FW, Ganem B (1993) Mass spectrometric studies on noncovalent dimers of leucine zipper peptides. J Am Chem Soc 115:8409–8413
62. Li YT, Hsieh YL, Henion JD, Ocain TD, Schiehser GA, Ganem B (1994) Analysis of the energetics of gas-phase immunophilin-ligand complexes by ion spray mass spectrometry. J Am Chem Soc 116:7487–7493
63. Darmanin C, Chevreux G, Potier N, Van Dorsselaer A, Hazemann I, Podjarny A et al (2004) Probing the ultra-high resolution structure of aldose reductase with molecular modelling and noncovalent mass spectrometry. Bioorg Med Chem 12:3797–3806
64. El Kabbani O, Rogniaux H, Barth P, Chung RP, Fletcher EV, Van Dorsselaer A et al (2000) Aldose and aldehyde reductases: correlation of molecular modeling and mass spectrometric studies on the binding of inhibitors to the active site. Proteins 41:407–414
65. Rogniaux H, Van Dorsselaer A, Barth P, Biellmann JF, Barbanton J, van Zandt M et al (1999) Binding of aldose reductase inhibitors: correlation of crystallographic and mass spectrometric studies. J Am Soc Mass Spectrom 10:635–647
66. Lofgren JA, Dhandapani S, Pennucci JJ, Abbott CM, Mytych DT, kaliyaperumal A et al (2007) Comparing ELISA and surface plasmon resonance for assessing clinical immunogenicity of panitumumab. J Immunol 178:7467–7472
67. Orosz F, Ovadi J (2002) A simple method for the determination of dissociation constants by displacement ELISA. J Immunol Methods 270:155–162

Chapter 17

Conformational Analysis of Recombinant Monoclonal Antibodies with Hydrogen/Deuterium Exchange Mass Spectrometry

Damian Houde and John R. Engen

Abstract

Understanding the conformation of antibodies, especially those of therapeutic value, is of great interest. Many of the current analytical methods used to probe protein conformation face issues in the analysis of antibodies, either due to the nature of the antibody itself or due to the limitations of the method. One method that has recently been utilized for conformational analysis of antibodies is hydrogen/deuterium exchange mass spectrometry (H/DX MS). H/DX MS can be used to probe the conformation and dynamics of proteins in solution, requires small sample quantities, is compatible with many buffer systems, and provides peptide-level resolution. The application of H/DX MS to immunoglobulin gamma 1 (IgG1) recombinant monoclonal antibodies can provide information about IgG1 conformation, dynamics, and changes to conformation as a result of protein modification(s), changes in storage conditions, purification procedures, formulation, and many other parameters. In this article we provide a comprehensive H/DX MS protocol for the analysis of an antibody.

Key words Immunoglobin, Deuterium, Conformation, Protein structure, Antibody, Hydrogen exchange

1 Introduction

Recombinant monoclonal antibodies (mAbs) are valuable biopharmaceuticals that can be used to treat a wide range of illnesses (1–5). The specificity and diversity of antibodies, along with manufacturing compatibility, have made mAbs dominant players in the development pipelines and clinical programs of many biopharmaceutical companies (5, 6). A major class of therapeutic mAbs is immunoglobulin gamma 1 (IgG1), a large globular glycoprotein with approximately 1,330 amino acids and a molecular weight approaching 150 kDa. Intact IgG1 is a homodimer containing two heavy chains (~440 residues each) and two light chains (~215 residues each) tethered together with interchain disulfide linkages.

Alain Beck (ed.), *Glycosylation Engineering of Biopharmaceuticals: Methods and Protocols*, Methods in Molecular Biology, vol. 988, DOI 10.1007/978-1-62703-327-5_17, © Springer Science+Business Media New York 2013

In total, there are 16 disulfide bonds in an IgG1 (4 interchain and 12 intrachain), in addition to one conserved but heterogeneous N-linked glycan on each heavy chain (7, 8).

Characterizing the conformation and structural dynamics of an antibody (or any protein for that matter) can be a major analytical challenge. Many available structural techniques are either extremely sophisticated, requiring very specialized skills and large amounts of sample (>μM quantities), or are of low resolution, making detailed structural analysis difficult. As a result, it is highly desirable to have techniques available that can probe protein structure with low sample requirements, good resolution, and relatively fast turnaround time. We have explored (9, 10) the suitability of hydrogen/deuterium exchange mass spectrometry (H/DX MS) for this purpose and found that it provides a great number of analytical advantages for the conformational analysis of antibodies.

H/DX is a phenomenon whereby hydrogen atoms, at labile positions in proteins, spontaneously change places with hydrogen atoms in the surrounding solvent (11). Backbone amide hydrogens are particularly of interest in this process and due to variations in their chemical and physical environment induced by protein structure, exchange rates of these hydrogens in a folded protein can vary over many orders of magnitude (11–13). Information about protein conformation and, most importantly, differences in protein conformation between two or more forms of the same protein can be extracted by monitoring the exchange reaction. An analytical method sensitive to the differences between the isotopes of hydrogen is required to observe hydrogen exchange. Nuclear magnetic resonance (NMR), infrared spectroscopy (FTIR), and mass spectrometry (MS) have been all utilized to make the measurement; hydrogen exchange measured by mass spectrometry is described here.

The combination of hydrogen exchange with mass spectrometry has been extensively reviewed (e.g., refs. 12–20). Upon introduction of >95% D_2O to a protein in an all H_2O buffer at physiological pH (7.0–8.0), the exchange reaction itself is catalyzed primary by a base-driven mechanism, but is dramatically slowed (by at least four orders of magnitude) when the pH is reduced to 2.5 (11). The exchange reaction is also temperature dependent: by lowering the temperature to 0°C, the rate of exchange is slowed by another order of magnitude. Coupling low pH with low temperature (quench conditions) reduces the rate of exchange such that the incorporation of deuterium can be measured with modern liquid chromatography and mass spectrometry. As hydrogen has a mass of 1.008 Da and deuterium (the second isotope of hydrogen) has a mass of 2.014 Da, hydrogen exchange can be followed by measuring the mass of a protein with a mass spectrometer. By incorporating proteolytic digestion between the quench step and the mass analysis (21), the location of the deuterium in the labeled protein can be resolved to short stretches of the protein backbone.

While an H/DX MS experiment can be applied to mAbs, particularly an IgG1, this experiment is not without its challenges. On the following pages, we describe our detailed protocol for making these measurements and provide helpful tips we have discovered in the process of optimizing the protocol for H/DX MS analysis of an IgG1.

2 Materials

In general, approximately 50 μL of a 3 mg/mL protein solution is required to carry out these experiments for a whole, intact IgG1. The protein concentration required may be less if the protein being investigated, such as scFV's, antibody fragments (i.e., Fc's and Fab's), nanobodies, etc., is <50 kDa. In the case of smaller proteins, at least 100 μL of a 1 mg/mL protein solution would be ideal. As an example, the following protocol focuses on the analysis of a recombinant monoclonal IgG1 antibody. This example has been specifically optimized for the final amount of protein (in pmoles) that is needed for a high quality MS signal. Other volumes, concentrations and amounts of protein can be used, but the key variable is the amount loaded into the mass spectrometer. Should deviation from the following be used, the volumes and concentrations at each step must be determined.

2.1 Antibody Sample

1. The recombinant monoclonal IgG1 antibody sample used in this protocol was formulated at approximately 3 mg/mL in 50 mM sodium phosphate, 100 mM NaCl, pH 6.0 (see Note 1).

2.2 Buffers and Solutions

1. Sodium phosphate (Sigma, St. Louis, MO).
2. Sodium chloride (NaCl) (Sigma, St. Louis, MO).
3. Hydrochloric acid (HCl) (Sigma, St. Louis, MO).
4. Sodium hydroxide (NaOH) (Arcos, Morris Plains, NJ).
5. Acetonitrile (ACN) optima (Fisher, Pittsburgh, PA).
6. Water, optima W6 (Fisher, Pittsburgh, PA).
7. Formic Acid ~98% (Fluka, St. Louis, MO).
8. Deuterium oxide (D_2O), (D 99.96%), low paramagnetic (Cambridge Isotope Laboratories, Andover, MA) (see Note 2).
9. Deuterium chloride (DCl), 20% in D_2O (D 99.96%) (Cambridge Isotope Laboratories, Andover, MA).
10. Sodium deuteroxide (NaOD) 40% in D_2O (D 99.5%) (Cambridge Isotope Laboratories, Andover, MA).
11. Guanidine hydrochloride (GndHCl) (Pierce, Rockford, IL).

12. TCEP (Tris(2-carboxyethyl)phosphine) (Pierce, Rockford, IL).
13. pH 4.00, 7.00, and 10.00 standard buffer solutions (Fisher, Pittsburgh, PA).
14. (Glu[1])-Fibrinopeptide B human (GFP), product #F3261 > 90% (HPLC) (Sigma, St. Louis, MO).
15. Mobile phase A: Water and 0.05% formic acid.
16. Mobile phase B: ACN and 0.05% formic acid.

    Mobile phase C (digestion buffer): Water and 0.05% formic acid.
18. Lockmass solution: 400 fmol/μL GFP, 50% ACN and water, 0.1% formic acid.
19. Equilibration buffer: 50 mM sodium phosphate, 100 mM NaCl, H_2O, pH 6.00 (see Note 3).
20. Labeling buffer: 50 mM sodium phosphate, 100 mM NaCl, D_2O, pD 6.00 (see Note 4).
21. Quench buffer: 200 mM sodium phosphate, 0.5 M TCEP, 4 M GndHCl, H_2O, pH 2.35 (see Note 5).

2.3 Equipment

1. pH meter capable of accuracy to 0.01 units and 3-point calibration (e.g., Accumet Basic, AB15 plus, Fisher, Pittsburgh, PA).
2. pH micro probe capable of pH measurements in 50 μL solution (e.g., Mettler Toledo, Schwerzenbach, Switzerland).
3. Timer (Fisher, Pittsburgh, PA).
4. 10 K MWCO Amicon Biomax centrifugal membrane filters (0.5, 4, or 15 mL) (Millipore, Billerica, MA).
5. Porozyme pepsin digestion column (Applied Biosystems, Carlsbad, CA).
6. UPLC system able to control at least three mobile phase solutions simultaneously, along with a cooled chamber capable of maintaining 0 ± 0.5°C or equivalent (see Note 6).
7. UPLC separation column, nanoACQUITY BEH C18, 1.7 μm, 1 × 100 mm (Waters, Milford, MA), or equivalent.
8. UPLC guard column (peptide trap), nanoACQUITY BEH C18, 1.7 μm, 2.1 × 5 mm (Waters, Milford, MA), or equivalent (see Note 7).
9. Mass spectrometer with electrospray ionization (ESI) capable of tandem MS, Waters Synapt MS (Milford, MA) or equivalent.
10. Excel-based macro HX Express (22), freely available from www.hxms.com.
11. Syringe for sample injection, 50 μL gastight syringe or equivalent (Hamilton, Reno, NV).

3 Methods

Localizing conformational information for an antibody by H/DX MS involves several steps: labeling the protein with deuterium, quenching the reaction, digesting the protein, chromatographically separating the peptides, and mass analysis. Figure 1 shows the workflow of a typical continuous labeling H/DX MS experiment. The details of each step are described below in Subheading 3.1. At the start of each experiment, an undeuterated control is often analyzed several times (at least three times) to ensure that the proteolytic digestion and peptide identification are reproducible. While H/DX MS experiments can be conducted on essentially any chromatographic system, we have found that UPLC offers many benefits over traditional HPLC (23, 24), including lower sample requirements, higher sensitivity, and better reproducibility (9). Figure 2 shows a representative chromatographic separation of a pepsin digested IgG1 using HPLC (Fig. 2a, b) and UPLC (Fig. 2c). To promote the efficient pepsin digestion of the protein, the quench buffer contains a denaturant (guanidine hydrochloride) and a reducing agent (TCEP) (25) to break disulfide bonds

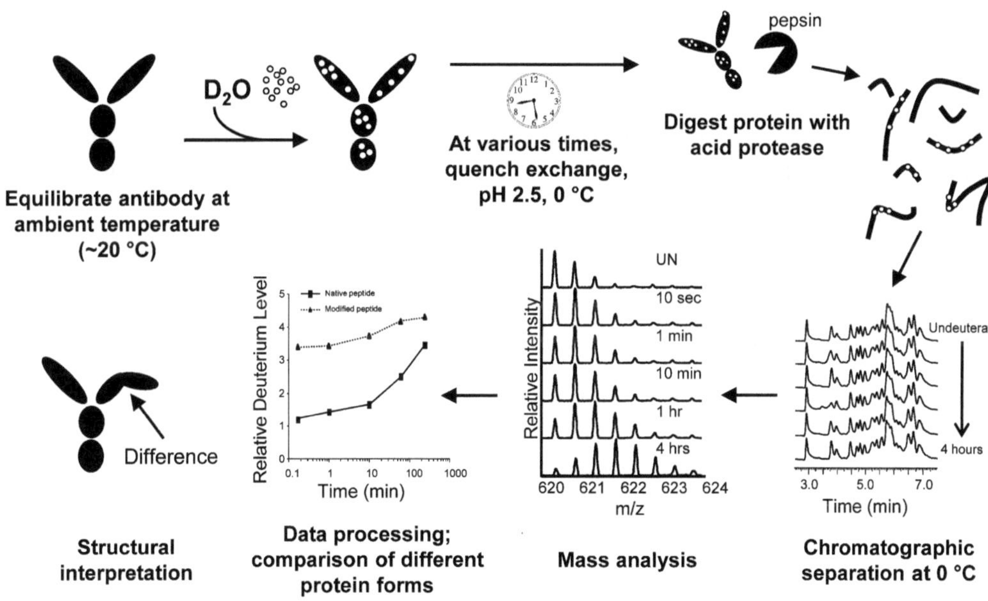

Fig. 1 Workflow of a hydrogen exchange mass spectrometry experiment. This schematic represents a continuous labeling experiment, with arrows indicating the direction of the experiment. The antibody is incubated at ambient temperature and in formulation buffer. The protein is diluted ~20-fold with deuterated formulation buffer. The antibody is incubated for predetermined amounts of time before the reaction is quenched by dropping the pH to ~2.5 and the temperature to 0°C. The quenched antibody is digested with an acid protease (pepsin) and peptides are separated by UPLC at 0°C before being introduced into the mass spectrometer. The mass of each peptide is determined for each deuterium time point and the deuterium incorporation is plotted versus time. A structural interpretation can then be made if the structure is known or a structural model is available

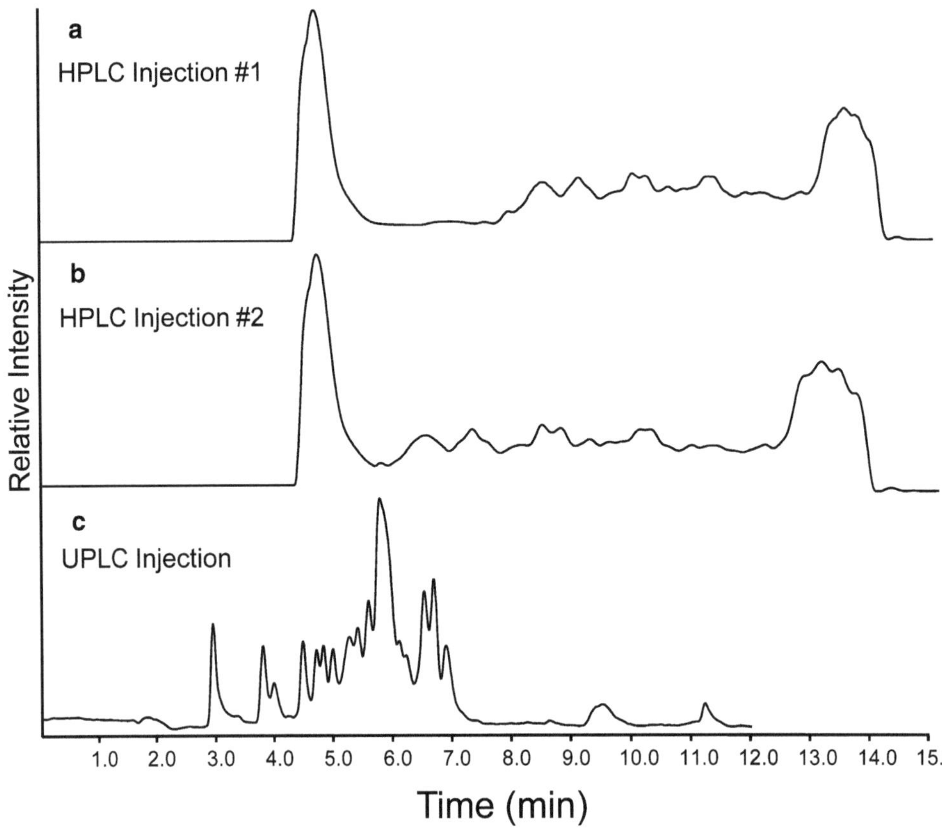

Fig. 2 Total ion chromatograms of IgG1 peptic peptides separated by HPLC and UPLC. (**a**) HPLC injection #1, ~100 μg of pepsin digested IgG1. (**b**) HPLC injection #2, a replicate using the same sample as in HPLC injection #1. (**c**) UPLC separation of ~15 μg of pepsin digested IgG1. See Subheading 3.6.3 for gradient program

(see Notes 8 and 9). The reducing agent is necessary only if the protein contains disulfide bonds, which is the case for IgG1.

3.1 UPLC and MS System Set Up

1. For on-line digestion the LC system must be capable of handling three mobile phases (see Note 6). Mobile phases A and B are for separating peptides and mobile phase C is for the on-line digestion (26, 27) and desalting the peptides. The system should also contain a cooling apparatus capable of maintaining 0 ± 0.5°C and valves able to divert solvent flow for antibody digestion, desalting, and separation (23, 26).
2. Before preparing antibody samples for analysis, ensure that the UPLC system mobile phases are in place and all lines have been purged of air bubbles.
3. The UPLC system should be equilibrated at initial gradient conditions for at least 30 min before sample analysis. For the Waters nanoACQUITY HDX system, mobile phases A and B operating at 40 μL/min should produce a back pressure

between 8,500 and 9,500 psi when connected to the nanoACQUITY BEH C18 1.7 μm guard and separation columns and running at 0°C. At room temperature the back pressure is approximately 5,000–6,000 psi.

4. Mobile phase C, the digestion buffer, is most often set at a flow rate of 100 μL/min. At this flow rate, the operating back pressure should be 700–1,100 psi at 10°C. The pressure of the column under non-injection conditions will be approximately 700 psi, while during an injection the pressure will increase to approximately 1,000 psi. These pressures primarily result from the UPLC lines (see Note 10).
5. The cooling system should be turned on and set to 0°C. Let the system run for at least 60 min to equilibrate the temperature before analysis. This will ensure that the entire thermal mass of the system has reached 0 ± 0.5°C.
6. Before sample analysis begins, the mass spectrometer should be tuned and calibrated with an appropriate standard, such as Glu-fib (GFP) (see Note 11). GFP is a common peptide standard and recommended for calibration by MS/MS fragmentation. GFP will fragment into several ions, at least ten of these ions should be found to make the calibration curve such that the average mass accuracy is at least 3 ppm.

3.2 Antibody Sample Preparation

1. Measure antibody concentration. Concentrations can be calculated from the absorbance measured at 280 nm using an experimentally determined or theoretical extinction coefficient (e.g., $\varepsilon = 218{,}292\ M^{-1}\ cm^{-1}$) (27, 28) or with a Bradford or similar assay (29).
2. Bring the antibody to the right starting concentration, in this example 3 mg/mL (see Note 1).

3.3 Label Antibody and Quench Each Reaction

The steps outlined below describe labeling of individual time points, one at a time. If multiple time points are desired, an alternative is to make them all from a larger "master solution" by multiplying the volumes in step 1 by the number of samples desired. Aliquots are then removed from the master labeling solution at each labeling timepoint and quenched as described in steps 3 and 4.

1. Add 19 μL of labeling buffer to 1 μL of 3 mg/mL antibody solution (the resulting solution is ~95% D_2O). Adding a larger volume to a smaller volume will result in better sample mixing. Allow labeling to proceed at ambient temperature (20 ± 1°C), or any desired temperature, for 10 s, 1, 10, 60, and 240 min (or for any desired time course).
2. An undeuterated control (0 time point) is created by substituting the 19 μL of labeling buffer with 19 μL of equilibration buffer.

3. Once each labeling time has been reached, quench the reaction by diluting the deuterated antibody sample 1:1 with quench buffer (20 μL of quench buffer +20 μL of deuterated antibody sample). Other volumes of quench buffer could also be used. Vortex or mix the solution for 20–30 s, some antibodies may require a longer mixing time.
4. If quenched samples are not to be analyzed immediately, they can be flash-frozen and stored at −80°C for less than 14 days before mass analysis (see Note 12).

3.4 Sample Injection

1. For this example, the UPLC system should be equipped with a sample loop to accommodate up to 40 μL of injected sample (see Note 13).
2. Quenched samples (~20 pmol or 3 μg) are injected directly onto the pepsin column. If quenched samples were frozen prior to analysis, these samples should take no more than 20 s to thaw (see Note 14).

3.5 Antibody Digestion

1. Antibody samples are digested with the on-line pepsin column in water and 0.05% formic acid. 0.05% formic acid has a pH of approximately 2.5, which is optimal for preserving the deuterium label (11).
2. Here, the valve system is set up (Fig. 3) so that after the antibody sample is injected onto the sample loop, it is transferred into the pepsin column for digestion (26). The digestion typically lasts 30 s for a 50 μL sample loop with mobile phase C flowing at 100 μL/min. The antibody peptides produced in the pepsin column are collected on the UPLC BEH C18 peptide trap and subsequently desalted by mobile phase C for 2–3 min before separation begins.
3. A digestion and trapping time of 3 min is often sufficient to completely digest the antibody and desalt the antibody peptides (see Note 15).

3.6 Chromatographic Separation and Mass Analysis

1. During the antibody digestion, the valve system (Fig. 3) isolates the analytical separation column from the online digestion. Once the digestion is complete, a switching valve switches, enabling mobile phases A and B to access the UPLC BEH C18 peptide trap.
2. The acetonitrile gradient from pumps A and B elutes the peptic peptides from the peptide trap onto the separation column.
3. A representative chromatographic gradient (see Note 16) is below, and the resulting separation of IgG1 peptic peptides is shown in Fig. 4.

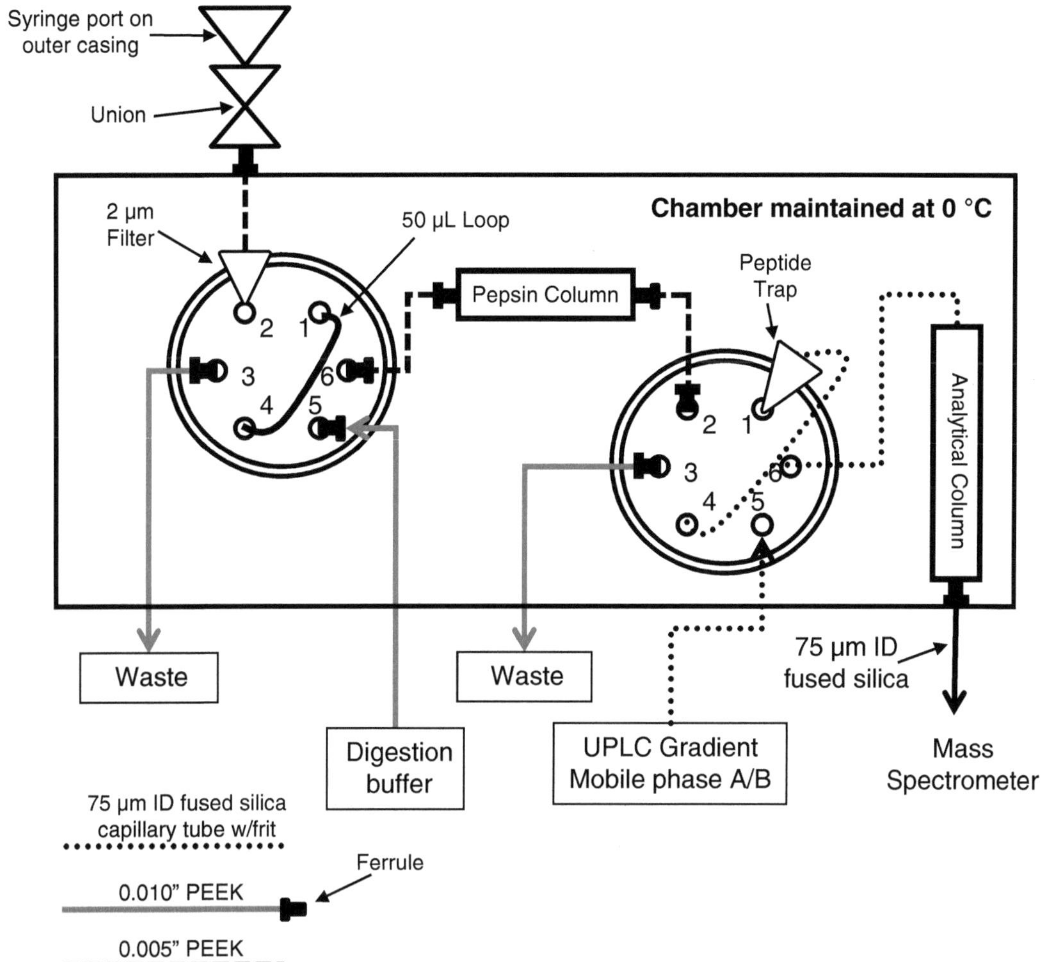

Fig. 3 Schematic representation of UPLC valve setup for an H/DX MS experiment (based on ref. 27). The injection valve is on the *left* in this diagram and the switching valve is on the *right*. The valves and columns are cooled to 0°C for digestion, desalting and separation prior to MS analysis. The UPLC solvents are typically held at ambient temperature but are chilled by the Waters HDX system before entering the cold chamber

Time (min)	Flow (µL/min)	%A (water)	%B (acetonitrile)
0.0	40	95	5
8.0	40	70	30
10.0	40	60	40
10.5	40	5	95
13.0	40	5	95
14.0	40	95	5

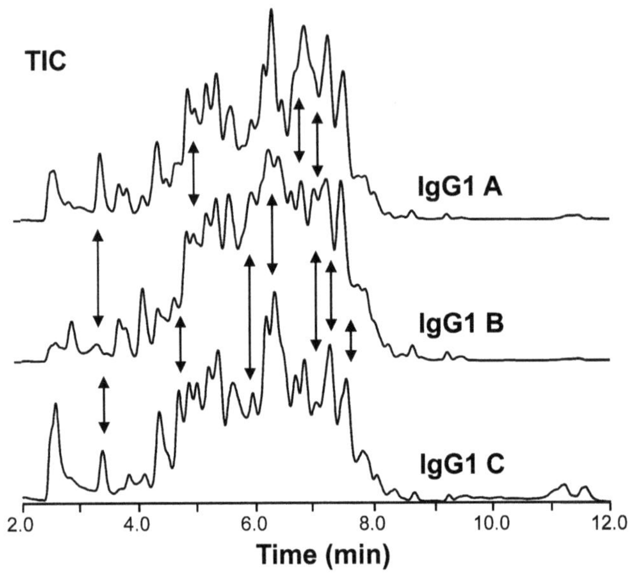

Fig. 4 Total ion chromatograms of the peptic digests of three different IgG1s. Separation was performed at 0°C on a Waters UPLC system designed for hydrogen exchange (24). The three IgG1s share >97% amino acid sequence similarity. The *black arrows* indicate differences in the chromatographic profiles

4. Chromatographically separated antibody peptides are eluted directly into the electrospray source of a mass spectrometer for mass analysis. For the Waters Synapt HDMS, the following mass spectrometer conditions are appropriate, but this will vary depending on the instrument used.

Instrument parameter	Value
Polarity	ESI+
Analyzer	V mode
Capillary (kV)	3.2
Sampling cone (V)	40
Source temp (°C)	80
Desolvation temp (°C)	175
Survey start m/z	255
Survey end m/z	1,800
Product high CE (V)	30
Parent survey low CE (V)	6
Cone gas flow (L/h)	0
Desolvation gas flow (L/h)	600

(continued)

Instrument parameter	Value
Survey scan time (s)	0.25
Survey interscan time (s)	0.02
Survey data format	Continuum

5. Figure 5a shows the total ion chromatogram of the pepsin fragments from digestion of an IgG1, as obtained from a typical H/DX MS experiment. Mass spectra of the peptide ions at elution time ~4.5 min are shown in Fig. 5b. A zoomed in trace of Fig. 5b is shown in Fig. 5c, illustrating the *m/z* region from approximately 1,030–1,060.

3.7 Data Processing

A labor intensive component of H/DX MS is data processing. While there are currently various software programs (30–35) and specialized applications (22) available that semi-automate the processing steps and help to accelerate data processing, some of the analysis still relies on manual processing and intervention, at least at this time. When analyzing the deuterium levels of peptides generated from proteins as large as antibodies, there are a lot of peptides. Maintaining high chromatographic resolution for all these peptides is challenging under the constraints of the quench conditions. Overlapping peptides can confuse some automatic processing algorithms and ions of different intensities may require manual processing.

Processing peptide level exchange data involves identifying each pepsin peptide, extracting the raw chromatographic profile for each peptide, determining the centroid mass for each peptide at each deuterium labeling time point, and determining the deuterium incorporation per labeling time. In the example shown here, all peptides in an undeuterated control sample were identified using CID-based tandem mass spectrometry (see Notes 17 and 18). In our experience, an unambiguous peptide identification requires mass accuracy of <3 ppm on the accurate mass measurement of the precursor ion and identification of 4 product ions for peptides under 12 residues in length and at least 6 product ions for peptides over 12 residues in length. We do not trust peptide maps and conclusions derived from maps when the MS quality is poor (i.e., unit mass resolution, accuracy of >10 ppm). In addition, the same peptide identification must be made in two out of three replicate experiments. It is for this reason that we perform at least three replicates of the undeuterated control sample.

During peptide identification, the retention time of each peptide is recorded either automatically through the software or manually. Since incubation with deuterium changes the peptide mass, knowing the retention time is a critical parameter that helps locate each peptide in the digestions of deuterated samples. Deuterium

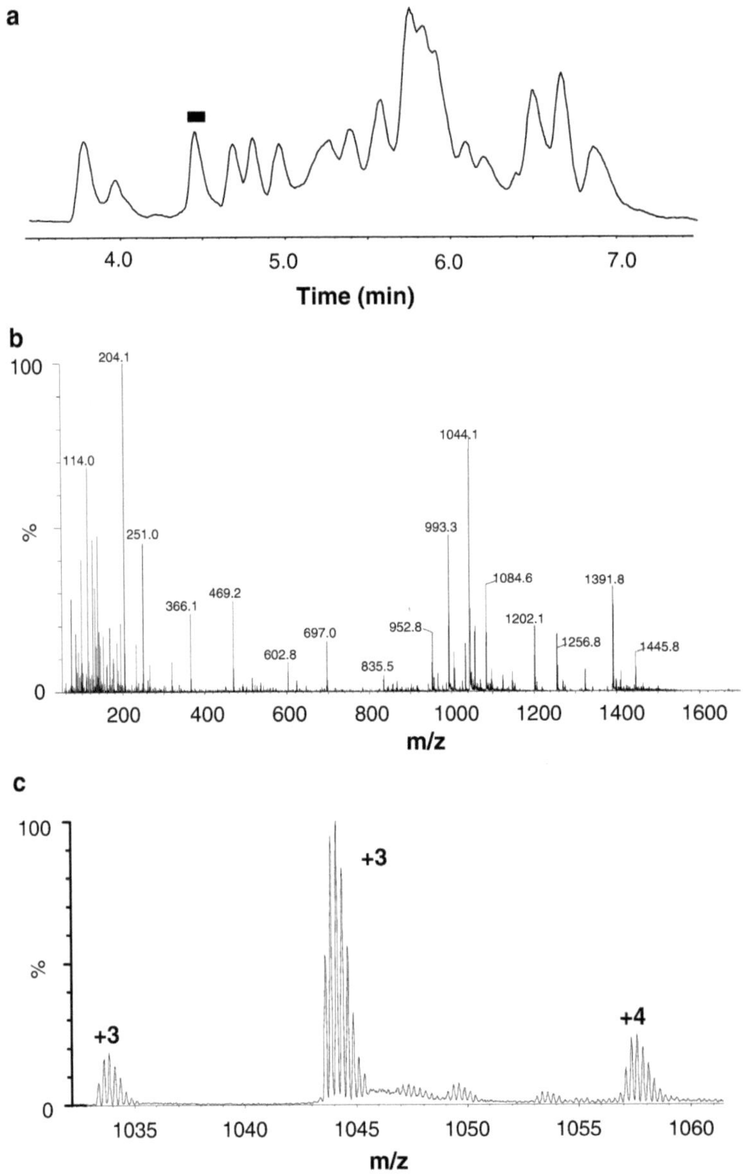

Fig. 5 Representative data from a hydrogen exchange run. (**a**) Total ion chromatogram of the peptic peptides from the pepsin digestion of an IgG1. (**b**) The mass spectrum under the black bar at ~4.5 min shown panel (**a**). (**c**) A zoom in of the mass spectrum in panel (**b**), showing the *m/z* range from 1,030 to 1,060

incorporation does not change chromatographic retention time for IgG1 peptides (or peptides from any other proteins we have worked with). Figure 6 shows the total ion chromatograms for the IgG1 undeuterated control sample and five exchange time points. The reproducibility of the chromatograms is again typical of what is seen using the setup described (see also Figs. 4 and 5a). As an example, the +2 charge state of the peptide representing residues 242–253 from the heavy chain of the IgG1 is shown in Fig. 6b.

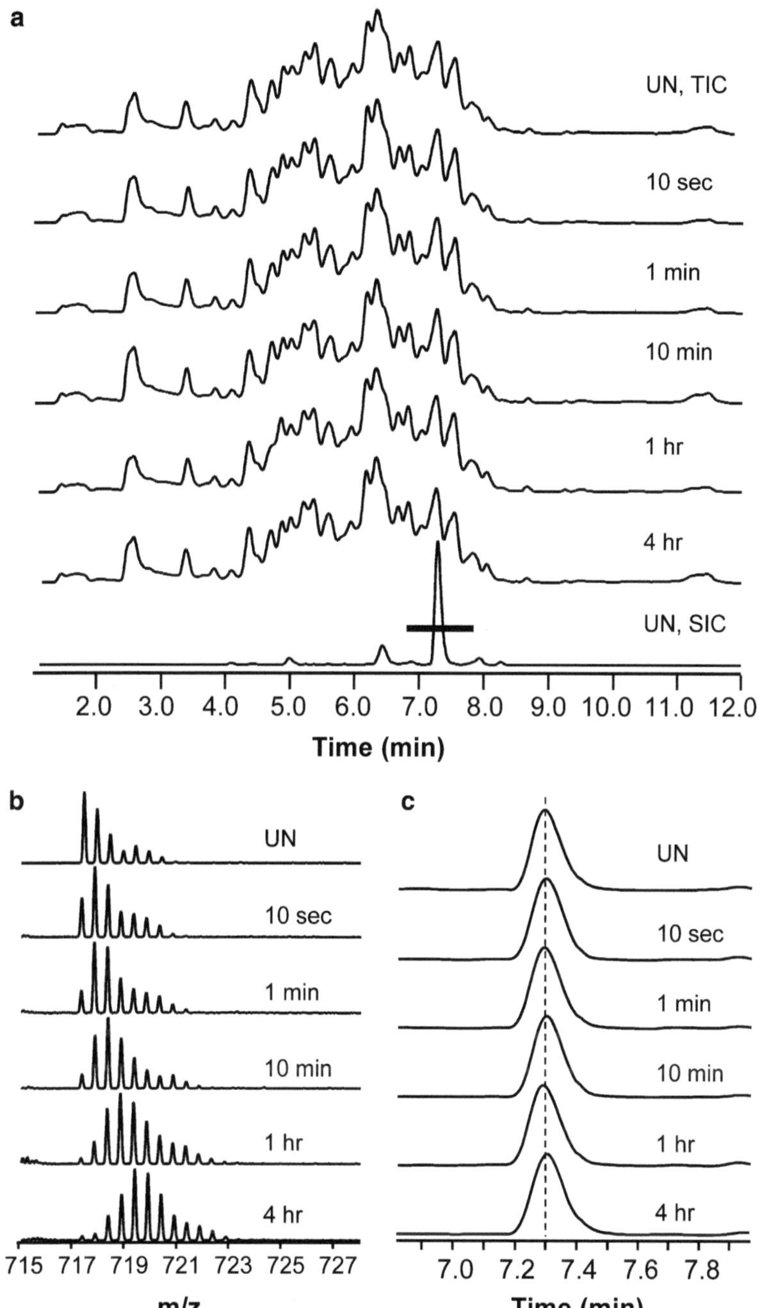

Fig. 6 Deuterium incorporation does not change the retention time of peptides. (**a**) UPLC separations of peptides from an IgG1 digestion are shown, with the total ion chromatogram for the unlabeled sample on top and the total ion chromatograms (TICs) for five deuterium exchange time points just below the undeuterated sample. The *bottom trace* in panel (**a**) is the selected ion chromatogram (SIC) for the +2 charge state ($m/z = 740$) of the IgG1 peptide that covers heavy chain residues 242–253. The mass spectra for each exchange time point for this ion are shown in (**b**). The two most *intense peaks* of the isotope distributions for this ion, in each sample, were used to generate the selected ion chromatograms shown in (**c**). The elution time of this ion does not change with increasing deuterated (*dotted line*). The *black bar* at ~7.0–8.0 min on the *bottom trace* of panel (**a**) indicates the retention time window that is shown in panel (**c**)

The selected ion chromatograms of the two most intense peaks in the isotope distributions for the peptide shown in Fig. 6b are shown in Fig. 6c and indicate that there is no change to retention time as a result of deuterium incorporation.

Since each deuterium incorporation time point is a separate chromatographic trace, data for each peptide must be extracted from each trace to generate a deuterium incorporation curve. Knowing the retention time of each peptide ion makes it easier to search through the chromatograms and extract the necessary data; many automated H/DX processing applications rely on this property. For each peptide, we generally select a representative ion (typically a +1 through +4 charge state) to follow for each peptide. Sometimes nearby ions dictate which charge state to select, and often we will process two or more charge states for the same peptide. The mass spectrometer processing software, in our case Waters MassLynx, is used to visualize and extract the data for each ion (Fig. 7). For each peptide ion, all deuterium incorporation time points are shown in order of increasing deuterium exposure (top to bottom), see Fig. 7, Step 1. The Excel application HX Express (22) is used to extracted the spectral list (*x*, *y* data where *x* is *m*/*z* and *y* is intensity) for each ion at each exchange time point. Other software packages (30–35) can also be used for this purpose, but HX Express was designed to interface with Waters MassLynx (see Note 6). The HX Express settings, which include charge state and centroid distribution width, have been previously described (22); shown below is what was used for IgG1 analysis.

1. The HX Express settings were input: Charge state for each ion and centroid distribution width of 30% peak height.
2. Check the box labeled, "Use isotopic peak detection." This will enable the macro to detect and identify the individual isotopic peaks related to each peptide ion.
3. Peak tolerances can be left at HX Express default settings.
4. The Output tab can be used to change the data reporting format as desired.
5. Once all preferred settings are in place, select OK and run the application.
6. HX Express determines the centroid mass of every peptide for each deuterium time point (Fig. 7, Step 2), based on the set distribution width, step 1 above.
7. The centroid mass at each deuterium time point is then plotted versus deuterium labeling time (on a log scale) (Fig. 7, Step 3).
8. This process is repeated for each peptide and sometimes for all peptide charge states that are observed. Again, software can automate this part of the experiment. We always manually inspect the processing to ensure quality.

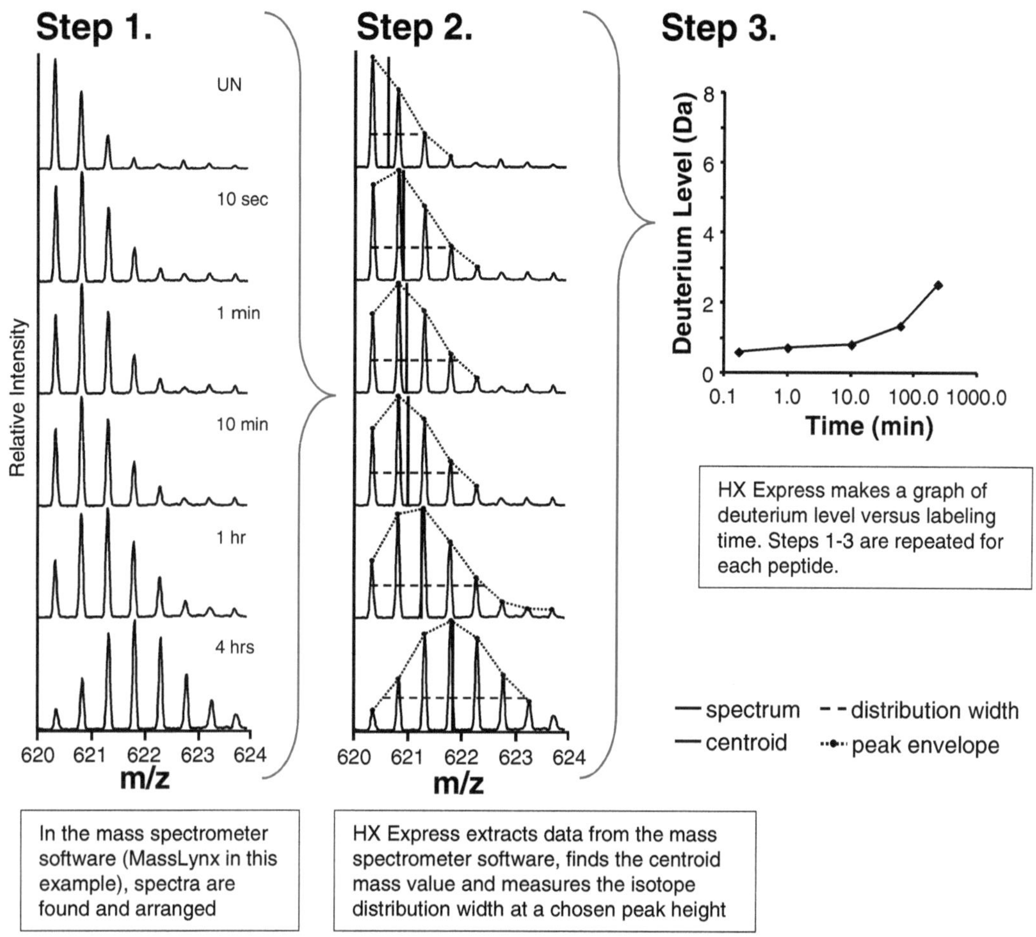

Fig. 7 Processing peptide level hydrogen exchange data. (*Step 1*) Mass spectra from a single peptide ion charge state are found and displayed using the MS instrument software (e.g., MassLynx). The spectra are arranged vertically with increasing deuterium exposure time going from *top* to *bottom*. (*Step 2*) Peptide specific processing parameters are entered into HX Express. HX Express determines the center of mass of each distribution. (*Step 3*) The deuterium incorporation graph for the peptide is made, based on the values from *Step 2*

3.8 Data Interpretation

Deuterium content in each peptide can then be compared for another species of the same protein, (i.e., an IgG1 with and without carbohydrates). Recall that all the information about protein conformation of the fully active and folded molecule was captured in the deuterium pattern found in each peptide. The main feature monitored in a comparison study is the location of where differences in deuterium uptake occur and at what rate. Some peptides will have the same amount of deuterium in normal and modified forms of the protein, while others will have more or less deuterium upon modification. Figure 8 shows the effect of deuterium incorporation on IgG1 heavy chain residues 279–294 as a result of deglycosylation. Such results provide

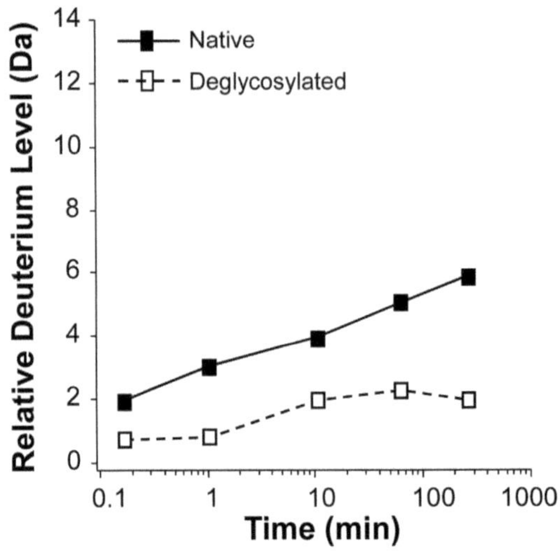

Fig. 8 Hydrogen exchange data for IgG1 heavy chain residues 279–294. See Note 18

information concerning changes to the conformation as a result of a modification (i.e., deglycosylation, oxidation, deamidation, etc.). In this example, deglycosylation of the IgG1 resulted in a decrease to the amount of deuterium that was incorporated into residues 279–294. The meaning of this can then be deciphered in light of the structure of the protein, the location of the deglycosylation relative the region that was affected, etc., can be determined.

4 Notes

1. If the protein sample concentration is below 3 mg/mL, use a centrifugal MWCO filter to concentrate the sample. Be sure to prewet and wash the membrane with at least two volumes of protein buffer before concentrating the protein.
2. There are several commercial suppliers of deuterated solvents. In our experience Cambridge Isotope Laboratories delivers the purest and most consistent product and is what we recommend.
3. The experimental procedure will be demonstrated using PBS buffer (50 mM sodium phosphate, 100 mM sodium chloride, pH 6.00). Any buffer that is free from high concentrations of detergents, which are not MS friendly in later steps of the protocol, will work. We have had success with HEPES, TRIS, histidine, arginine, succinate, and citrate buffer systems.
4. The pH of all solutions was determined with a hydrogen glass electrode that is designed for reporting H^+-ion concentrations (pH). Since the average lab does not have a deuterium

electrode, the concentration of D^+ ion (pD) in the labeling solution was determined with the correction equation pD = pH* + 0.40 where pH* is the reading obtained with a hydrogen electrode (36).

5. The strength of the quench buffer is critical. The quench buffer must be able to reduce the pH of the solution from the labeling pH (usually around 6.00–8.30) to the quench pH (2.50–2.60) reliably and very reproducibly. We recommend buffers for this practice as they are more reproducible. Some experiments have been performed where concentrated acid is used for the quench step, but this not as reliable (in terms of pH) a method for quenching as is a quench buffer. The rate of backbone amide hydrogen exchange is at a minimum at pH 2.5–2.6 (11). The effectiveness of lowering the pH of the quench buffer should be tested by diluting the quench buffer 1:1 with deuterated labeling buffer. The resultant pH should read between 2.5 and 2.6, adjust accordingly if the criteria are not met. Quench buffer pH may need to be varied as much as ±0.3 pH units depending on the buffer strength and pH of the labeling and sample buffers. These pH tests can be made in solutions of the buffer that do not contain protein.
6. The specific instrumentation discussed (i.e., Waters nanoACQUITY UPLC as described first in ref. 23 and now fully commercialized) is not required, but is recommended due to the superior performance of UPLC separation, the integrated cooling chamber for temperature control, the ease of peptic peptide identification using MS^E and the integration of Waters data analysis software (MassLynx) with tools designed for processing H/DX MS data.
7. For optimal peptide retention and separation, the column and trap should consist of the same stationary phase.
8. Other denaturants could be used such as urea or guanidine isothiocynate but guanidine hydrochloride has proven to be very effective. Other reducing agents could be also used, such as dithiothreitol (DTT); however, TCEP is a more effective reducing agent at low pH. It should be noted that TCEP is not stable in phosphate buffers for more than 24 h and should be prepared fresh before the start of a labeling experiment.
9. TCEP gives a very strong molecular ion peak at ~251 *m/z*, thus the low *m/z* in the acquisition *m/z* range is 255 *m/z* in both the precursor (survey) and product ion experiments.
10. During pepsin digestion, if the pressure of the digestion column rises over 1,100 psi during the non-injection phase of the analysis, it is likely that pepsin column may be getting clogged. The pepsin column can be cleaned with 1% formic acid and no more than 5% acetonitrile. If this does not resolve the issue a

new pepsin column may be required. The life of an individual pepsin column will vary. In our experience, a column should last for more than 600 injections; however, this will depend on how the column was cared for and what samples the column was exposed to. Never expose the pepsin column to pH above 5.0 as pepsin will be irreversibly denatured and the column will be ruined. We typically perform the digestion at 10°C, which in the Waters HDX system is facilitated by a small column heater inside the cooled chamber.

11. Other mass calibrants can be used. We prefer to calibrate on the MS/MS fragments of GFP. GFP theoretical fragment ion list:

Fragmention #	*m/z*
1	72.0813
2	120.0813
3	175.1195
4	187.0719
5	246.1566
6	333.1886
7	480.2570
8	627.3254
9	684.3469
10	813.3895
11	942.4321
12	1,056.4750
13	1,171.5020
14	1,285.5448

12. Many proteins, including some antibodies, can be sensitive to multiple freeze–thaw cycles. Avoid multiple freeze–thaw cycles if the stability of the antibody is not known. We prefer to prepare samples and immediately analyze them without a freeze–thaw cycle.
13. Depending upon the set up of the sample loop, there may be a dead volume of approximately 10 μL before the sample loop. If needed, ensure that the sample loop is full by injecting a sample volume equal to the size of the sample loop +10 μL.
14. Avoid large final volumes of quenched protein if possible. If samples have been frozen, thawing large volumes (i.e., >100 μL) may take too long and will result in inconsistency and more experimental variability.

15. Pepsin is a nonspecific enzyme; however, given the same experimental conditions pepsin digestion is reproducible. Because of pepsin's lack of specificity, predicting what peptides will be produced based solely on the amino acid sequence of a protein is not possible. All peptides produced during digestion must be identified, see Note 17 below.
16. Chromatography in H/DX MS is typically performed in less than 10 min. The chromatographic gradient should be optimized for each protein. The gradient shown in Subheading 3.6, point 3 is a good representative gradient and is sufficient for differentiating the peptides of an IgG1. Figure 4 shows an overlay of representative TICs from three different IgG1 antibodies with >97% amino acid sequence similarity. Chromatographic carryover can become an issue in these experiments (37). To avoid misinterpretation of hydrogen exchange data, blank injections or wash methods may be required before each sample injection to ensure that all columns and lines are clean and free of material from prior injections.
17. All peptides in our work are identified using a combination of exact mass (<3 ppm accuracy) and Waters MS^E technology. MS^E is CID (collision-induced dissociation) fragmentation that works by alternating the energy in the collision cell between a low (~6 V) to a high (~30 V) energy (alternating every 0.25 s) (38). All ions in the cell at the time the high energy cycle starts are fragmented and then algorithms and software are used to determine which product ions go with which precursor ions. The average peak width for each peptide identified was approximately 6 s wide, which results in about 12 MS and 12 MS/MS points per peptide peak. MS^E is, in our experience, ideal for identifying peptic peptides in H/DX MS experiments. Because all of the ions can be identified under the identical separation conditions that were used for the analysis of deuterium incorporation, there is no need for data-dependent switching and precursor ion selection. The Waters PLGS software used to interpret peptide fragmentation has been optimized for the analysis of nonspecific pepsin digestion. However, we always visually inspect all MS/MS spectra to verify the quality of the fragmentation and validity of each peptide assignment.
18. The error of each data point should be determined with replicate analyses. In our experience, the greatest source of error in determination of the deuterium level comes from inconsistencies in the pH of labeling and quenching. The overall error in a well controlled experiment is on the order of ±0.5 Da per data point, but this can partially depend on the experimental conditions. At least three replicates of an experiment should be performed to determine this error and to allow addition of error bars or tolerance values to each deuterium incorporation graph. See also ref. 39.

Acknowledgments

We would like to thank Dr. Steven A. Berkowitz for his advice and encouragement, Dr. Rohin Mhatre, Dr. Helena Madden, and Dr. Geoff Gerhardt for their support of H/DX MS, and Dr. Thomas E. Wales, Dr. Keith Fadgen, Dr. Martha Stapels, and Dr. Michael Eggertson for their technical assistance. This work was supported in part by funding from the NIH (GM-086507) and a research collaboration with Waters Corporation. This is contribution 974 from the Barnett Institute.

References

1. Chan AC, Carter PJ (2010) Antibody therapeutics for autoimmunity and inflammation. Nat Rev Immunol 10:301–316
2. Reichert JM, Valge-Archer VE (2007) Development trends for monoclonal antibody cancer therapeutics. Nat Rev Drug Discov 6:349–356
3. Weiner LM, Surana R, Wang S (2010) Antibody-targeted cancer immunotherapy: magic bullets with memory. Nat Rev Immunol 10:317–327
4. Brekke OH, Sandlie I (2003) Therapeutic antibodies for human diseases at the dawn of the twenty-first century. Nat Rev Drug Discov 2:52–62
5. Walsh G (2010) Biopharmaceutical benchmarks 2010. Nat Biotechnol 28:917–924
6. Leavy O (2010) Therapeutic antibodies: past, present and future. Nat Rev Immunol 10:297
7. Davies DR, Padlan EA, Segal DM (1975) Three-dimensional structure of immunoglobulins. Annu Rev Biochem 44:639–667
8. Saphire EO, Parren PW, Pantophlet R, Zwick MB, Morris GM, Rudd PM et al (2001) Crystal structure of a neutralizing human IGG against HIV-1: a template for vaccine design. Science 293:1155–1159
9. Houde D, Arndt J, Domeier W, Berkowitz S, Engen JR (2009) Characterization of IgG1 conformation and conformational dynamics by hydrogen/deuterium exchange mass spectrometry. Anal Chem 81:2644–2651
10. Houde D, Peng Y, Berkowitz SA, Engen JR (2010) Post-translational modifications differentially affect IgG1 conformation and receptor binding. Mol Cell Proteomics 9:1716–1728
11. Englander SW, Kallenbach NR (1983) Hydrogen exchange and structural dynamics of proteins and nucleic acids. Q Rev Biophys 16:521–655
12. Smith DL, Deng Y, Zhang Z (1997) Probing the non-covalent structure of proteins by amide hydrogen exchange and mass spectrometry. J Mass Spectrom 32:135–146
13. Wales TE, Engen JR (2006) Hydrogen exchange mass spectrometry for the analysis of protein dynamics. Mass Spectrom Rev 25:158–170
14. Hoofnagle AN, Resing KA, Ahn NG (2003) Protein analysis by hydrogen exchange mass spectrometry. Annu Rev Biophys Biomol Struct 32:1–25, Epub;%2003 Feb 18., 1–25
15. Brier S, Engen JR (2008) Protein analysis by hydrogen exchange mass spectrometry. In: Chance M (ed) Mass spectrometry analysis for protein–protein interactions an dynamics. Wiley-Blackwell, New York, pp 11–43
16. Engen JR (2009) Analysis of protein conformation and dynamics by hydrogen/deuterium exchange MS. Anal Chem 81:7870–7875

    Maier CS, Deinzer ML (2005) Protein conformations, interactions, and H/D exchange. Methods Enzymol 402:312–360
18. Yan X, Maier CS (2009) Hydrogen/deuterium exchange mass spectrometry. Methods Mol Biol 492:255–271
19. Konermann L, Tong X, Pan Y (2008) Protein structure and dynamics studied by mass spectrometry: H/D exchange, hydroxyl radical labeling, and related approaches. J Mass Spectrom 43:1021–1036
20. Morgan CR, Engen JR (2009) Investigating solution-phase protein structure and dynamics by hydrogen exchange mass spectrometry. Curr Protoc Protein Sci. 2009 Nov; Chapter 17: Unit 17.6.1–17. doi: 10.1002/0471140864.ps1706s58
21. Zhang Z, Smith DL (1993) Determination of amide hydrogen exchange by mass spectrometry: a new tool for protein structure elucidation. Protein Sci 2:522–531

22. Weis DD, Engen JR, Kass IJ (2006) Semi-automated data processing of hydrogen exchange mass spectra using HX-Express. J Am Soc Mass Spectrom 17:1700–1703
23. Wales TE, Fadgen KE, Gerhardt GC, Engen JR (2008) High-speed and high-resolution UPLC separation at zero degrees Celsius. Anal Chem 80:6815–6820
24. Wu Y, Engen JR, Hobbins WB (2006) Ultra performance liquid chromatography (UPLC) further improves hydrogen/deuterium exchange mass spectrometry. J Am Soc Mass Spectrom 17:163–167
25. Woods VL, Jr, Hamuro Y (2001) High resolution, high-throughput amide deuterium exchange-mass spectrometry (DXMS) determination of protein binding site structure and dynamics: utility in pharmaceutical design. J Cell Biochem Suppl 37:89–98
26. Wang L, Pan H, Smith DL (2002) Hydrogen exchange-mass spectrometry: optimization of digestion conditions. Mol Cell Proteomics 1:132–138
27. Gill SC, von Hippel PH (1989) Calculation of protein extinction coefficients from amino acid sequence data. Anal Biochem 182:319–326
28. Pace CN, Vajdos F, Fee L, Grimsley G, Gray T (1995) How to measure and predict the molar absorption coefficient of a protein. Protein Sci 4:2411–2423
29. Noble JE, Bailey MJ (2009) Quantitation of protein. Methods Enzymol 463:73–95
30. Pascal BD, Chalmers MJ, Busby SA, Mader CC, Southern MR, Tsinoremas NF et al (2007) The Deuterator: software for the determination of backbone amide deuterium levels from H/D exchange MS data. BMC Bioinformatics 8:156
31. Pascal BD, Chalmers MJ, Busby SA, Griffin PR (2009) HD desktop: an integrated platform for the analysis and visualization of H/D exchange data. J Am Soc Mass Spectrom 20:601–610
32. Slysz GW, Baker CA, Bozsa BM, Dang A, Percy AJ, Bennett M et al (2009) Hydra: software for tailored processing of H/D exchange data from MS or tandem MS analyses. BMC Bioinformatics 10:162
33. Hotchko M, Anand GS, Komives EA, Ten Eyck LF (2006) Automated extraction of backbone deuteration levels from amide H/2H mass spectrometry experiments. Protein Sci 15:583–601
34. Nikamanon P, Pun E, Chou W, Koter MD, Gershon PD (2008) "TOF2H": a precision toolbox for rapid, high density/high coverage hydrogen–deuterium exchange mass spectrometry via an LC-MALDI approach, covering the data pipeline from spectral acquisition to HDX rate analysis. BMC Bioinformatics 9:387
35. Lou X, Kirchner M, Renard BY, Kothe U, Boppel S, Graf C et al (2010) Deuteration distribution estimation with improved sequence coverage for HX/MS experiments. Bioinformatics 26:1535–1541
36. Glasoe PK, Long FA (1960) Use of glass electrodes to measure acidities in deuterium oxide1,2. J Phys Chem 64:188–190
37. Fang J, Rand KD, Beuning PJ, Engen JR (2011) False EX1 signatures caused by sample carryover during HX MS analyses. Int J Mass Spectrom 302:19–25
38. Silva JC, Gorenstein MV, Li GZ, Vissers JP, Geromanos SJ (2006) Absolute quantification of proteins by LCMSE: a virtue of parallel MS acquisition. Mol Cell Proteomics 5:144–156
39. Houde D, Berkowitz SA, Engen JR (2011) The utility of hydrogen/deuterium exchange mass spectrometry in biopharmaceutical comparability studies. J Pharm Sci 100:2071–2086

Chapter 18

Epitope Mapping of Antibodies by Mass Spectroscopy: A Case Study

Victor H. Obungu, Valentina Gelfanova, and Lihua Huang

Abstract

Epitope mapping of antibodies is the identification and characterization of binding sites of monoclonal antibodies (mAbs) on target antigens. This knowledge can be useful in generating novel antibodies to a particular target as well as elucidating an antibody mechanism of action. Several techniques are available to identify antibody epitopes among which are preliminary and simple ones like sequence homology analysis ELISA and Western blotting. However, the more widely used robust methods typically involve the use of mass spectrometry to fully analyze and interpret the data and accurately identify the binding site. Such methods include epitope extortion/excision, hydrogen deuterium exchange.

Key words Epitope mapping, Mass spectrometry, Epitope extraction, Epitope excision, Hydrogen deuterium exchange

1 Introduction

Antibody epitope mapping has emerged as an extremely useful tool in determining the uniqueness and mechanism of action of neutralizing monoclonal antibodies. This characterization of binding sites has been used in generation, selection, and optimization of new and unique therapeutic antibodies as well molecules to be used in diagnostics and vaccines. Additionally, epitope mapping information is increasingly being relied upon to bolster intellectual property claims by drug companies as well as by academic institutions.

In identifying binding site of an antibody, several techniques can be used either individually or orthogonally. Since an increasing number of targets have complex structures, it has become common to use multiple approaches to clearly identify and characterize antibody epitopes. These approaches can broadly be divided into two, i.e., those that use mass spectrometry of some kind for characterization, analysis, and interpretation of data and those that do not.

Alain Beck (ed.), *Glycosylation Engineering of Biopharmaceuticals: Methods and Protocols*, Methods in Molecular Biology, vol. 988, DOI 10.1007/978-1-62703-327-5_18, © Springer Science+Business Media New York 2013

Some of the commonly used methods include those that do not necessarily rely on the use of mass spectrometry such as ELISA, X-ray crystallography, peptide scanning, site-directed mutagenesis, surface plasmon resonance (SPR), domain swapping, and yeast display among others. Some of these non-MS analytical techniques can be used in conjunction with SPR. However, a good number of epitope mapping techniques rely on mass spectrometry to fully identify the binding site. These include limited proteolysis, epitope foot-printing, epitope extraction, excision, and hydrogen deuterium exchange. Decision on what method to use is informed by whether the epitope is linear or conformational and complexity of the target, i.e., whether it is soluble, membrane bound, etc.

Mass spectrometry (MS) as a technique to characterize peptides and proteins has been employed for the analysis of molar masses of these polymeric molecules for the past 50 years (1). Over this period it has evolved greatly due to its high sensitivity, high mass accuracy, rapid analysis time, and low sample consumption (2).

This chapter will limit itself to those techniques that rely on mass spectrometry as a tool in identifying an antibody epitope, and specifically the use of epitope extraction/excision and hydrogen deuterium exchange with mass spectrometry. The two mass spectrometry methods described are MALDI-TOF and liquid chromatography MS (LC/MS). Further, we will provide methods and procedures to perform epitope mapping using a case study of identifying the epitope of an anti-FasL antibody.

1.1 Mass Spectrometric Techniques for Epitope Mapping

Antibody epitope mapping methods which incorporate MS analysis typically use two of the most common primary ionization techniques for biological molecules, i.e., matrix laser assisted desorption ionization (MALDI) and electrospray ionization (ESI). Another technique used is liquid chromatography MS (LC/MS). We describe here MALDI and LC/MS as analytical techniques for epitope extraction/excision and hydrogen deuterium exchange.

1.1.1 MALDI-MS

In MALDI-MS, a sample is dissolved with an excess of light absorbing compound (a matrix) dissolved in a volatile organic solvent. This mixture is spotted onto a metal plate and allowed to crystallize. When a laser beam is applied to the mixture, the matrix crystals absorb light and transfer energy thereby facilitating desorption and ionization of the matrix and analyte. In majority of MALDI analyses, the mass of the analyte is determined by a time-of-flight (TOF) analyzer tube. This relies on the principle that if ions are accelerated with the same potential at a fixed point and a fixed initial time and are allowed to drift, the ions will separate according to their mass to charge ratios. During the flight in this tube, different molecules are separated according to their mass to charge ratio and reach the detector at different times. In this way, each molecule yields a distinct signal. The time required between the triggering electrical pulse and the detection event correlates to the mass and charge of the analyte (2).

This method is capable of detecting and characterizing biomolecules with molecular masses ranging between 400 and 350,000 Da.

1.1.2 LC-MS

Liquid chromatography/mass spectroscopy (LC/MS) is a technique which combines high performance liquid chromatography (HPLC) with mass spectroscopy. In LC/MS, compounds are separated chromatographically before they are introduced to the ion source and mass spectrometer. The most commonly used ionization method in LC/MS for protein and peptide analysis is ESI. ESI is soft ionization technique and compatible with most chromatographic separations. Both MALDI-TOF and LC/MS are applied for hydrogen/deuterium exchange (HDX) experiments, but LC/MS for HDX experiments are the most common due to high recovery of the proteolytic peptides.

1.2 Epitope Extraction and Excision

Two approaches have been used for the mass spectrometric identification of functional epitopes on antigens bound to antibodies: epitope extraction and epitope excision (3). In epitope extraction, an antigen protein is digested by a proteolytic enzyme to produce an appropriate set of peptide fragments, then peptide fragments containing the linear epitope are selected and separated from the pool of peptide fragments by immunoprecipitation with the monoclonal antibody (Fig. 1). The immunoprecipitated peptides are identified by MALDI-MS (4). In epitope excision, an antigen protein bound to monoclonal antibody is subjected to enzymatic digestion. Because an antibody bound to a protein antigen decreases the rate of proteolytic cleavage of the antigen, especially of those regions involved in antibody contact, an epitope can be identified by the ability of the antibody to protect one region of the antigen more than others from proteolysis (5).

1.3 Hydrogen Deuterium Exchange (Fig. 2)

Proteins contain three types of hydrogen, i.e., very fast exchange hydrogen, such as amine, hydroxyl, carboxyl hydrogen; very slow exchange hydrogen, such as aryl, aliphatic hydrogen; and amide hydrogen with a characteristic exchange rate (pH dependent). Protein hydrogen exchange studies principally focus on the main chain amide NH hydrogens. These hydrogens are distributed uniformly at every amino acid, except proline, in every protein molecule. Besides extreme pH effect, amide hydrogen exchange rate of protein is not only dependent on protein higher order structures but also on solvent accessibility. When an antibody binds an antigen, the exchange of amide hydrogens at or near the protein/protein interaction site (epitope) is slower, due to solvent inaccessibility. Hence, if we compare the exchange rate of hydrogen to deuterium between antigen/antibody complex and antigen only,

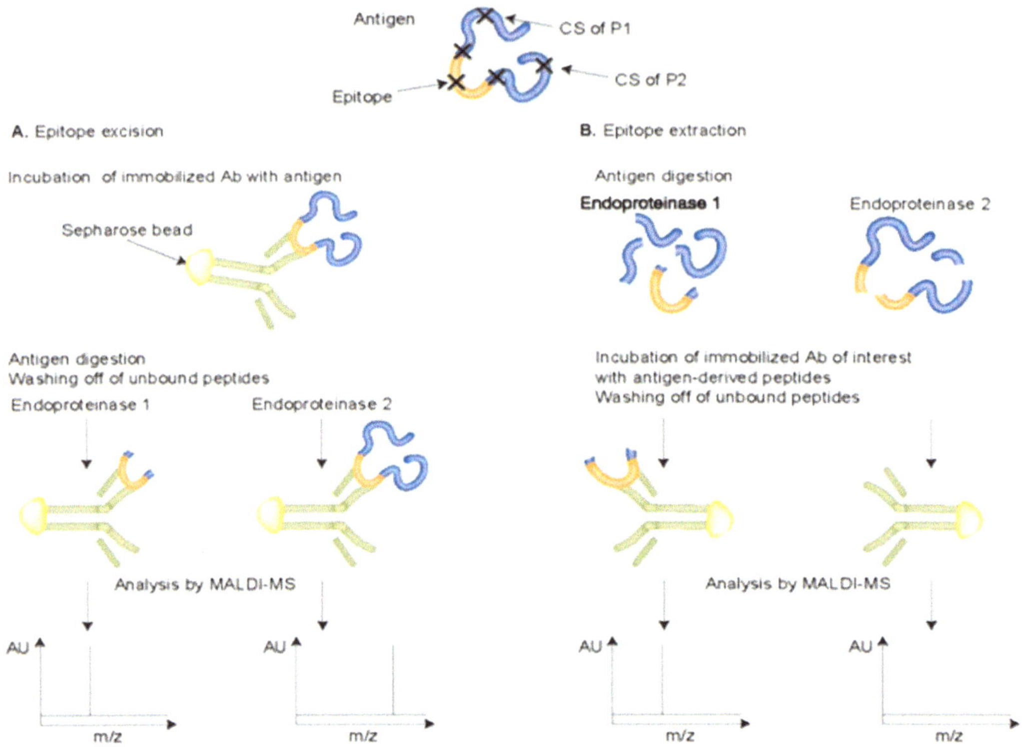

Fig. 1 Identification of epitopes by epitope excision and extraction. (**a**) For epitope excision, the antigen is bound to the immobilized Ab and digested with various endoproteinases. Unbound peptides are washed off and the affinity-bound peptides are analyzed by MALDI-MS. (**b**) In epitope extraction, the antigen is digested in solution, passed over the immobilized Ab and subsequently analyzed for the presence of affinity-bound antigen fragments. *Ab* antibody, *AU* arbitrary units, *CS* cleavage site, *MALDI* matrix-assisted laser desorption/ionization, *MS* mass spectrometry, *m/z* mass-to-charge ratio, *P1* proteinase 1, *P2* proteinase 2. Reproduced from *Expert Review of Proteomics* (2005) 2(5), 745–756 with permission of Expert Reviews Ltd

antigen protein sequence at the epitope or near epitope can be deduced.

To perform H/D exchange, complex solutions of antigen/antibody, antigen/control protein (no interaction), or antigen only is diluted with D_2O (to 80–90% D for the final solution) at one pH (high pH) for some duration. The exchange is quenched by adding low pH buffer (pH, ~2.5). Typically amide hydrogen has slowest exchange rate at pH 2.5 (6). The quenched solution is immediately analyzed by mass spectrometry or subjected to proteolytic digest before mass spectrometry. The most common enzyme for H/DXMS is pepsin. Peptic peptides derived from the antigen will gain less deuterium atoms (smaller masses) when in complex with the antibody than when free, if the peptides are at or near the antigen/antibody binding site (epitope).

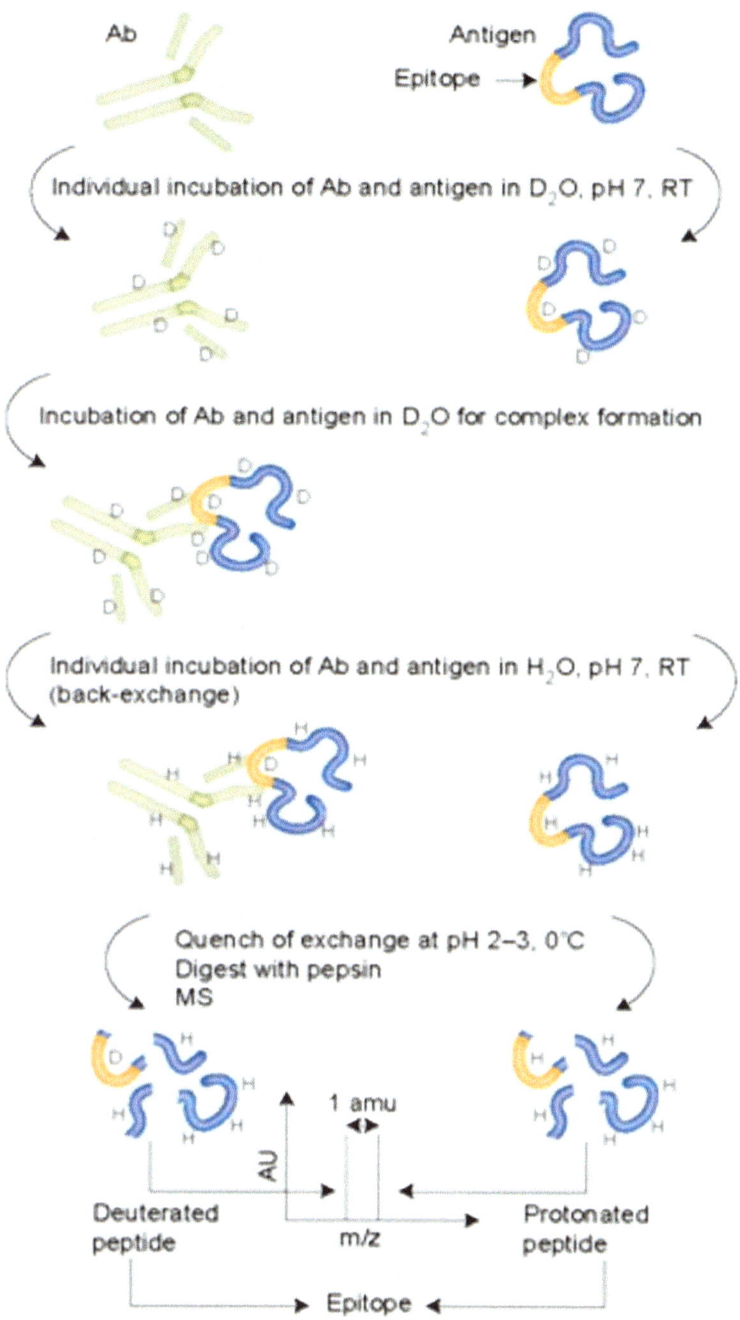

Fig. 2 Epitope mapping by H/D exchange. Antibody (Ab) and antigen are incubated in deuterium (D_2O) at neutral pH at RT for a specified amount of time to allow the deuteration of the sample. Ab and antigen are then combined for the formation of a complex. Subsequently, the solvent is changed to H_2O to allow back exchange from deuterium to proton. By reducing the pH to approximately 2–3 and cooling the sample to 0°C, the back exchange is significantly decreased, which traps the deuterium label. The proteins are analyzed by MS or immediately digested with pepsin, an acid-stable proteinase, and analyzed by MS for the presence of deuterated peptides. The mass difference between deuterated peptide and protonated peptide is 1 amu per deuterium. *Ab* antibody, *amu* atomic mass unit, *AU* arbitrary units, *D* deuterium, *H* proton, *MS* mass spectrometry, *m/z* mass-to-charge ratio, *RT* room temperature. Reproduced from *Expert Review of Proteomics* (2005) 2(5), 745–756 with permission of Expert Reviews Ltd

The anti-FasL antibody, LA296, used in this test case is an IgG4 monoclonal antibody with a binding affinity of 100 pM for soluble FasL. LA296 binds to and neutralizes the activities of both soluble and membrane-associated human and cynomolgus monkey FasL, but not that of mouse or rabbit, as determined by ELISA and BIAcore measurements. In order to further understand the epitope binding specificity of this antibody and its mechanism of action, we have mapped critical residues involved in its recognition of FasL using orthogonal approaches among which was hydrogen–deuterium exchange mass spectrometry (H/DXMS) and mass spectrometry. However, the methods described below can be broadly applied to any antibody that binds to a soluble target.

2 Materials

2.1 Biotinylation Anti-FasL Antibody in Solution

1. EZ-Link NHS_LC-Biotin, succinimidyl-6-(biotinamido) hexanoate (Pierce, Rockford, IL). Upon receipt store desiccated at 4°C (see Note 1).
2. Phosphate-buffered saline (PBS) or other amine-free buffer having pH 7–8 for use as reaction buffer (see Note 2).
3. Centricon Centrifugal Filter devices with YM-50 MW membrane (Millipore (Bedford, MA)) (see Note 3).
4. DMSO.

2.2 Epitope Extraction and Excision

1. Recombinant human Fas Ligand/TNFSF6 (FasL) carrier-free (R&D Systems) is the N-terminal 6× histidine-tagged human Fas Ligand that was expressed in CHO cells. Store lyophilized protein at −20°C (see Note 4).
2. LA296 is a recombinant antibody of Eli Lilly and Co. LA296 was biotinylated with NHS-LC-biotin (Pierce, Rockford, IL).
3. Dynabeads M-280 Streptavidin (Dynal, Lake Success, NY).
4. Trypsin (see Note 5).

2.3 MALDI-Mass Spectrometry

1. Acetonitrile (ACN) (Honeywell B&J, Muskedon, MI).
2. Ammonium Bicarbonate, 100 mM, pH 8. Store at room temperature.
3. MassPREP MALDI Matrix, α-cyano-4-hydroxycinnamic acid (CHCA) (Waters (Milford, MA)). Saturated matrix solution contains CHCA in 50% ACN and 0.1% TFA in water.
4. ZipTip (uC18) (Millipore, Bedford, MA).

2.4 Hydrogen Deuterium Exchange

1. Acetonitrile (ACN) and water (Honeywell B&J, Muskedon, MI).
2. Formic Acid.
3. 100% D_2O (ACROS).

4. Neuraminidase solution (10 U/1 ml) (Roche Diagnostics).
5. Pepsin.

3 Methods

3.1 Biotinylating Antibody in Solution

1. Remove vial of NHS-LC-Biotin from freezer and equilibrate it to room temperature before opening. Immediately before use, prepare a 10 mM solution of the NHS-LC-Biotin by dissolving 2.3 mg reagent in 500 μl of DMSO (see Note 6).
2. Add the appropriate volume (see Note 7) of 10 mM biotin reagent solution to the protein solution.
3. Incubate reaction at room temperature for 30 min (or on ice for 2 h).
4. To remove excess non-reacted and hydrolyzed biotin reagent from solution:
   (a) Pre-rinse two Centricon Centrifugal Filter devices with YM-50 MW membrane to remove glycerin: fill the devices with 2 ml of PBS, spin at 5,000 ×*g* for 10 min at room temperature; then invert the devices and spin at 300 ×*g* for 2 min to remove the remaining rinse. Discard filtrate vials and insert prerinsed sample reservoirs into fresh filtrate vials.
   (b) Dilute labeled antibody sample to 2 ml with PBS and transfer solution to two prerinsed Centricon Centrifugal Filter devices.
   (c) Spin at 5,000 ×*g* for 10 min at room temperature.
   (d) Remove centrifugal filter assembly from centrifuge, and add 1 ml of PBS to each device.
   (e) Spin at 5,000 ×*g* for 10 min at room temperature.
   (f) Remove centrifugal filter assembly from centrifuge, remove and discard 1–1.5 ml of filtrate from filtrate vial and add 1 ml of PBS to sample reservoir of each device.
   (g) Spin at 5,000 ×*g* for 10 min at room temperature.
   (h) Remove centrifugal filter assembly from centrifuge, place retentate vials over sample reservoirs, invert units to recover the retentates. Centrifuge at 300 ×*g* for 2 min to transfer concentrate into retentate vials. Combine recovered antibody and adjust volume to desired concentration (1–2 mg/ml) with PBS. Store antibody sample at 4°C.

3.2 Epitope Extraction

3.2.1 Preparation of rhFasL Tryptic Digest

rhFasL (R&D, carrier-free) is reconstituted at 0.1 mg/ml in 100 mM NH_4HCO buffer, pH 8 containing 0.1 μg/ml of trypsin. Equal volume of trypsin containing buffer is aliquoted in separate vial and assigned as control. Samples are incubated at 37°C for 6 h (see Note 8).

3.2.2 Preparation of LA296 Immobilized to Dynabeads M-280 Strepavidin Beads

1. Label three tubes as follows: (1) FasL/LA296 (protein digest/antibody sample), (2) -/LA296 (trypsin/antibody control), and (3) FasL/- (protein digest/no-antibody control).
2. Resuspend the Dynabeads M-280 Strepavidin by gently shaking the vial to obtain a homogeneous suspension. Add 100 μl (1 mg) of Dynabeads M-280 Strepavidin to all tree tubes.
3. Place the tubes in the magnet (Dynal MPC) for 1–2 min. Do not remove the tubes from the magnet during the separation process.
4. Remove the supernatant by aspiration with a pipette while the tubes remain on the Dynal MPC.
5. Remove the tubes from Dynal MPC. Add 1 ml of 100 mM NH_4HCO buffer, pH 8. Resuspend gently.
6. Repeat steps 3–5 twice.
7. After the last wash add 100 μl of 0.3 mg/ml antibody in PBS to two tubes (1 and 2) of the Dynabeads M-280 Strepavidin and 100 μl of PBS to tube 3.
8. Incubate at room temperature with gentle rotation of the tube (see Note 9).
9. Separate the Dynabeads M-280 Strepavidin now coated with biotinylated antibody using a Dynal MPC for 2–3 min.
10. Wash 4–5 times in PBS/0.1% BSA using a Dynal MPC.
11. Resuspend beads in 100 μl of PBS/0.1% BSA.

3.2.3 Peptide(s) Capturing Procedure

1. Dilute 20 μl (2 μg) of the rhFasL digest from step 1 in 200 μl of PBS/0.1% BSA. Add 100 μl of rhFasL solution to tube 1 and tube 3. Dilute 20 μl of trypsin containing buffer from step 1 in PBS/0.1% BSA. Add 100 μl of trypsin containing solution to tube 2 (see Note 10).
2. Incubate samples overnight at 4°C.
3. Separate the Dynabeads M-280 Strepavidin now coated with biotinylated antibody-peptide(s) complexes using a Dynal MPC for 2–3 min. Put aside supernatants if desired (see Note 11).
4. Wash beads three times in PBS and two times with 100 mM NH_4HCO buffer, pH 8 using a Dynal MPC.
5. After last wash, spin down tubes with beads shortly to remove remaining liquid and immediately resuspend in 10 μl of 10% formic acid by gently shaking the vial for 10–30 s to obtain a homogeneous suspension. Place the tubes in the magnet and collect beads-free eluates. Eluates are desalted with ZipTip C18, and peptides are analyzed by MALDI-TOF.

3.3 Epitope Excision

1. Label three tubes as follows: (1) FasL/LA296 (protein/antibody sample), (2) -/LA296 (no-FasL control), and (3) FasL/-(no-antibody control).

2. rhFasL (R&D, carrier-free) is reconstituted at 0.1 mg/ml in 100 mM NH_4HCO buffer, pH 8.
3. Add 10 μl (1 μg) of rhFasL solution to two tubes (tubes 1 and 3). Add 10 μl of 100 mM NH_4HCO buffer (pH 8) to tube 2.
4. Add 4 μl (5 μg) of LA296 antibody to tubes 1 and 2. Add 4 μl of PBS to tube 3. Incubate samples overnight at 4°C.
5. Briefly spin down samples to collect liquid on the bottom of tubes.
6. Add 0.12 μg of trypsin (in 10 μl of 100 mM NH_4HCO buffer) to all three samples and incubate for 4–6 h at 37°C (see Note 5).
7. Briefly spin down samples to collect liquid on the bottom of tubes and stop reaction by adding 10 μl of 10% formic acid.
8. Desalt samples with ZipTip C18 and analyze peptides by MALDI-TOF.

3.4 MALDI-Mass Spectrometry

3.4.1 Desalting

1. Prewet tips by aspirating 10 μl of 100% ACN three times.
2. Equilibrate tips by washing with 10 μl of 0.1%TFA three times.
3. Bind samples by slowly aspirating them through the tips up and down for 2–5 min.
4. Wash tips with 10 μl of 0.1%TFA for 10–15 times.
5. Elute samples with 2 μl of 50% ACN/0.1% TFA.
6. Mix eluates (1:1) with matrix solution (saturated α-cyano-4-hydroxycinnamic acid in 50% ACN/0.1% TFA) and place on MALDI plate.

3.4.2 MALDI-MS Analysis

1. All spectra are acquired with a 4,700 MALDI-TOF–TOF mass spectrometer (PerSeptive Biosystems, Framingham, MA, USA) in the reflector or linear delayed extraction mode. Spectra are acquired in 12 separate areas of each spot with 200 laser shots per acquisition and all acquisitions are averaged for each spot.
2. The following criteria are used to assign the FasL peptides: (1) the measured molecular mass of the peak matched the calculated molecular mass of FasL peptide based on the rhFasL amino acid sequence, (2) the peak are not found in no-FasL controls samples.

3.5 Hydrogen Deuterium Exchange

1. FasL:LA296 complex was prepared by mixing 10 μl aliquot of 1.0 mg/ml FasL in 1× PBS buffer with 5 μl aliquot of 10 mg/ml LA296 in 1× PBS buffer and 20 μl of 1× PBS buffer to yield an approximate molar ratio (FasL/LA296) of 1:1.5.
2. Free FasL solution was obtained by mixing 10 μl aliquot of 1.0 mg/ml FasL in 1× PBS buffer with 25 μl of 1× PBS buffer.

3. Both solutions were treated with 1 µl of 10 U/ml neuraminidase solution at 37°C for 90 min.
4. 5 µl aliquot of free FasL or FasL:LA296 complex were mixed with 20 µl of 100% D_2O, resulting in 80% D_2O in the final sample.
5. The solution was incubated at ambient temperature for 1–10 min (see Note 12).
6. The exchange was quenched by adding 20 µl of 0.15% formic acid aqueous solution and immediately treated by adding 2 µl of 2 mg/ml pepsin aqueous solution and incubating at ambient temperature for 30 s (see Note 13).
7. The digest was immediately injected onto a C18 reversed-phase column, manually.

3.6 LC/MS Analysis

1. A Waters 2795 HPLC was coupled to a Micromass LCT premier mass spectrometer.
2. The HPLC stream from the HPLC pump was connected via metal tubing (approximately 1 ml) to the manual injector and to a Zorbax C18 column (2.1 × 50 mm) and then to the mass spectrometer.
3. The metal tubing, injector loop and column were continuously submerged in ice water.
4. HPLC mobile phase A: 0.15% formic acid aqueous solution and mobile phase B: 0.12% formic acid in acetonitrile.
5. HPLC gradient was set as follows:

Time (min)	0	2	13	14	16.5	17.0	19
% Mobile phrase B:	1	15	40	90	90	1	1
Flow rate (ml/min)	0.2	0.2	0.2	0.2	0.2	0.4	0.4

6. The mass spectrometry was performed with the following settings:

Electrospray/mode	Positive/V
Capillary voltage (kV)	3.0
Cone voltage (V)	80
Aperture 1 voltage (V)	15
Mass scan range	300–3,500
Desolvation temperature	300°C
Desolvation gas flow (l/h)	700

7. Each peptic peptide of FasL was assigned from LC/MS analysis (see Note 14).
8. Mass spectrum of each peptic peptide was after H/D exchange with or without LA296 present.

9. The average mass of each peptic peptide was calculated (see Note 15).
10. Mass difference for each peptic peptide between free FasL and FasL/LA296 was calculated.

4 Notes

1. NHS-LC-Biotin is moisture-sensitive. Store the vial of biotin reagent at 4°C with desiccant. To avoid moisture condensation onto the product, equilibrate vial to room temperature before opening.
2. Avoid buffers containing primary amines (e.g., Tris or glycine) as these will compete with the intended reaction. If necessary, dialyze or otherwise desalt to exchange the antibody sample into an amine-free buffer such as phosphate-buffered saline.
3. Other types of desalting columns or dialysis units for buffer exchange may be used.
4. Carrier-free protein is used, because high concentration of BSA (as carriers) prevents identification of FasL peptides in tryptic digests. Store lyophilized protein at –20°C. The carrier-free protein should be used immediately upon reconstitution to avoid losses due to nonspecific binding to the inside surface of the vial.
5. Other proteases could/should be considered based on the sequence of the studied protein. Trypsin was chosen in this study, because we were able to detect all but C-terminal tryptic peptide in the FasL digest (C-terminal of the FasL has two potential glycosylation sites and could have been glycosylated) by MALDI-TOF.
6. As directed in the procedure, dissolve the biotin reagent immediately before use. The NHS-ester moiety readily hydrolyzes and becomes non-reactive; therefore, do not prepare stock solutions for storage. Discard any unused reconstituted reagent. The NHS-Biotin Reagent is not directly water soluble but must be dissolved in organic solvents such as DMSO or DMF before adding to aqueous solutions at the final concentration for the labeling reaction. Alternatively water soluble Sulfo-NHS-Biotin reagents could be used for antibody labeling.
7. The extent of biotin labeling depends on the distribution of amino groups on the protein, protein concentration, and the amount of reagent used. We followed manufacturer suggested protocol and used a 20-fold molar excess of biotin reagent to label 0.6 mg antibody (in 0.5 ml).
8. Efficiency of digestion was monitored by analyzing of portion (1 μg) of tryptic digest by MALDI-Mass Spectrometry.
9. According to manufacturer protocol, 5–10 μg antibodies per mg (100 μl) Dynabeads would saturate the Strepavidin, assuming

100% biotinylation of the antibody, at room temperature in 30 min. We used longer incubation times (up to 5 h). Efficiency of immobilization procedure was monitored by ability of immobilized antibody to precipitate FasL (SDS-PAGE).

10. Due to sensitivity of MALDI and for economy, as well as to minimize nonspecific binding of protein to the beads, we only used 0.5–1 μg of FasL or FasL digest per mg of LA296 immobilized or control beads.
11. Collect supernatants and label as flow-through. Flow-through samples can be analyzed to confirm that FasL tryptic peptides are still present (in solutions) or were depleted from samples.
12. Since antigen epitope for antibody is on protein surface, amide hydrogen exchange is very faster at 1× PBS buffer (pH 7.4). The exchange for epitope amide hydrogen will reach equilibrium in very short time when antibody does not exist.
13. In order to minimize back exchange during pepsin digest, quenching solution and digest can be conducted in water ice bath. Of course, the digest time need longer (2–10 min depending protein).
14. Since pepsin has a broad specificity, Tandem MS may be required to assist complete peptic peptide assignment.
15. When peptic peptide is very large and there is no resolution among isotopic ions, average mass of the peptide was directly obtained. When mass spectrometry resolution is high enough, the average mass of peptic peptide was obtained by the following equation.

$$\text{Average mass peptide} = \left(\frac{\sum_{i} m_i A_i}{\sum_{i} A_i} - 1.008 \right) \times z$$

where m_i is isotopic ion, A_i is intensity or area of isotopic ion m_i, and z is charge state of the ion.

References

1. Burlingame AL, Baillie TA, Russell DH (1992) Mass spectrometry. Anal Chem 64: 467R–502R
2. Hager-Braun C, Tomer KB (2005) Determination of protein-derived epitopes by mass spectrometry. Expert Rev Proteomics 2: 745–756
3. Parker CE, Tomer KB (2000) Epitope mapping by a combination of epitope excision and MALDI-MS. Methods Mol Biol 146:185–201
4. Zhao Y, Chalt BT (1994) Protein epitope mapping by mass spectrometry. Anal Chem 66:3723–3726
5. Jemmerson R, Paterson Y (1986) Mapping epitopes on a protein antigen by the proteolysis of antigen-antibody complexes. Science 232: 1001–1004
6. Bai Y, Milne JS, Mayne L, Englander SW (1993) Primary structure effects on peptide group hydrogen exchange. Proteins 17:75–86

Part IV

PK/PD Assays for Therapeutic Antibodies

Chapter 19

Evaluation of Antibody-Dependent Cell Cytotoxicity Using Lactate Dehydrogenase (LDH) Measurement

Matthieu Broussas, Lucile Broyer, and Liliane Goetsch

Abstract

In order to improve therapeutic antibodies efficacy in cancer patients, several strategies were developed. One of these strategies consists in the enhancement of effector functions. Antibody-dependent cellular cytotoxicity (ADCC) was shown to mediate the activity of several therapeutic antibodies through interaction of the constant fragment (Fc) with immune cells. The interactions of Fc fragment can be modulated by engineering through modifications of the carbohydrate moieties or through modifications of some critical amino acids for its binding. Such modifications have to be studied in an in vitro assay to evaluate their impact on the regulation of effector functions. Here, we described a method to evaluate ADCC using a nonradioactive assay based on the measurement of lactate dehydrogenase (LDH) release. NK cells were purified by negative immunomagnetic selection and used as effector cells to trigger ADCC against specific target tumor cells. The LDH release measurement from lysed cells is performed after 4 h incubation. This method can replace the ^{51}Cr release assay since it is less restrictive and highly sensitive.

Key words ADCC, Antibody-dependent cell cytotoxicity, LDH release, Non-colorimetric method, Herceptin

1 Introduction

Therapeutic monoclonal antibodies represent one of the most promising innovative molecules for cancer treatment. Selection of antibodies in oncology is mainly based on their capability to reduce tumor growth. This efficacy depends on their ability to directly kill tumor cells via degradation of the targeted protein, programmed cell death (PCD) induction (1–3), cell cycle arrest (CCA), sensitization of cells to chemotherapy (3, 4), and/or effector functions activation. The latter functions include complement dependent cytotoxicity (CDC) (5), antibody-dependent cellular cytotoxicity (ADCC) (3, 5, 6), and antibody dependent cellular phagocytosis (ADCP) (7).

Alain Beck (ed.), *Glycosylation Engineering of Biopharmaceuticals: Methods and Protocols*, Methods in Molecular Biology, vol. 988, DOI 10.1007/978-1-62703-327-5_19, © Springer Science+Business Media New York 2013

Human presents five class of immunoglobulin that are classified on the basis of the constant region of the heavy chain: IgA, IgD, IgE, IgG, and IgM. Two IgA (IgA1 and IgA2) and four IgG isotypes (IgG1, IgG2, IgG3, and IgG4) are identified (8). Despite the large repertoire of Ig, the IgG isotype is the main selected for cancer therapy maybe due to its in vivo stability and its capability to induce effector functions. The intact molecule is composed of four polypeptide chains, including two identical light chains and two identical heavy chains (9). The variable fragment (Fv) of an antibody is involved in target recognition and specificity. The constant fragment (Fc) is responsible of interactions with immune cells that express Fc receptors (FcR) on their surface mediating ADCC and ADCP and of complement binding triggering CDC (10).

Fc-receptors are expressed constitutively on the majority of hematopoietic cells (NK cells, monocytes, granulocytes, B cells). ADCC and ADCP are mediated through interaction of mAb targeting cells with FcγR-bearing cells of the innate immune system (i.e., NK cells, monocytes, and granulocytes) (11). Recognition of the Fc domain of antibody, bound to target cells, by FcγR triggers effector cells to release cytosolic perforin, granzymes, and granulysin to induce opsonized-cell lysis (12). In human, the Fcγ receptor family is composed of three structurally homologous cellular FcγR types: FcγRI (CD64), FcγRII (CD32), and FcγRIII (CD16) which differ in their antibody affinities due to their different molecular structure (13). FcγRI is the receptor eliciting the highest affinity for IgG1, whereas FcγRII and FcγRIII have a relative weak affinity for this Ig. FcγRs are classified into activating receptors (FcγRI, FcγRIIa, and FcγRIII) and inhibiting receptors (FcγRIIb). In addition, FcγRs differ in their IgG affinity according to genetic factors. The three classes of FcγR are encoded by eight genes, all located on chromosome one. Some of them display a functional allelic polymorphism generating allotypes with different receptor properties (13–15). Genomic polymorphism of the FcγRIIIA corresponds to phenotype expression of valine (V) or phenylalanine (F) at amino acid 158. The FcγRIIIa-V^{158} variant binds to IgG1 with a great affinity and mediates a higher ADCC than the FcγRIIIa-F^{158} variant (13, 16). NK cells, which are the major effector cells in ADCC, only express the FcγRIIIa receptor (17). Based on clinical correlative studies using anti-CD20 antibody, Rituximab, an association has been established between polymorphism on FcγRIIIa and clinical response to single-agent Rituximab. Patients with follicular lymphoma with the VV genotype have a better clinical response to Rituximab than patients with the VF or FF genotype (18). These data highlighted the importance of Fc–FcγR interactions in the antitumor effect of Abs and led to Fc Fragment engineering, in order to increase their ability to mediate effector functions such as ADCC (19, 20).

Among strategies developed in this field, modifications of Fc glycosylation are extensively studied. For example, it was demonstrated that low-fucosylated IgG1 has a higher ADCC than high fucosylated-IgG1 (21, 22).

To test all constructs in this optimization process, a robust, sensitive and nonradioactive method is needed to replace the original chromium-release assay. We here describe such a method using purified NK cells.

2 Materials

2.1 Cell Culture

1. *A*431, epidermoïd human cells are obtained from the American Type Culture Collection (ATCC) and grown in culture conditions as recommended by ATCC.
2. Fetal Bovine Serum (FBS) (Sigma Aldrich).
3. Trypsin–EDTA (Gibco/Invitrogen).
4. Trypan blue solution (0.4%) used at 0.04% (Sigma Aldrich).
5. Phosphate Buffer Saline (PBS) (w/o Ca^{2+} and Mg^{2+}) (Lonza).
6. Assay medium, RPMI 1640 (Lonza), supplemented with 1× L-Glutamine 100× (Gibco/Invitrogen) and 1% Bovine Serum Albumin (BSA) (Sigma Aldrich).

2.2 Mononuclear Cells and NK Separation

1. Blood obtained from healthy volunteers.
2. Dilution buffer (PBS w/o Mg^{2+} and Ca^{2+} supplemented in 2% FBS and 1 mM EDTA).
3. Ficoll-Paque PLUS (GE Healthcare).
4. EasySep® Multistand with four EasySep® Magnets (STEMCELL Technologies).
5. RoboSep® Human NK Cell Enrichment kit (STEMCELL Technologies).
6. RoboSep® buffer (STEMCELL Technologies).

2.3 PBMCs and NK Staining

1. BD Tritest™ CD3/CD16+CD56/CD45 (BD Bioscience).
2. Flow cytometer BD FACS Calibur (BD Bioscience).

2.4 Target Cells and Antibodies Preparation

1. 96-well Microplate, clear, tissue-culture-treated, flat-bottom with lid (BD Falcon).
2. Deepwell plate 96/2,000 μl (VWR).
3. hIgG1 isotype control (Binding Site).
4. Positive control Ab (see Note 1).

The other necessary reagents or solutions or cell suspension have already been described or prepared in the previous sections.

2.5 Seeding of ADCC Assay Plates and LDH Release Measurement

1. Cytotoxicity Detection KitPLUS (LDH) (Roche).
2. Mithras LB 940 (Berthold Technologies).

3 Methods

3.1 Isolation of Peripheral Blood Mononuclear Cells

1. Blood samples were obtained from healthy volunteers and are manipulated few hours after their collection (see Note 2). Blood is considered as potentially infected and should be carefully manipulated according to safety rules (Gloves, laboratory coat, laminar flow hood, etc.).
2. Prepare two 50 ml Falcon tubes/donor by adding 20 ml of PBS supplemented with 2% of heat-inactivated FBS.
3. Disinfect scissors and blood bags with 70% alcohol solution under a laminar flow hood.
4. Cut the cannula of the blood bag and drain 20 ml of blood into the two 50 ml Falcon tubes/donor containing 20 ml PBS supplemented with FBS. Mix gently by inverting the tubes.
5. In parallel, prepare four 50 ml Falcon tubes/donor by adding 15 ml of Ficoll.
6. Carefully layer 20 ml of the diluted blood over the Ficoll minimizing mixing of the two phases.
7. Centrifuge the tubes at 400 × *g* for 30 min at 20°C without breaking the rotor. Blood cells are separated according a gradient density.
8. After centrifugation, handle tubes with caution. A white ring appears in the middle of each tube containing lymphocytes and monocytes. The upper layer contains platelets, whereas pellet is rich in red blood cells and granulocytes.
9. Slowly, aspirate the upper layer with a 10 ml pipette leaving the mononuclear cell layer undisturbed at the interface. Carefully transfer the mononuclear cell layer with a 5 ml pipette in a new 50 ml Falcon tube.
10. Collect two rings/Falcon tube and complete the tubes with RPMI 1640 media to 40 ml and mix by inverting.
11. Centrifuge at 250 × *g* for 10 min at 20°C with brake.
12. Carefully remove supernatant.
13. Resuspend the pellet in 10 ml of RPMI 1640 media and pool the mononuclear cell fractions from the same donor in one 50 ml Falcon tube. Repeat the wash steps 10 and 11.
14. Resuspend the pellet in 5 ml of Dilution buffer at 4°C and keep at 4°C overnight.

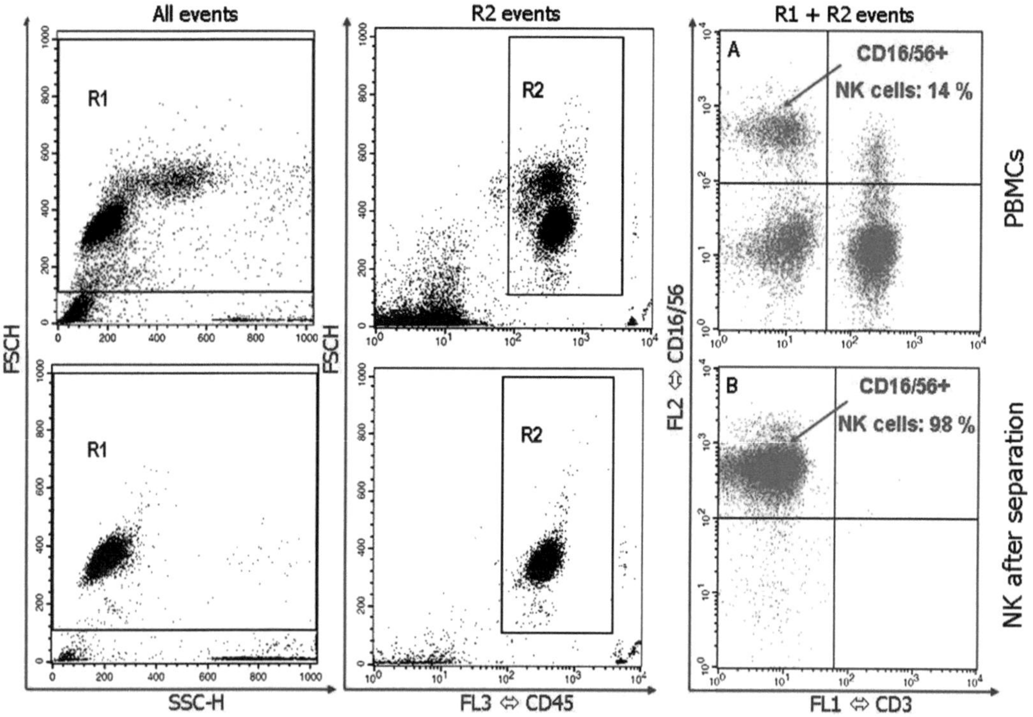

Fig. 1 PBMCs and NK cells staining for flow cytometry analysis. PBMCs and NK cells staining for flow cytometry analysis. A FSC/SSC *dot plot* is created to identify all events. Define a large gate containing all PBMCs and some debris (Gate R1). A second *dot plot* is created to separate leukocytes from cell debris using PerCP-labelled anti-CD45 antibody. The gate R2 is then defined. A third *dot plot* is created. This *dot plot* is gated on the two gates R1 and R2. (**a**) Represents the analysis of PBMCs before cell separation. In this example, the % of NK cells is 14. (**b**) Represents the analysis of purified NK cells. In this example, purity was 98%

3.2 NK Cells Separation

1. Mix the mononuclear cell pellet by gently pipetting up and down and count the suspension on a hemocytometer chamber to estimate total viable cell number using trypan blue.
2. Keep one tube containing 2×10^5 PBMCs in 400 μl of Dilution buffer at 4°C for PBMCs staining (see Subheading 3.3).
3. Adjust the concentration of each mononuclear cell suspension to 50×10^6 PBMCs/ml with the dilution buffer.
4. Pipette 2 ml of this suspension into four Falcon 5 ml polystyrene tubes.
5. Add 100 μl of EasySep® Human NK Cell Enrichment Cocktail in each tube. Mix with a 1 ml pipette and incubate at room temperature for 10 min. The cocktail contains tetrameric antibody complexes which target cells for depletion.
6. Vortex EasySep® Magnetic Particles for 30 s. Make sure that the particles form a uniform suspension with no visible aggregates. Add 200 μl of the magnetic particles and mix gently by

pipetting up and down. Incubate at room temperature for 5 min. Magnetic particles bind to labelled cells.

7. Add 200 μl of the RoboSep® buffer and mix the cells by gently pipetting up and down.
8. Place the four tubes without cap in the four magnets of the Multistand. Set aside for 2.5 min and in one constant motion invert magnets and tubes, pouring off the NK cell desired fraction into new 50 ml tubes. Leave magnets and tubes in inverted position for 2–3 s before returning to upright position. The magnetic field pulls the unwanted cells to the inner surface of the tube. NK cells are poured off and left untouched since they are not labelled with either antibodies complexes or magnetic particles.
9. Pool the four NK negative fractions from the same donor into one 50 ml tube. Keep the NK fraction at 4°C after separation.
10. Repeat separation steps 5–9 for each donor.
11. Recover the four NK fractions from each donor and centrifuge at $450 \times g$ for 5 min. Resuspend each NK cells pellet in 1 ml of assay media (RPMI + 1% BSA + 1% l-Glutamine) with a 200–1,000 μl pipette.
12. Dilute 50 μl of cell suspension in PBS and count the NK cell suspension on a hemocytometer chamber to estimate total viable effector cells number (see Note 3).
13. Keep one tube containing 2×10^5 NK cells in 400 μl for NK cell staining (see Subheading 3.3).
14. NK cell concentration is adjusted to plate the optimal effector: target (E:T) ratio (see Note 4). For a 50:1 E:T ratio, adjust NK cell concentration of each suspension to 5×10^6 NK/ml in assay media to allow each 50 μl to give 2.5×10^5 NK/well. Target cells suspension will be adjust to 1×10^5 cells/ml to allow each 50 μl to give 5×10^3 (see Subheading 3.4.6).

3.3 PBMCs and NK Staining

The staining of PBMCs and NK cells is performed to evaluate the yield and the purity of NK cell separation. NK cells are identified as $CD3^-$ and $CD16^+$ and/or $CD56^+$. CD45 is used to clearly separate leukocytes from cells debris (see Note 5).

1. Take the tubes containing 2×10^5 PBMCs in 400 μl/donor and the tubes that contain 2×10^5 NK cells in 400 μl/donor.
2. Add 20 μl of TriTEST CD3/CD16 + CD56/CD45 in each tube (BD Bioscience) and vortex gently to mix.
3. Incubate in the dark for 15 min.
4. Do not wash.
5. Analyze the samples with a flow cytometer by creating dot plots and monoparametric histogram as illustrated in Fig. 1 to evaluate the yield and NK cell purity (see Note 6).

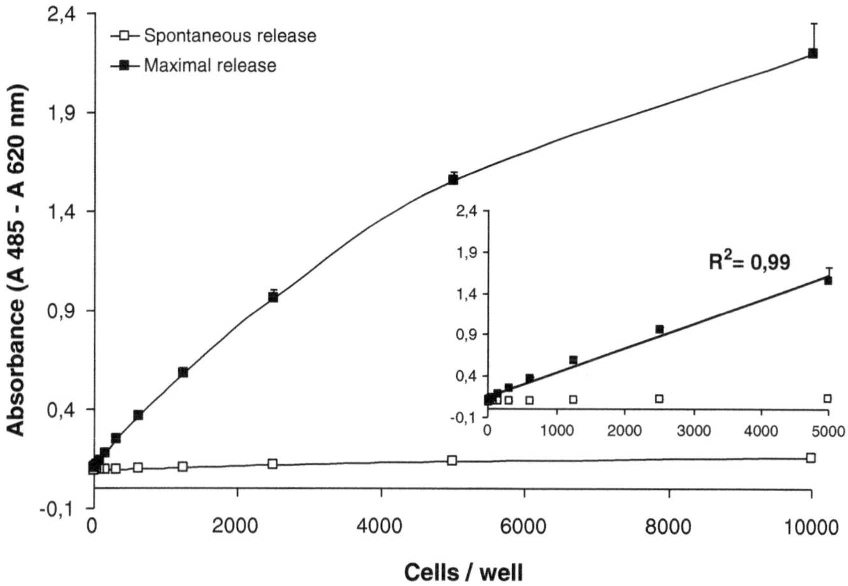

Fig. 2 Determination of the optimal target cell quantity/well. A dose range of *A*431 cells were plated in 96-multiwell plate and spontaneous and maximal LDH releases were determined. The optimal cell quantity/well should correspond to the highest difference between maximal release and spontaneous release. For this cell type, the optimal quantity was 5,000 cells/well (see *inset graph*). Linear regression analysis of the data and Pearson correlation are shown in the *inset* graph

3.4 Target Cells and Antibodies Preparation

1. Target cells are seeded at least 2 days before ADCC assay to reach 70–80% confluence.
2. On experiment day, aspirate culture medium from adherent cells growing in 162 or 75 cm^2 flask. Wash twice with 10 ml PBS, and add 2 ml of Trypsin–EDTA. Let the cell layer detach in the incubator at 37°C.
3. Inhibit the trypsin by adding 10 ml of complete media containing 10% FBS and collect cells into a 50 ml Falcon tube.
4. Spin cell suspension at 450 × *g* for 5 min at room temperature.
5. Resuspend cell pellet in 10 ml of assay media and dilute 50 μl of cell suspension in PBS and count the cell suspension on a hemocytometer chamber to estimate total viable target cell number using the trypan blue solution.
6. Add assay media to bring target cells to the optimum concentration defined (see Note 7 and Fig. 2). For *A*431 cells, optimal quantity to seed is defined at 5,000 cells/well, i.e., a 100,000 cells/ml optimal concentration.
7. Split *A*431 cell suspension in five wells of 1 ml in a 96-deep well plate.
8. In each well containing *A*431 cells, add either assay media or isotype control or specific antibody until they are twice as concentrated as the optimal concentration determined (see Note 8).

Table 1
Overview of the experimental and control conditions

	Target cells + antibodies (20 μg/ml)	Target cells + media	NK (ratio 50:1)	Assay media	Lysis solution (added after incubation)
Background control	X	X	X	100 μl	X
$SR_{\text{target cells}}$	X	50 μl	X	50 μl	X
$MR_{\text{target cells}}$	X	50 μl	X	50 μl	5 μl
$SR_{\text{NK+target cells}}$	X	50 μl	50 μl	X	X
Exp	50 μl	X	50 μl	X	X

Table 2
Experimental setup for four different blood samples

NK from blood sample 1	NK from blood sample 2	NK from blood sample 3	NK from blood sample 4
Background control	Background control	Background control	Background control
$SR_{\text{target cell}}$	$SR_{\text{target cell}}$	$SR_{\text{target cell}}$	$SR_{\text{target cell}}$
$MR_{\text{target cell}}$	$MR_{\text{target cell}}$	$MR_{\text{target cell}}$	$MR_{\text{target cell}}$
$SR_{\text{NK+target cells}}$	$SR_{\text{NK+target cells}}$	$SR_{\text{NK+target cells}}$	$SR_{\text{NK+target cells}}$
Exp herceptin antibody	Exp herceptin antibody	Exp herceptin antibody	Exp herceptin antibody
Exp isotype control antibody	Exp isotype control antibody	Exp isotype control antibody	Exp isotype control antibody

In this test, a final optimal concentration of 10 μg/ml is defined. So antibodies are diluted in deep wells to reach a 20 μg/ml concentration. Let antibodies to opsonize cells during 10 min at room temperature.

3.5 Seeding of ADCC Assay Plates and LDH Release Measurement

3.5.1 Seeding ADCC Assay Plates

1. In a sterile 96-well flat bottom plate, seed 250,000 NK in 50 μl/well. Prepare all test samples in triplicate.
2. To each well containing effector cells, add 50 μl of target cell suspension treated with either assay media, isotype control, or tested antibodies.
3. All following conditions have to be performed in triplicate (see Tables 1 and 2):
   (a) Background control (BC): Measure the LDH activity contained in the assay media (see Note 9). The absorbance

value of background control has to be subtracted from all other values.

(b) Spontaneous release ($SR_{target\ cells}$): Measure the spontaneous LDH release from untreated target cells.

(c) Maximal release ($MR_{target\ cells}$): Measure the maximum releasable LDH activity in the target cells.

(d) Spontaneous release of the mix ($SR_{target\ cells+NK}$): Measure the spontaneous LDH release from untreated target cells and effector cells.

(e) Experimental conditions (Exp): Measure the LDH release of effector cells and treated target cells with different antibodies. For each tested antibody, negative and positive antibody controls have to be added to the assay.

(f) Another control condition could be added to determine LDH release from NK cells incubated with antibodies. It will be notably interesting in the case of the antibody targeted antigen is expressed on NK cells.

4. Incubate the plates for 4 h in an incubator (37°C, 5% CO_2).

3.5.2 LDH Release Measurement

The Cytotoxicity Detection KitPLUS (LDH) marketed by Roche is a colorimetric assay for quantify cytotoxicity by measuring LDH activity released from damaged cells. LDH activity is determined in an enzymatic test and is proportional to the number of lysed cells.

1. Thaw all the reagents of the kit: The catalyst, the dye solution, the lysis solution and the stop solution. Reconstitute the lyophilized catalyst (blue cap) by adding 1 ml of distilled water with the 200–1,000 μl pipette. Let it dissolve for 10 min, then mix thoroughly. Shortly before use, mix 250 ml of reconstituted catalyst with 11.25 ml of Dye Solution (red cap) for 100 conditions. Do not store this reaction mixture.
2. Recover plates from the incubator.
3. To wells that contain Maximal release controls, add 5 μl of Lysis solution (white cap) with the multichannel pipette (5–50 μl) and homogenize.
4. Incubate the plate for 15 min in the incubator.
5. To determine LDH activity, add 100 μl of freshly prepared reaction mixture by reverse pipetting with the multichannel pipette (50–300 μl).
6. Incubate plates protected from light for 30 min at room temperature.
7. Add 50 μl of Stop solution (green cap) to each well by reverse pipetting with the multichannel pipette (50–300 μl).
8. Measure absorbance of samples at 485 and 620 nm with a microplate reader.

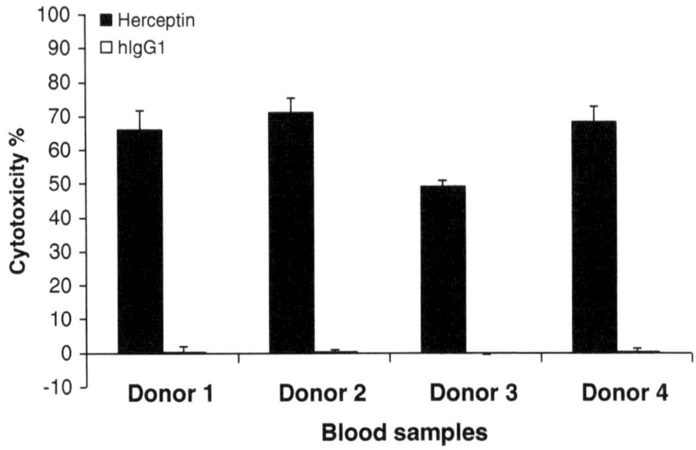

Fig. 3 Antibody-dependent cell cytotoxicity was induced by Herceptin. *A*431 cells were incubated with either hIgG1 isotype control or Herceptin in the presence of purified NK cells (E:T = 50:1) for 4 h at 37°C. No ADCC was observed with the isotype control hIgG1, whereas Herceptin® induced a strong ADCC on the four donors. Each condition was made in triplicate. Results represent cytotoxicity mean ± SD calculated from triplicate

3.6 Data Processing

1. The measurement at 620 nm is used as a reference wavelength. Therefore, to all 485 nm measurements, corresponding 620 nm measurement is subtracted.
2. To determine the cytotoxicity percentage, calculate the average absorbance values of the triplicate samples and controls. Then subtract from each of these, the absorbance value obtained in the background control. Finally, substitute the resulting values in the following equation:

$$\text{Cytotoxicity}(\%) = \frac{\text{Exp} - \left(\text{SR}_{\text{NK+target cells}}\right)}{\text{MR}_{\text{target cells}} - \text{SR}_{\text{target cells}}} \times 100.$$

3. An example of the cytotoxic effect of Herceptin® performed with NK cells isolated from four healthy volunteer donors is represented in Fig. 3. The variation between triplicates was less than 5% and no cytotoxic effect of the isotype control hIgG1 was obtained.

4 Notes

1. Here Herceptin® was used as a positive control. However, another mAb should be selected depending on cells used as target cells. An IgG4 isoform of the same Ab could be used as a negative control. Another elegant alternative should be the same IgG1 isotype mutated in position L234 or L235.

2. Blood samples are obtained from Blood Center of Geneva. Buffy coats are prepared from healthy donor blood taken in the morning, and are delivered in the afternoon. Isolation of mononuclear cells is immediately performed after the receipt to get good mononuclear cell isolation.
3. Since dead effector cells release LDH in the medium, high levels of dead NK cells bring an enhanced background and reduce the efficacy of ADCC. The viability of effector cells should be greater than 90% to avoid any interference with the assay.
4. The optimal effector:target (E:T) ratio is determined by using different E:T ratios of 100–6.25 keeping constant the number of target cells and antibody concentration. The optimal E:T ratio is chosen when a plateau is reached. Optimal E:T ratio defined for *A*431 cells was 50:1.
5. The TriTEST CD3/CD16+CD56/CD45 reagent from Becton Dickinson company contains a FITC-labelled an anti-CD3, a PE-CD16/CD56 antibody, and a PerCP-labelled anti-CD45 antibody. This TriTest reagent can be used with a standard flow cytometer only equipped with an argon laser. However, before flow cytometry analysis, it is important to check instrument sensitivity and determine the best instrument settings before use and the best setting for fluorescence compensation. When appropriate settings are defined, isotype control reagents are no longer needed.
6. NK cells were always used when purity was greater than 95% to avoid any interference of other leukocytes with ADCC. For example, monocytes were described to significantly reduce ADCC due to shaving of the targeted receptor by the mAb (23).
7. For each target cell type, the optimal cell concentration should be determined since the amount of LDH depends on the cell type. Prepare a dose range of target cells and plate each dilution into six wells. For each cell dilution, designate three wells as spontaneous release and three wells as maximal release. To the cell dilutions designated as maximal release, add 5 μl/well of Lysis solution. Incubate the plate for 15 min in an incubator and determine the LDH activity as described in Subheading 3.6. The assay has to show a linear relationship between cell number and maximal LDH released (see Fig. 2). The optimal cell concentration presents the greatest difference between the maximal and the spontaneous LDH release. The optimal quantity of target cells has to be lower than 10,000 cells/well to limit the NK cell quantity to be added. Finally, the maximal LDH release of the optimal concentration should be between 1 and 2 absorbance to keep the same test sensitivity whatever cell line used.

8. A titration curve with the antibody should be determined by flow cytometry since the expression level of the target receptor depends on the cell type. Mostly, a 10 μg/ml final concentration is used in order to add antibodies in excess and thus saturate specific cell binding sites.
9. Both human and animal sera contain various amounts of LDH, which may increase background absorbance in the assay. Therefore, the assay is performed in media supplemented with 1% BSA. Verify also the interference of the culture media on the assay since some culture media inhibit LDH reaction. For instance do not use culture media that contains pyruvate.

References

1. Glennie MJ, French RR, Cragg MS, Taylor RP (2007) Mechanisms of killing by anti-CD20 monoclonal antibodies. Mol Immunol 44:3823–3837
2. Cragg MS, Walshe CA, Ivanov AO, Glennie MJ (2005) The biology of CD20 and its potential as a target for mAb therapy. Curr Dir Autoimmun 8:140–174
3. Oflazoglu E, Audoly LP (2010) Evolution of anti-CD20 monoclonal antibody therapeutics in oncology. MAbs 2:14–19
4. Galizia G, Lieto E, De Vita F, Orditura M, Castellano P, Troiani T et al (2007) Cetuximab, a chimeric human mouse anti-epidermal growth factor receptor monoclonal antibody, in the treatment of human colorectal cancer. Oncogene 26:3654–3660
5. Pescovitz MD (2006) Rituximab, an anti-cd20 monoclonal antibody: history and mechanism of action. Am J Transplant 6:859–866
6. Ravandi F, O'Brien S (2005) Alemtuzumab. Expert Rev Anticancer Ther 5:39–51
7. Watanabe M, Wallace PK, Keler T, Deo YM, Akewanlop C, Hayes DF (1999) Antibody dependent cellular phagocytosis (ADCP) and antibody dependent cellular cytotoxicity (ADCC) of breast cancer cells mediated by bispecific antibody, MDX-210. Breast Cancer Res Treat 53:199–207
8. Correia IR (2010) Stability of IgG isotypes in serum. MAbs 2:221–232
9. Yoo EM, Yu LJ, Wims LA, Goldberg D, Morrison SL (2010) Differences in N-glycan structures found on recombinant IgA1 and IgA2 produced in murine myeloma and CHO cell lines. MAbs 2:320–334
10. Liu XY, Pop LM, Vitetta ES (2008) Engineering therapeutic monoclonal antibodies. Immunol Rev 222:9–27
11. Ravetch JV, Bolland S (2001) IgG Fc receptors. Annu Rev Immunol 19:275–290
12. Bryceson YT, March ME, Ljunggren HG, Long EO (2006) Activation, coactivation, and costimulation of resting human natural killer cells. Immunol Rev 214:73–91
13. Nimmerjahn F, Ravetch JV (2008) Fcgamma receptors as regulators of immune responses. Nat Rev Immunol 8:34–47
14. Cartron G, Dacheux L, Salles G, Solal-Celigny P, Bardos P, Colombat P et al (2002) Therapeutic activity of humanized anti-CD20 monoclonal antibody and polymorphism in IgG Fc receptor FcgammaRIIIa gene. Blood 99:754–758
15. Bruhns P, Iannascoli B, England P, Mancardi DA, Fernandez N, Jorieux S et al (2009) Specificity and affinity of human Fcgamma receptors and their polymorphic variants for human IgG subclasses. Blood 113:3716–3725
16. Jefferis R (2009) Recombinant antibody therapeutics: the impact of glycosylation on mechanisms of action. Trends Pharmacol Sci 30:356–362
17. Bryceson YT, Chiang SC, Darmanin S, Fauriat C, Schlums H, Theorell J et al (2011) Molecular mechanisms of natural killer cell activation. J Innate Immun 3:216–226
18. Weng WK, Levy R (2003) Two immunoglobulin G fragment C receptor polymorphisms independently predict response to rituximab in patients with follicular lymphoma. J Clin Oncol 21:3940–3947
19. Beck A, Wurch T, Bailly C, Corvaia N (2010) Strategies and challenges for the next generation of therapeutic antibodies. Nat Rev Immunol 10:345–352
20. Lazar GA, Dang W, Karki S, Vafa O, Peng JS, Hyun L et al (2006) Engineered antibody Fc variants with enhanced effector function. Proc Natl Acad Sci U S A 103:4005–4010
21. Okazaki A, Shoji-Hosaka E, Nakamura K, Wakitani M, Uchida K, Kakita S et al (2004)

Fucose depletion from human IgG1 oligosaccharide enhances binding enthalpy and association rate between IgG1 and FcgammaRIIIa. J Mol Biol 336:1239–1249
22. Shields RL, Lai J, Keck R, O'Connell LY, Hong K, Meng YG et al (2002) Lack of fucose on human IgG1 N-linked oligosaccharide improves binding to human Fcgamma RIII and antibody-dependent cellular toxicity. J Biol Chem 277:26733–26740
23. Beum PV, Lindorfer MA, Taylor RP (2008) Within peripheral blood mononuclear cells, antibody-dependent cellular cytotoxicity of rituximab-opsonized Daudi cells is promoted by NK cells and inhibited by monocytes due to shaving. J Immunol 181:2916–2924

Chapter 20

Evaluation of Complement-Dependent Cytotoxicity Using ATP Measurement and C1q/C4b Binding

Lucile Broyer, Liliane Goetsch, and Matthieu Broussas

Abstract

One of the major issues for antibody treatment is enhancement of efficacy. Recent studies have highlighted the important role of effector functions in improvement of antibody therapy. Among effector functions, complement-dependent cytotoxicity (CDC), which induces cell lysis by a cascade of activation triggered by the binding of C1q subunits to the Fc regions of antibodies bound to the cell surface, is part of the mechanism of several antibody therapies. CDC can be modulated by either Fc isotype engineering or Fc genetic mutations or Fc glycosylation profile modifications. To evaluate the impact of such modifications on CDC, we describe a luminescence method based on ATP measurement to estimate tumor damaged cells and a flow cytometry method to evaluate the binding of C1q on the Fc region and the binding of C4b on cell surface. The luminescence method coupled with complement protein analysis by flow cytometry encompasses all needed methods to evaluate antibody ability to trigger CDC.

Key words CDC, Complement-dependent cytotoxicity, ATP measurement, C1q, C4b

1 Introduction

Monoclonal antibodies (mAbs) represent a third of all new pharmaceutical treatments and could become prevalent products over the next decade. At that time, about 30 mAbs were approved by the Food and Drug Administration (FDA) while over 300 new drugs are currently in clinical trials. mAbs are largely used for cancer therapy and treatment of autoimmune and inflammatory diseases. The clinical efficacy of mAbs often results from their ability to recognize specific molecular targets on tumor cells or in tumor microenvironment (1). This antibody specificity triggers direct effects either on tumor growth, such as tumor cells apoptosis, or on inhibition of proliferation. mAbs can also induce indirect effects driven by activation of immune-mediated effector functions including antibody dependent cell cytotoxicity (ADCC), antibody dependent cell phagocytosis (ADCP), and complement-dependent cytotoxicity (CDC) (2).

Alain Beck (ed.), *Glycosylation Engineering of Biopharmaceuticals: Methods and Protocols*, Methods in Molecular Biology, vol. 988, DOI 10.1007/978-1-62703-327-5_20, © Springer Science+Business Media New York 2013

CDC is a cytolytic cascade mediated by a series of complement proteins C1 to C9 that activate binding of C1q to the constant fragment (Fc) region bound to the cell surface. Antibody ability to bind complement varies with the Ig isotypes. Interactions with IgM, IgG1, and IgG3 are strong while IgG2 is a poor inductor of complement activation. IgG4 is devoid of any effector function. Furthermore, IgA, IgD, and IgE are not able to induce CDC (3–5). The complement classical pathway is initiated by the interaction of at least two Fc antibodies-opsonized cells with the C1q, first complement component. This causes conformational changes permitting to C1r and C1s components to bind C1q and cleave C4 molecules resulting in C4a and C4b fragments. C4a diffuses away while C4b covalently binds either to cell membrane or to antibody through an ester or amide linkage. C1 complex also cleaves C2 resulting in C2b and C2a, able to fix C4b to form C3 convertase. This enzyme catalyzes degradation of C3 into C3a and C3b. This latter component binds to C4bC2a complex and form C5 convertase to induce C5 cleavage into C5a and C5b. C5b binds to the cell membrane and initiates the formation of the membrane attack complex (MAC) composed of the terminal complement components C5b to C9. This last complex is able to insert into the cell lipid bilayer leading to pore formation in cell membrane and cell destruction.

Many studies support the role of complement activation in patients antibody immunotherapy (6–12). Optimization of IgG1 antibodies for complement becomes an important key to improve efficacy of the next generation of therapeutic antibodies. Different approaches are described to enhance CDC activity of mAbs. A first approach to enhance CDC deals with the engineering of the heavy chain by shuffling Fc fragment portions of both IgG1 and IgG3 to form IgG1/IgG3 chimeric isotype antibodies. Several variants expressing a set of IgG1/IgG3 mixed sequences show a highest CDC activity than the parental wild-type antibodies (3, 13). In addition, mutations inserted into the polypeptide chain of either CH2 domain, that constitutes the C1q binding epicenter, or hinge region could result in antibodies with enhanced C1q binding (4, 9). The complement activation could also be modulated by changing the glycosylation profile of Fc region. The presence of terminal galactose has been demonstrated to affect C1q binding. Degalactosylation of antibody Fc fragment could significantly reduce antibody complement activation. (14, 15).

Effector functions, particularly CDC, could easily be controlled either by selection of the immunoglobulin isotype or by engineering of Fc fragment or by mutations in CH2 domain or hinge region polypeptide chains. Optimization of Fc fragment to enhance effector functions is a promising way to improve efficiency of antibody therapies (13). Here, we first describe a nonradioactive method to evaluate the activation of complement by measuring viable cells and a flow cytometry protocol to analyze the binding of both C1q and C4b on mAbs bound on cell surface.

2 Materials

2.1 Cell Culture

1. Human cancer cells obtained from the American Type Culture Collection (ATCC) and grown in culture conditions as recommended by ATCC.
2. Fetal Bovine Serum (FBS) (Sigma Aldrich).
3. Trypsine-EDTA (Gibco/Invitrogen).
4. Trypan blue solution (0.4%) used at 0.04% (Sigma Aldrich).
5. Phosphate Buffer Saline (PBS) (w/o Ca^{2+} and Mg^{2+}) (Lonza).

2.2 Complement-Dependent Cytotoxicity Assay

1. Culture plate—96-well white opaque (Perkin Elmer).
2. Deepwell plate 96/2,000 μl (VWR).
3. Human Serum from human male AB (Sigma Aldrich).
4. CellTiter Glo® Luminescent Cell Viability Assay (Promega).
5. Microplate reader: Mithras LB 940 (Berthold Technologies).
6. Assay medium: RPMI 1640 (Lonza), supplemented with 1× l-Glutamine 100× (Gibco/Invitrogen) and 1% Bovine Serum Albumin (BSA) (Sigma Aldrich).

2.3 C1q and C4b Binding Assays

1. 96-Well Microplate, round-bottom, tissue culture, sterile, lid (BD Falcon).
2. FACS Buffer: PBS Dulbecco w/o Ca^{2+} w/o Mg^{2+} 1× (Biochrom), 1% BSA, 0.01% sodium azide (Sigma Aldrich).
3. C1q protein (CompTech).
4. Human Serum from human male AB (Sigma Aldrich).
5. Polyclonal rabbit anti-human C1q/FITC (Dako).
6. Polyclonal rabbit anti-human C4c/FITC (Reacts with C4, C4b, and C4c) (Dako).
7. Propidium iodide (1 mg/ml) used at 20 μg/ml (Interchim).

The other necessary materials, reagents, or cell suspensions are already described or prepared in the previous sections.

3 Methods

3.1 Complement-Dependent Cytotoxicity Assay

3.1.1 Antibodies Preparation

1. For each assay, a negative control (usually an irrelevant antibody) and a positive control have to be included in the test in parallel with the tested antibody.
2. For each antibody, prepare 1 ml of a stock solution at 33 μg/ml in assay media. This solution will be diluted later to reach a 10 μg/ml concentration in the assay.

Table 1
Overview of the experimental and control conditions

	Antibodies (33 μg/ml)	Assay media	Target cells	AB human serum
Untreated target cells	*X*	30 μl	50 μl	20 μl
Exp antibody	30 μl	*X*	50 μl	20 μl

3. Perform a series of dilution by using halving dilutions of each antibody. Make this 1:2 dilution series in assay media with either the antibody to be tested, an irrelevant antibody, or a positive antibody. Use a 96-deep well plate to perform the dilutions.
4. Transfer each antibody concentration by plating 30 μl of solution in a white 96-well flat-bottom plate. Perform all the samples and controls conditions in triplicate (see Tables 1 and 2).

3.1.2 Target Cells Preparation

1. Target cells are seeded at least 2 or 3 days before CDC assay (see Note 1).
2. On the day of experiment, aspirate medium from adherent cells growing in 162 or 75 cm^2 flasks. Wash twice with 10 ml PBS and add 2 ml of pre-warmed trypsin. Let the cell layer detach in the incubator at 37°C.
3. Inhibit the trypsin by adding 10 ml of complete media containing 10% FBS and collect cells into a 50 ml Falcon tube.
4. Spin cell suspension at 450 × *g* for 5 min at room temperature. Discard the supernatant.
5. Resuspend cell pellet in 10 ml of assay media.
6. Count cell suspension on a hemocytometer chamber to estimate total viable target cell number using the trypan blue solution.
7. Adjust the cell concentration in assay media to 2×10^5 cells/ml. Plate 50 μl of the cell suspension (1×10^4 cells/well) in each well containing either antibodies or assay media (see Note 2). Prepare wells with media and without cells to determine the luminescence background.
8. Incubate cells with antibodies or assay media for 10 min at 4°C (see Note 3).

3.1.3 Seeding of CDC Assay Plates

1. Thaw AB human serum on ice and keep on ice until use. Avoid repeated freezing and thawing cycles.
2. AB human serum from either male or female is used at 20% final concentration (see Note 4). Add 20 μl of serum in each

Table 2
Experimental setup

Ab (µg/ml)	Isotype control antibody	Positive control antibody	Antibody of interest
10			
5			
2.5			
1.25			
6.25E−01			
3.13E−01			
1.56E−01			
0	Untreated target cells		
Background luminescence	Media without cells		

well containing untreated or antibody treated target cells and homogenize (see Note 5). The final volume of the assay is then 100 μl (30 μl of antibody + 50 μl of cells + 20 μl of serum).

3. All following conditions have to be performed in triplicate (see Tables 1 and 2):
   (a) Untreated target cells: Measure the amount of ATP naturally presents in target cells.
   (b) Experimental conditions (Exp): Measure the ATP amount remaining in antibody treated target cells after incubation.
   (c) Background: Measure luminescence in wells containing 100 μl of the media assay.
4. Incubate the plate for 1 h in an incubator (37°C, 5% CO_2).

3.1.4 ATP Measurement

The CellTiter-Glo® Luminescent assay from Promega is used to determine the number of viable cells. This assay is based on the quantification of ATP, which signals the presence of metabolically active cells. The mono-oxygenation of luciferin is catalyzed by the luciferase in the presence of notably ATP, resulting in the production of light. This signal is detected and directly proportional to the viable cell number.

1. Thaw CellTiter-Glo® buffer and substrate, and equilibrate at room temperature before use.
2. Transfer 10 ml of the buffer in the lyophilized substrate reagent to reconstitute enzyme mixture. Mix gently by inversion of the tube.
3. Recover the plate from the incubator.
4. Add 100 μl of CellTiter Glo® reagent by reverse pipetting with a multichannel pipette (50–300 μl).
5. Agitate contents for 2 min to induce cell lysis.
6. Allow the plate to incubate at room temperature for 10 min to stabilize luminescent signal.
7. Record luminescence with the luminometer.

3.1.5 Data Processing

1. To determine the cytotoxicity percentage, calculate the average bioluminescence values of the triplicate samples and controls.
2. Calculations are performed using:

$$\text{Cytotoxicity }(\%) = 100 - \frac{(\text{Experimental condition} - \text{Background})}{(\text{Untreated target cells} - \text{Background})} \times 100$$

3. An example of complement-dependent cytotoxic effect induced by a dose range of Rituximab® is presented in Fig. 1.

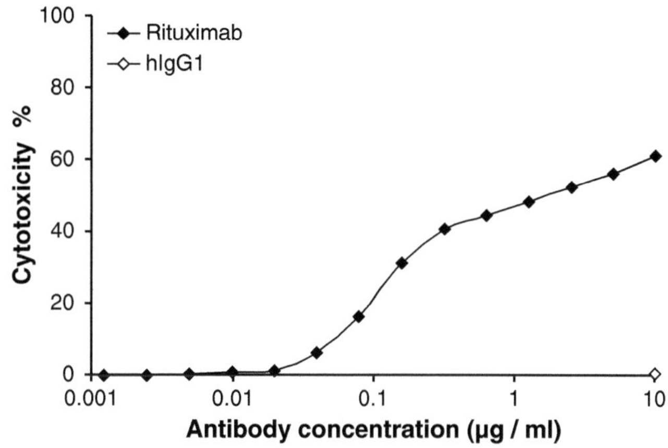

Fig. 1 Complement-dependent cytotoxicity of anti-CD20 antibody was induced by AB human serum (20%) against RAMOS cell line. After 1 h incubation, ATP amount in living cells is measured. % of cytotoxicity was calculated as described in the Subheading 3.1.5

3.2 C1q and C4b Binding Assays

C1q is the first component of the cascade that binds to the Fc fragment of antibody. Once bound to the antibody, C1q leads to C4b formation that binds the cell surface. The binding of either C1q or C4b can be evaluated by flow cytometry.

3.2.1 Antibodies Preparation and Target Cell Preparation

1. For each assay, a negative control (usually an irrelevant antibody) and a positive control have to be included in the test in parallel with the tested antibody.
2. For each antibody, prepare 1 ml of a stock solution at 40 µg/ml in assay media. Perform a serial dilution of the antibodies as described in Subheading 3.1.1.
3. Transfer each antibody concentration by plating 25 µl in a 96-multiwell plate.
4. Prepare a cell suspension at 4×10^5 cells/ml as described in Subheading 3.1.2.
5. Plate 25 µl of treated target cells in each well containing antibodies and homogenize.
6. Let antibodies opsonize target cells for 10 min on ice (see Note 3).

3.2.2 C1q Binding Assay

1. Thaw exogenous C1q on ice and keep on ice until use. Avoid repeated freezing and thawing cycles.
2. Exogenous C1q is used at 70 µg/ml final concentration (see Note 6). Dilute exogenous C1q in assay media to reach a 140 µg/ml concentration. Add 50 µl of exogenous C1q/well in all wells containing antibody-opsonized target cells and homogenize. At this stage, antibodies used for targeting cells are diluted fourfold.

3. Incubate plates at 37°C for 15 min in the incubator.
4. Recover plates and centrifuge at 450×*g* for 2 min at 4°C in a centrifuge. Throw supernatant away by inversing the plate and wash each with 100 μl of cold FACS buffer.
5. Repeat step 6 once.
6. Resuspend pellets in 100 μl of FACS buffer containing 10 μg/ml of the anti-C1q antibody (see Note 7).
7. Incubate the plate at 4°C for 20 min.
8. Recover plates and centrifuge at 450×*g* for 2 min at 4°C in a centrifuge. Throw supernatant away by inversing plate and wash each well with 100 μl of cold FACS buffer.
9. Repeat step 6 twice.
10. Resuspend pellets in 100–200 μl cold FACS buffer.
11. Leave at 4°C until cytometry analysis.

3.2.3 C4b Binding Assay

1. Prepare antibodies and cells as described in Subheading 3.2.1.
2. The source of C4b comes from AB serum. AB human serum is used at 20% final concentration. Dilute serum in assay media to reach 40% concentration.
3. Add 50 μl of the diluted serum in each well containing antibodies and homogenize.
4. Incubate plates at 37°C for 10–15 min maximum in the incubator.
5. Proceed as described in steps 4 and 5, Subheading 3.2.2.
6. Resuspend pellets in 100 μl of FACS buffer containing 10 μg/ml of the anti-C4b antibody (see Note 7).
7. Proceed as described in steps 7–11, Subheading 3.2.2.

3.2.4 Flow Cytometry Analysis

1. Add 10 μl of propidium iodide (PI) at 20 μg/ml in each well.
2. Analyze each condition with a flow cytometer to evaluate FITC fluorescence intensity directly correlated to the C1q or C4b binding. Create dot plots and monoparametric histogram as illustrated in Fig. 2 (see Note 8).

Mean of fluorescence intensity is noted for each well and reported in a graph as illustrated in Fig. 3. As previously described, C1q or C4b binding is not maximal with 10 μg/ml of anti-CD20 antibody (3, 16).

4 Notes

1. The expression of the targeted protein could vary during cell culture. For example, CD20 expression is higher after 2 days of culture than after 3–4 days. Since expression level of the receptor

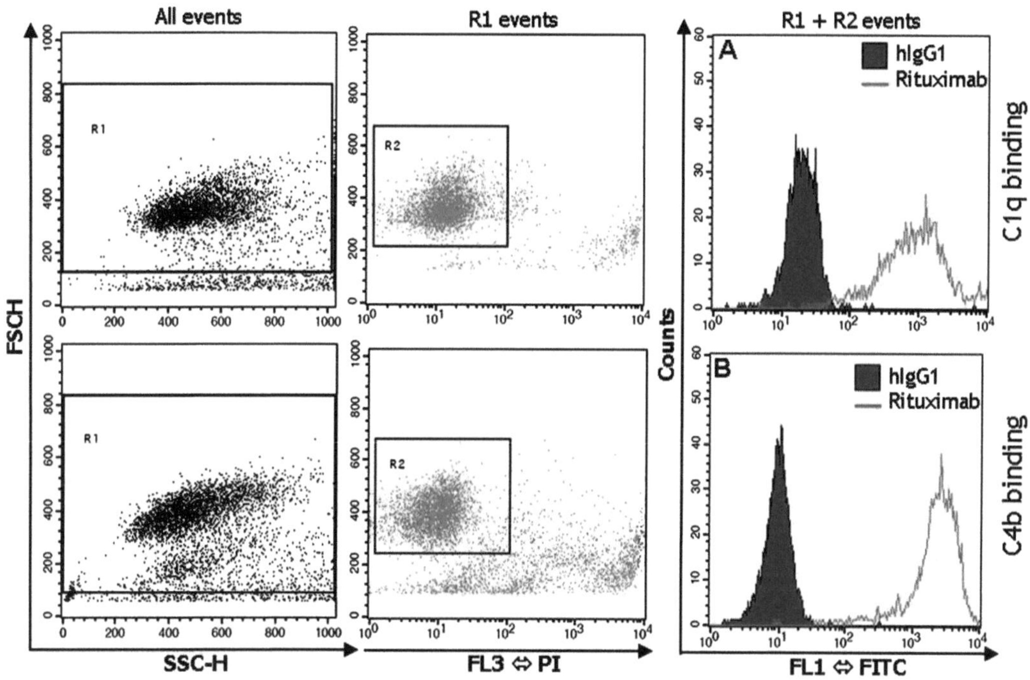

Fig. 2 C1q and C4b binding were analyzed by flow cytometry. Cells are opsonized with antibodies, Exogenous C1q or AB serum is added, and binding of both C1q and C4b is detected with the corresponding antibody. A first FSC/SSC *dot plot* is created to identify all events and a R1 gate is defined to select cells and eliminate some debris. The second *dot plot* FSC/FL3 gated on R1 and R2 is created to separate living cells from dead cells and debris. Finally, Fluorescence intensity was evaluated on gate R1 + R2. The Mean of Fluorescence Intensity (MFI) represents binding of either C1q (**A**) or C4b (**B**) on target cells. Antibodies hIgG1 and Rituximab are used at 10 μg/ml

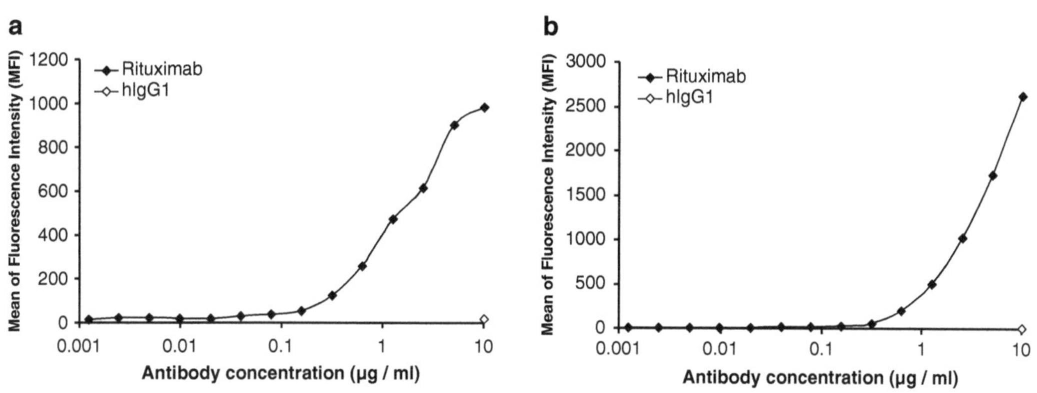

Fig. 3 C1q (**a**) and C4b (**b**) binding assays. Anti-CD-20 antibody, Rituximab is tested in dose range. hIgG1, isotype control, is used at 10 μg/ml. Antibodies are opsonized on RAMOS target cells, after addition of exogenous C1q or AB serum, C1q and C4b bindings are detected by flow cytometry

can influence the % of cytotoxicity, the assay has to be performed each time with the same cell culture conditions to minimize variations between experiments.

2. For each target cell type, the optimum cell concentration should be determined since the amount of ATP depends on the cell type. Prepare a dose range of target cells and plate each dilution. The assay has to show a linear relationship between cell number and the ATP content.
3. Some antibodies can induce receptor internalization during the opsonization step. Thus, this step has to be performed on ice to avoid internalization and a reduced cell cytotoxicity.
4. The percentage of AB serum has to be determined in a previous experiment. Use a series of serum dilution to determine the percentage of AB serum that gives the highest cell cytotoxicity in presence of the positive control. No cytotoxicity should be observed in the well that contains the irrelevant antibody.
5. In order to validate the role of the complement in the assay, human AB serum can be decomplemented by heating it at 56°C for 30 min. After filtration, use it as a negative control.
6. The optimal concentration of C1q has to be evaluated in a preliminary experiment using serial dilutions. In our hands, the optimal concentration of exogenous C1q is 70 μg/ml.
7. Anti-C1q and anti-C4b antibody concentration has to be previously determined using a serial dilution of the antibody. Usually, a 10 μg/ml antibody concentration is a good starting point.
8. Create a first FSC/SSC dot plot analyzing all events. A G1 gate is made to eliminate most of debris. The second dot plot FSC/FL3 is gated on R1. FL3 canal detects PI incorporated in damaged cells, this plot allows selecting only living cells with the gate R2. PI stained cells are not considered for the analysis as they can bind C1q and C4b nonspecifically. Finally create an histogram, gated on R1 and R2, to estimate FITC fluorescence intensity. Mean of fluorescence intensity (MFI) detected is directly proportional to C1q or C4b binding.

References

1. Strome SE, Sausville EA, Mann D (2007) A mechanistic perspective of monoclonal antibodies in cancer therapy beyond target-related effects. Oncologist 12:1084–1095
2. McEarchern JA, Oflazoglu E, Francisco L, McDonagh CF, Gordon KA, Stone I et al (2007) Engineered anti-CD70 antibody with multiple effector functions exhibits in vitro and in vivo antitumor activities. Blood 109:1185–1192
3. Natsume A, In M, Takamura H, Nakagawa T, Shimizu Y, Kitajima K et al (2008) Engineered antibodies of IgG1/IgG3 mixed isotype with enhanced cytotoxic activities. Cancer Res 68:3863–3872
4. Moore GL, Chen H, Karki S, Lazar GA (2010) Engineered Fc variant antibodies with enhanced ability to recruit complement and mediate effector functions. MAbs 2:181–189

5. Daha NA, Banda NK, Roos A, Beurskens FJ, Bakker JM, Daha MR et al (2011) Complement activation by (auto-) antibodies. Mol Immunol 48:1656–1665
6. Cragg MS, Morgan SM, Chan HT, Morgan BP, Filatov AV, Johnson PW et al (2003) Complement-mediated lysis by anti-CD20 mAb correlates with segregation into lipid rafts. Blood 101:1045–1052
7. Di Gaetano N, Cittera E, Nota R, Vecchi A, Grieco V, Scanziani E et al (2003) Complement activation determines the therapeutic activity of rituximab in vivo. J Immunol 171: 1581–1587
8. Golay J, Zaffaroni L, Vaccari T, Lazzari M, Borleri GM, Bernasconi S et al (2000) Biologic response of B lymphoma cells to anti-CD20 monoclonal antibody rituximab in vitro: CD55 and CD59 regulate complement-mediated cell lysis. Blood 95:3900–3908
9. Idusogie EE, Wong PY, Presta LG, Gazzano-Santoro H, Totpal K, Ultsch M et al (2001) Engineered antibodies with increased activity to recruit complement. J Immunol 166: 2571–2575
10. Manches O, Lui G, Chaperot L, Gressin R, Molens JP, Jacob MC et al (2003) In vitro mechanisms of action of rituximab on primary non-Hodgkin lymphomas. Blood 101:949–954
11. Reff ME, Carner K, Chambers KS, Chinn PC, Leonard JE, Raab R et al (1994) Depletion of B cells in vivo by a chimeric mouse human monoclonal antibody to CD20. Blood 83: 435–445
12. Zent CS, Secreto CR, LaPlant BR, Bone ND, Call TG, Shanafelt TD et al (2008) Direct and complement dependent cytotoxicity in CLL cells from patients with high-risk early-intermediate stage chronic lymphocytic leukemia (CLL) treated with alemtuzumab and rituximab. Leuk Res 32:1849–1856
13. Kubota T, Niwa R, Satoh M, Akinaga S, Shitara K, Hanai N (2009) Engineered therapeutic antibodies with improved effector functions. Cancer Sci 100:1566–1572
14. Hodoniczky J, Zheng YZ, James DC (2005) Control of recombinant monoclonal antibody effector functions by Fc N-glycan remodeling in vitro. Biotechnol Prog 21:1644–1652
15. Boyd PN, Lines AC, Patel AK (1995) The effect of the removal of sialic acid, galactose and total carbohydrate on the functional activity of Campath-1H. Mol Immunol 32:1311–1318
16. Pawluczkowycz AW, Beurskens FJ, Beum PV, Lindorfer MA, van de Winkel JG, Parren PW et al (2009) Binding of submaximal C1q promotes complement-dependent cytotoxicity (CDC) of B cells opsonized with anti-CD20 mAbs ofatumumab (OFA) or rituximab (RTX): considerably higher levels of CDC are induced by OFA than by RTX. J Immunol 183:749–758

Chapter 21

Capture of the Human IgG1 Antibodies by Protein A for the Kinetic Study of h-IgG/FcγR Interaction Using SPR-Based Biosensor Technology

Thierry Champion and Alain Beck

Abstract

Surface plasmon resonance (SPR) is a key technology to evaluate IgG effector functions in vitro involving binding to Fcgamma receptors (FcγR). We describe here the chemical coupling of protein A to a sensor chip. IgGs prepared either from purified antibodies or from crude cell media supernatants are captured by the protein A and are used as the ligand. Soluble forms of FcγRs are injected at different concentrations and are used as the analyte. This setup allows the definition of the kinetic rates of binding of the FcγRI (high affinity) or the FcγRIIIa (low affinity) on the Fc domain of IgGs.

Key words ADCC, Biacore X100, Fcgamma receptor, IgG effector functions, Protein A capture, Surface plasmon resonance

1 Introduction

Immune system effector cells actively lyse target cells that have been bound by specific antibodies. This mechanism of cell-mediated immunity is called Antigen Dependent Cell-mediated Cytotoxicity (ADCC). The effector cells express at their surface Fcgamma receptors (FcγR) which belong to the immunoglobulin superfamily (1, 2). This family includes several members such as FcγRI (CD64), FcγRIIA (CD32), FcγRIIB (CD32), FcγRIIIA (CD16a), and FcγRIIIB (CD16b). All, except the FcγRI, have an extracellular portion composed of two immunoglobulin (Ig)-like domains. FcγRI presents a third Ig-like domain and binds to IgGs more strongly than FcγRII or FcγRIII. This property allows activation of FcγRI by a single IgG molecule, while the latter two Fcγ receptors have to bind multiple IgG molecules within an immune complex to be activated. Typically ADCC involves activation of natural killer (NK) cells, these cells express at their surface the FcγRIIIa Fc receptors which play a central role in this activation.

Alain Beck (ed.), *Glycosylation Engineering of Biopharmaceuticals: Methods and Protocols*, Methods in Molecular Biology, vol. 988, DOI 10.1007/978-1-62703-327-5_21, © Springer Science+Business Media New York 2013

For many therapeutic antibodies used in oncology such as rituximab, trastuzumab, alemtuzumab, cetuximab, and pertuzumab, ADCC is an integral part of their mechanism of action (2, 3). For others, ADCC may be responsible for some undesirable side effects. It is thus important to document the effector functions in the early stages of the research and development of therapeutic antibodies. Moreover, the affinity of the FcγRIIIa for human IgG is not only affected by the subclass nature of the IgG isotype but also by the glycan attached to the Asn 297 of the heavy chains (4) and by the aggregation tendency of the antibody. The glycan composition depends on the cell system and the culture conditions used for the production of the therapeutic antibody. The aggregation rate of an antibody depends on its primary sequence (5), the purification process and the storage condition of the final drug product. Therefore, it is mandatory to test for modifications to the effector potential of an antibody after any changes in the process to its production.

Easy to handle and low protein consuming SPR based experiments represent valuable surrogates to in vitro and in vivo ADCC assays to test any step of the production process for such potential modifications.

For this purpose, protein A is chemically coupled to a sensor chip. A solution of antibody (a purified solution or even a crude cell media supernatant) is captured by the protein A and is used as ligand. The soluble forms of FcγR are injected at different concentrations and are used as analyte. This setup allows the definition of the kinetic rates of binding of the FcγRI (high affinity) but also the FcγRIIIa (low affinity) on the Fc domain of IgGs.

2 Materials

1. The experiments were run on a Biacore X100.
2. CM5 sensor chips (Biacore).
3. HBS-EP + as the running buffer:
   - 10 mM HEPES [*N*-(2-hydroxyetyl) piperazine-*N'*-(2-ethanesulfonic acid)].
   - 150 mM NaCl.
   - 3 mM EDTA (Ethylenediaminetetraacetic acid).
   - 0.05% (v/v) surfactant P20 (Polyoxyethylene sorbitan) pH 7.4.
4. Protein A from *Staphylococcus aureus* (Sigma ref. P7837).
5. Trastuzumab (Herceptin), Roche.
6. *N*-hydroxysuccinimide (NHS).

7. 1-Ethyl-3-(3-dimethylaminopropyl)carbodiimide hydrochloride (EDC).
8. Sodium acetate.
9. Ethanolamine hydrochloride.
10. Sodium hydroxide.
11. Glycine HCl.
12. Soluble recombinant h-FcγRI (R&D Systems, ref 1257-FC-050).
13. Soluble FcγRIIIa (R&D Systems, ref 4325-FC-050).

3 Methods

3.1 Coupling of Protein A on the CM5 Sensor Chip

1. Protein A was coupled on both flowcells using the amine coupling kit.
2. The two flowcells were prepared successively as follow at a flow rate of 5 μl/min.
3. The carboxymethyldextran matrix was first activated by a 7 min injection of a 50/50 (v/v) mix of *N*-hydroxysuccinimide (NHS) 0.1 M and 1-ethyl-3-(3-dimethylaminopropyl)carbodiimide hydrochloride (EDC) 0.39 M.
4. A 200 μg/ml solution of protein A was prepared in a 10 mM sodium acetate pH 5.0 buffer and injected for 7 min on the activated flowcell.
5. At the end of this injection, the non-reacting activated carboxylic functions were quenched by a 7 min injection of a 1 M ethanolamine hydrochloride, NaOH pH 8.5 solution.
6. 2,984 RU and 2,935 RU of protein A were chemically coupled to the Flowcells 1 and 2, respectively.

3.2 Single Cycle Kinetic with Trastuzumab (Herceptin®) and the h-FcγRI

1. A 2 μg/ml solution of antibody was prepared by diluting the trastuzumab stock solution (21 mg/ml) in the running buffer.
2. The solution was injected for 1 min at a flow rate of 30 μl/min on the flowcell 2 solely.
3. This injection allows the capture of around 700 RU of h-IgG1 (see Note 1).
4. A solution of the soluble recombinant h-FcγRI was prepared at 200 nM (assuming a molecular weight of 50 kDa) in the running buffer.
5. A range of five concentrations from 40 to 1 nM was prepared by a 2.5-fold serial dilution scheme.

6. The five concentrations were injected in the growing order for 90 s each at a flow rate of 30 μl/min.
7. After the last injection, a delay of 600 s was left for the dissociation phase study.
8. As a second reference, the running buffer was injected five times in the same conditions as the h-FcγRI concentration range.
9. At the end of each cycle, the antibody/hFcγRI complexes were discarded by a 30 s injection of 10 mM Glycine HCl pH 1.7 buffer at a flow rate of 30 μl/min.
10. The sensorgrams correspond to the flowcell 2 signal corrected by the reference flowcell 1 signal (without any captured antibody).
11. The blank sensorgram was then subtracted from the sensorgram obtained with the recombinant hFcγRI range.
12. The corrected data were fitted using the Biacore X100 evaluation software with the Langmuir (1:1) model (see Notes 2 and 7).

3.3 Single Cycle Kinetic with Trastuzumab and the h-FcγRIIIa

1. Except for the concentration range, the experiment was identical to 3.2.
2. A solution of the recombinant soluble FcγRIIIa was prepared at 2,000 nM (assuming a molecular weight of 45 kDa) in the running buffer.
3. A range of five concentrations from 2,000 to 125 nM was prepared by a twofold serial dilution scheme.
4. The data were analyzed with the Biaevaluation software using the Langmuir (see Note 3), the heterogeneous ligand (see Notes 4, 7 and 8), and the two state reaction models (see Notes 5–8).

4 Notes

1. Protein A–CM5 Sensor Chip captured the h-IgG1 very efficiently. The capture of trastuzumab is perfectly reproducible, very stable during the five injections of HBS-EP + buffer plus the 6 min delay at the end of each cycle, the antibodies are efficiently removed by a single 30 s injection of a 10 mM Glycine pH 1.7 buffer (see Fig. 1). As expected, (6) this capture enables the binding of the recombinant soluble h-FcγR (I and IIIa) (see Note 9). This setup enables testing antibody batches before the purification step. Even, a crude cell culture supernatant may be used. This useful property would not be reached with the inverse setup where the FcγR are used as ligand and the antibodies as analyte. Moreover, this last setup presents some other drawbacks (see Note 10).

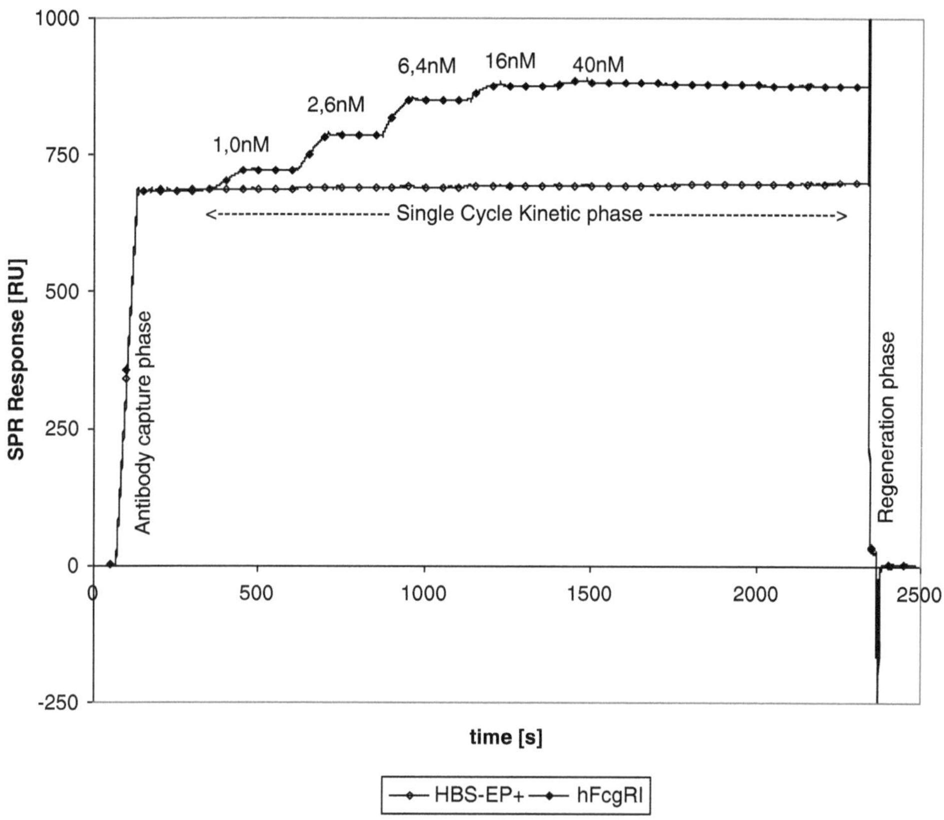

Fig. 1 Superposition of two sensorgrams corresponding to an experimental and the blank reference of a Single Cycle Kinetic experiment. IgG1 antibody (Trastuzumab) was captured on the flowcell 2 of a CM5–protein A grafted sensor chip (2,984 and 2,935 RU of protein A have been coupled, respectively, on the flowcells one and two). A growing range of five concentrations of h-FcγRI (from 1.0 to 40 nM) are injected during 1990s each after the capture phase. As a blank reference the running buffer was injected five times in the same conditions. After the fifth injection a 600 s delay was left to study the dissociation phase. At the end of each cycle the surface was regenerated by a 30 s injection of a 10 mM Glycine pH 1.7 buffer. The experiment was carried out on a Biacore X100 at a flow rate of 30 μl/min and at the temperature of 25°C

2. The kinetic binding of h-FcγRI on the h-IgG1 follows a Langmuir (1:1) model. The Langmuir model properly fits the single cycle kinetic experimental (see Fig. 2 and Table 1). The ratio $Chi^2/R_{max} = (1.41\ RU^2/186.4\ RU) = 7.6 \times 10^{-3}$ RU. This value is largely under the commonly used acceptable limit of 0.05 RU. The R_{max} value (186.4 RU) is in the range of the theoretical R_{max} value (229 RU) (see Note 11). The refractive index (RI) contribution is not significant: $-1.71\ RU < RI < 0.94\ RU$ (see Note 12). With these experimental conditions, the association kinetic rate k_a is equal to $(2.232 \pm 0.003) \times 10^6\ M^{-1}\ s^{-1}$ and the dissociation rate k_d is equal to $(6.82 \pm 0.13) \times 10^{-5}\ s^{-1}$. This latter value is close but above the $2 \times 10^{-5}\ s^{-1}$ limit. The dissociation

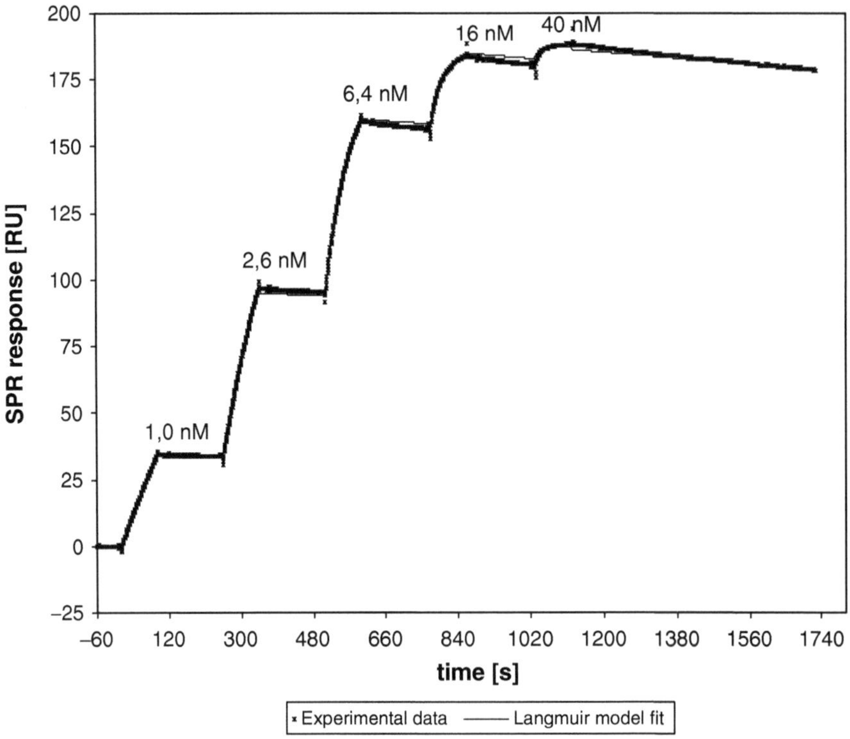

Fig. 2 Analysis of the single cycle kinetic experimental data corresponding to five injections in the growing range of concentrations (from 1 to 40 nM) of h-FcγRI on around 700 RU of captured h-IgG1 (Trastuzumab) corrected by the five HBS-EP + injections used as the double reference. These data are fitted with a Langmuir (1:1) model with the following parameters ($k_a = 2.23 \times 106\ M^{-1}\ s^{-1}$, $k_d = 6.8 \times 10^{-5}\ s^{-1}$, $R_{max} = 186.4$ RU, $Chi^2 = 1.14\ RU^2$). The experiment was carried out on a Biacore X100 at a flow rate of 30 μl/min and at the temperature of 25°C

constant K_D corresponding to the k_d/k_a ratio is equal to 30.6 ± 0.7 pM.

3. Kinetic study of binding of h-FcγRIIIa on the h-IgG1: The kinetic with h-FcγRIIIa do not follow the Langmuir model. The Langmuir model improperly fits the single cycle kinetic experimental data (see Fig. 3 and Table 1). The ratio $Chi^2/R_{max} = (23.4\ RU^2/141.6\ RU) = 0.165$ RU. This value is largely above the 0.05 RU acceptable limits. The contribution of the refractive index is high too ($-14.2\ RU < RI < 3.6\ RU$). The calculated dissociation rate (k_d) seems to be overestimated (quicker than the experimental data). With this model, the dissociation constant K_D is equal to 421 ± 64 nM.
4. Kinetic study of binding of h-FcγRIIIa on the h-IgG1: The heterogeneous model This model considers the ligand (i.e., the antibody) as a mix of two components and defines for each one a k_a, a k_d, and a R_{max}. This model improves at least the fitting of the dissociation phase (see Fig. 4 and Table 1). With this

Table 1
Kinetic parameters of the binding of the two soluble recombinant h-FcγR at 25°C on Trastuzumab captured on a Protein A–CM5 sensor chip

	h-FcγRI/h-IgG1	h-FcγRIIIa/h-IgG1		
Mathematical models	Langmuir	Langmuir	Heterogeneous ligand	Two state reaction
k_{a1}[a]	$2.232 \pm 0.003 \times 10^6$ M^{-1} s^{-1}	$2.33 \pm 0.18 \times 10^5$ M^{-1} s^{-1}	$3.36 \pm 0.34 \times 10^5$ M^{-1} s^{-1}	$7.12 \pm 0.90 \times 10^5$ M^{-1} s^{-1}
k_{d1}[a]	$6.82 \pm 0.13 \times 10^{-5}$ s^{-1}	$98.1 \pm 7.4 \times 10^{-3}$ s^{-1}	$13.8 \pm 0.7 \times 10^{-3}$ s^{-1}	$254 \pm 32 \times 10^{-3}$ s^{-1}
K_{D1}	–	–	41.2 ± 6.2 nM	–
R_{max1}	–	–	18.8 RU	–
k_{a2}	–	–	$2.89 \pm 0.10 \times 10^5$ M^{-1} s^{-1}	$3.33 \pm 0.35 \times 10^{-3}$ s^{-1}
k_{d2}	–	–	$145.9 \pm 4.6 \times 10^{-3}$ s^{-1}	$14.7 \pm 1.0 \times 10^{-3}$ s^{-1}
K_D[b]	30.6 ± 0.1 pM	421 ± 64 nM	505 ± 33 nM	291 ± 12 nM
R_{max}[b]	186.4 RU	141.6 RU	107.0 RU	118.6
Chi^2	1.14 RU^2	23.4 RU^2	12.4 RU^2	14.3 RU^2

[a] k_a and k_d for the Langmuir model
[b] K_{D2} and R_{max2} for "Heterogeneous ligand" model

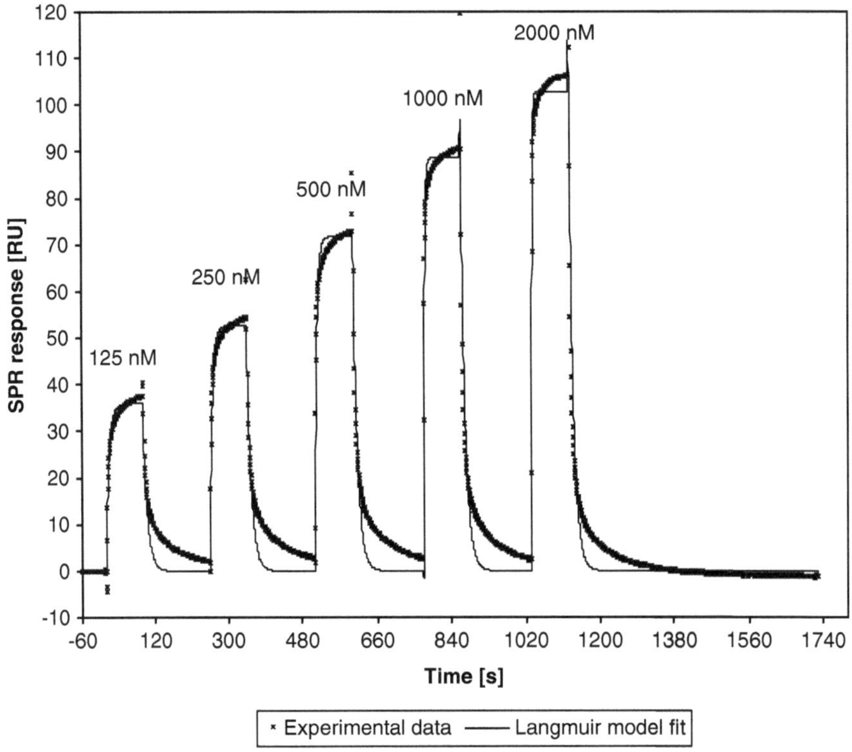

Fig. 3 Analysis of the single cycle kinetic experimental data corresponding to five injections in the growing range of concentrations (from 125 to 2,000 nM) of h-FcγRIIIa on around 700 RU of captured h-IgG1 (Trastuzumab) corrected by the five HBS-EP+ injections used as the double reference. These data are fitted with a Langmuir (1:1) model with the following parameters ($k_a = 2.3 \times 105\ M^{-1}\ s^{-1}$, $k_d = 98 \times 10^{-3}\ s^{-1}$, $R_{max} = 141.6$ RU, $Chi^2 = 23.4\ RU^2$). The experiment was carried out on a Biacore X100 at a flow rate of 30 μl/min and at the temperature of 25°C

model, the major (85.1%) and the minor (14.9%) component of the captured antibodies (see Note 13) present a dissociation constant K_D of 505 ± 33 and 41.2 ± 6.2 nM, respectively. The proportion defined by the kinetic analyses has to be compared with the published ratio of fucosylated (76%) and afucosylated (24%) glycan composition of trastuzumab (7) and the dissociation constant of a fully fucosylated (300 nM) and a fully afucosylated (7.2 nM) antibody with a soluble recombinant FcγRIIIa (4). So, this model makes sense even if the glycan heterogeneity is generally more complex than a simple binary component system (see Note 14).

Fig. 4 (**a**) Analysis of the same experimental data shown in Fig. 3 with the "Heterogeneous Ligand" model corresponding to the following parameters ($k_{a1} = 3.4 \times 105\ M^{-1}\ s^{-1}$, $k_{d1} = 14 \times 10^{-3}\ s^{-1}$, $R_{max1} = 18.8$ RU, $k_{a2} = 2.9 \times 105\ M^{-1}\ s^{-1}$, $k_{d2} = 146 \times 10^{-3}\ s^{-1}$, $R_{max2} = 107.0$, $Chi^2 = 12.4\ RU^2$). The Bulk and Drift (RI) component was fixed to 0. (**b**) Decomposition of the fitted signal in its two components (AB1: binding of h-FcγRIIIa on the minor kind of antibody, and AB2: binding of h-FcγRIIIa on the major kind of antibody)

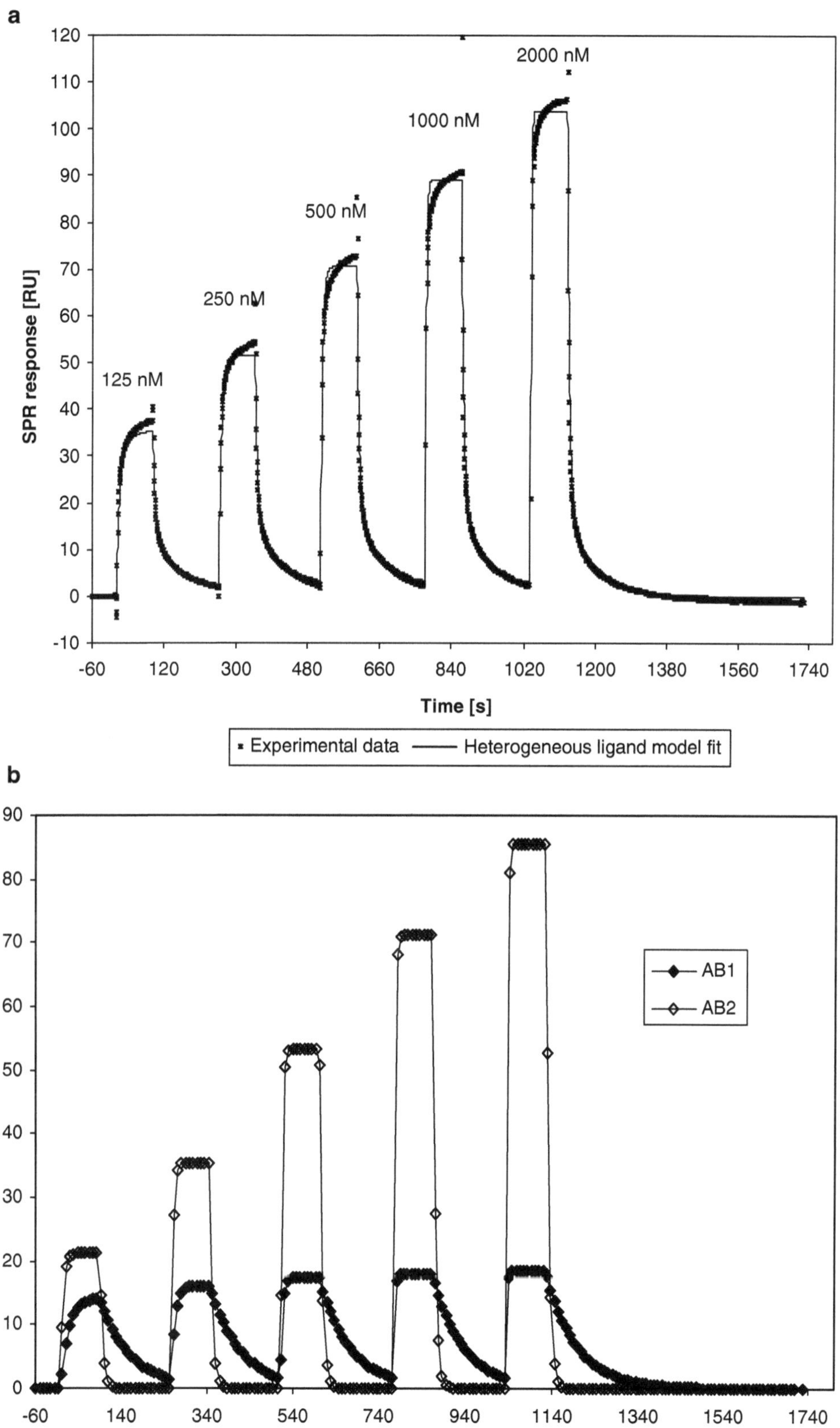
a
120
110
100
90
80
70
60
50
40
30
20
10
0
-10
SPR response [RU]
125 nM
250 nM
500 nM
1000 nM
2000 nM
-60
120
300
480
660
840
1020
1200
1380
1560
1740
Time [s]
Experimental data
Heterogeneous ligand model fit
b
90
80
70
60
50
40
30
20
10
0
-60
140
340
540
740
940
1140
1340
1540
1740
AB1
AB2

5. The two state reaction model supposes a conformational change induced by the binding of the FcγRIIIa to the Fc domain of the IgG1 antibody. This assumed mechanism is based on the 3D structure of the complex FcγRIIIa/Fc (PDB ID code 3SGJ) (4). This model, in accordance with other 3D models, shows that the FcγRIIIa makes close contact with both chains of the human Fc domain (see Note 15). At first, the Fc receptor may interact with one arm and then "lock" its position by slightly pushing aside to the second chain. Some glycan compositions (i.e., afucosylated structures) could facilitate this conformational change.
6. This model defined five kinetic parameters: two association rates, two dissociation rates and one R_{max} (see Table 1). The Chi^2 value of the fitting (14.3 RU^2) is slightly higher than the Chi^2 of the heterogeneous model (12.4 RU^2). Nonetheless this fitting is better than the fitting with the Langmuir model. The dissociation constant K_D, calculated with the four kinetic rates (see Note 16), is equal to 291 ± 12 nM.
7. The stable and easy to regenerate capture of the human IgG1 by protein A sensor chip is very efficient for the study of the binding of the Fcγ receptors. The high affinity FcγRI may be used as an assay to check the integrity of the antibody structure. The FcγRIIIa kinetic study is a little bit trickier. Therefore, the steady state analysis is generally used for the dissociation constant definition of the FcγRIIIa/Fc interaction. However, the kinetic rate definition gives more information than a sole K_D.
8. A kinetic study with different glycan compositions of a same antibody would be of great interest to understand how the carbohydrate portion of the Fc domain has this tremendous effect on the FcγRIIIa binding by finding a correlation between the glycan composition and the kinetic rates calculated at least by one of the two proposed models.
9. The protein A from *S. aureus* presents 4 IgG binding domains. Each of which is able to react with the "elbow" region between the CH2 and the CH3 domains of an IgG heavy chain (PDB ID code 1FC2). This interaction domain is apart from the binding domain of the FcγR.
10. There are at least two solutions for using the recombinant soluble FcγR as the ligand and the antibodies as the analyte. The first one is a chemical coupling of the soluble receptor. This approach needs to conjugate low quantities of protein to minimize the mass transport limit and to find a regeneration condition that does not unfold the coupled protein. When the device allows only two flowcells, one sensor chip per receptor is required. The second way is the capture of the receptor with an anti-Tag antibody (anti-6 His c-terminal Tag generally).

The capture allows assays with different receptors on the same sensorchip. Running the experiment, this capture seems very accurate: very stable capture and efficient regeneration. However, the h-FcγR interacts with the Fc domain of anti-Tag His antibody too (at least with mouse antibodies). This interaction complicates the analysis. NTA-sensor chip may be used to capture His-Tag protein to avoid this drawback. This surface has not been tested in house.

11. The theoretical R_{max} value is calculated with the following formula:

    Captured ligand level $(RU) \times MW_{analyte}/MW_{ligand}$ (MW = molecular weight in Da).

12. SPR based technology measures changes in the refractive index at the surface of the sensorchip. The analyte solutions may have a refractive index different from the refractive index of the running buffer. These changes are responsible for the bulk effect represented in the mathematical models by the Refractive Index value. The use of a reference flowcell is supposed to cancel the bulk effect on the sensorgrams. So generally the RI values have to tend to 0.

13. The proportion of the two components is estimated by the R_{max1} and R_{max2} ratio.

14. For instance the published oligosaccharide profile of Trastuzumab (6) is:

    -N-N-M[M]-M[M]-M: 5%, -N[F]-N-M[M]-M-N: 2%.

    -N-N-M[M-N]-M-N (G0): 8.5%, -N[F]-N-M[M-N]-M-N (G0F): 43%.

    -N-N-M[M-N]-M-N-G (G1): 8%, -N[F]-N-M[M-N]-M-N-G (G1F): 28%.

    -N-N-M[M-N-G]-M-N-G (G2): 3%, -N[F]-N-M[M-N-G]-M-N-G (G2F): 2.5%.

    With N = *N*-acetylglucosamine, M = Mannose, F = Fucose, G = Galactose, and [X] represents a branched oligosaccharide sequence.

    Each glycosylation composition may have an impact on the binding of the FcγRIIIa.

15. The FcγRIIIa makes close contacts (<5 Å) with the residues belonging to the following sequences of one chain: PELLGGPS^{239} + KALPAPI^{332} and with the following sequences of the other chain: PELLGGPS^{239} + DVSHE^{269} + YN[NAG-FUCα1-6]ST^{299} + NKA^{327}.

16. With the two state reaction model (Fig. 5), the K_D is calculated with the following formula:

$$K_D = \frac{(k_{d1}/k_{a1})}{(1 + (k_{a2}/k_{d2}))}.$$

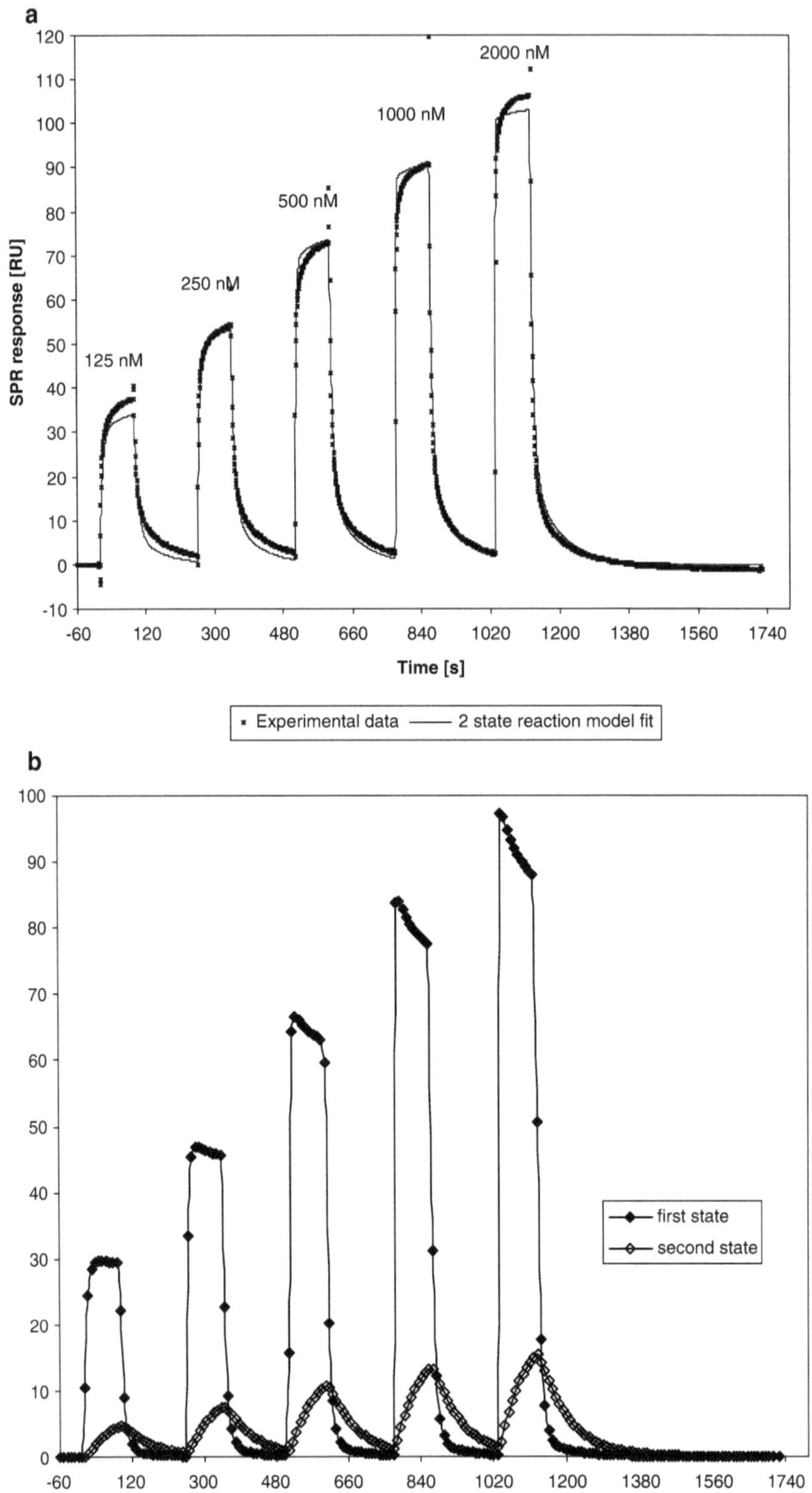

Fig. 5 Analysis of the same experimental data shown in Fig. 3 with the "Two state reaction" model corresponding to the following parameters ($k_{a1} = 7.1 \times 105\ M^{-1}\ s^{-1}$, $k_{d1} = 254 \times 10^{-3}\ s^{-1}$, $k_{a2} = 3.3 \times 0^{-3}\ s^{-1}$, $k_{d2} = 15 \times 10^{-3}\ s^{-1}$, $R_{max2} = 118.6$, $Chi^2 = 14.3\ RU^2$). The Bulk and Drift (RI) component was fixed to 0. (**b**) Decomposition of the fitted signal in its supposed two states of the binding of h-FcγRIIIa on the antibody

References

1. Nimmerjahn F, Ravetch JV (2008) Fcgamma receptors as regulators of immune responses. Nat Rev Immunol 8:34–47
2. Carton G, Watier H, Golay J, Solal-Celigny P (2004) From the bench to the bedside: ways to improve rituximab efficacy. Blood 104: 2635–2642
3. Jiang XR, Song A, Bergelson S, Arroll T, Parekh B, May K, Chung S, Strouse R, Mire-Sluis A, Schenerman M (2011) Advances in the assessment and control of the effector functions of therapeutic antibodies. Nat Rev Drug Discov 10:101–110
4. Ferrara C, Grau S, Jäger C, Sondermann P, Brünker P, Waldhauer I, Hennig M, Ruf A, Rufer AC, Sthile M, Umaňa P, Benz J (2011) Unique carbohydrate–carbohydrate interactions are required for high affinity binding between FcγRIII and antibodies lacking core fucose. Proc Natl Acad Sci U S A 108: 12269–12674
5. Wang X, Das TK, Singh SK, Kumar S (2009) Potential aggregation prone regions in biotherapeutics: a survey of commercial antibodies. MAbs 1:254–267
6. Sulica A, Medesan C, Laky M, Onică D, Sjöquist J, Ghetie V (1979) Effect of protein A of Staphylococcus aureus on the binding of monomeric and polymeric IgG to Fc receptor-bearing cells. Immunology 38:173–179
7. Juntilla TT, Parsons K, Olsson C, Lu Y, Xin Y, Theriault J, Crocker L, Pabonan O, Baginski T, Meng G, Totpal K, Kelley RF, Sliwkowski MX (2010) Superior in vivo efficacy of afucosylated Trastuzumab in the treatment of HER2-amplified breast cancer. Cancer Res 70:4481–4489

Chapter 22

Mass Spectrometry Protocol for the Absolute Quantification of a Monoclonal Antibody in Serum with Immunopurification

François Becher, Mathieu Dubois, François Fenaille, and Eric Ezan

Abstract

We present here an analytical protocol for the sensitive, specific, and accurate absolute quantification of cetuximab, a human:murine chimeric monoclonal antibody, using mass spectrometry. Extraction from human serum is performed with micrometric magnetized beads, functionalized with soluble epidermal growth factor receptor (sEGFR), the pharmacological target of cetuximab. This specific immunocapture step allows sample purification and, in parallel, assessment of the antibody's biological potency. The eluted mAb is digested with trypsin and specific peptides from light and heavy chains are monitored by liquid chromatography coupled with tandem mass spectrometry operated in the selected reaction monitoring (SRM) mode. The limit of quantification of the assay was 20 ng/mL in serum.

Key words Recombinant antibody, Mass spectrometry, Quantification, Serum, Immunopurification

1 Introduction

Discovery and development of monoclonal antibodies (mAbs) as therapeutic agents continue to grow rapidly within the pharmaceutical industry. Reliable bioanalytical methods are crucial to support the preclinical and clinical studies. Mass spectrometry is now an alternative to immunoassay methods for absolute quantification of mAbs. Recently, a few analytical approaches have proven applicable to clinical samples for pharmacokinetic evaluation (1–4). All these approaches are based on detection of proteotypic peptides, following trypsin digestion of the mAb, by liquid chromatography coupled with tandem mass spectrometry operated in the selected reaction monitoring (SRM) mode. Differences mainly relate to plasma/serum extraction, using major serum protein depletion, solid phase extraction, or immunopurification. The extraction step is mandatory to concentrate the sample and separate the mAb from the main endogenous proteins, especially immunoglobulins, before mass spectrometry analysis.

Alain Beck (ed.), *Glycosylation Engineering of Biopharmaceuticals: Methods and Protocols*, Methods in Molecular Biology, vol. 988,
DOI 10.1007/978-1-62703-327-5_22, © Springer Science+Business Media New York 2013

Using cetuximab (Erbitux®) as a model (1), we have recently illustrated the potential of immunopurification employing the mAb pharmacological target to enhance sensitivity and reach a detection limit at low ng/mL concentrations in human serum. It should also be noted that this approach makes the assay specific to the antibody active form, able to bind to the pharmacological target. We present here the detailed protocol of the assay with a description of each analytical step.

2 Materials

2.1 Proteotypic Peptide Identification by LC-MS/MS

1. Buffers and reagents:
   - Nano HPLC mobile phase A: Water containing 0.1% (v/v) formic acid.
   - Nano HPLC mobile phase B: Acetonitrile containing 0.1% (v/v) formic acid.
2. Lab equipment:
   - Nano LC-MS/MS: Ultimate 3,000 nanoLC (Dionex, Sunnyvale, CA) coupled with a QTOF-I (Waters, Milford, MA) equipped with a nanoelectrospray ion source and a PicoTip Emitter (FS360-20-10-D20-C7, New Objective Inc., Woburn, MA).
   - Maldi-TOF: Voyager DE-STR instrument (Applied Biosystems, Les Ulis, France) equipped with a pulsed nitrogen laser (337 nm). Samples were co-crystallized with 10 mg/mL alpha-cyano-4-hydroxycinnamic acid in acetonitrile/water (50/50, v/v) with 0.1% trifluoroacetic acid.

2.2 Immuno-purification from Serum

1. Buffers and reagents:
   - M-280 tosylactivated magnetic beads (Dynal Biotech, Oslo, Norway).
   - 0.1 M borate buffer (pH 9.5): 1.23 g/200 mL Milli-Q water.
   - PBS buffer (pH 7.4, 0.1% BSA): 0.15 g NaH_2PO_4, 1.45 g Na_2HPO_4, 4.39 g NaCl, 0.5 g BSA/500 mL Milli-Q water
   - 200 mM Tris buffer (pH 8.5, 0.1% BSA): 2.42 g Tris-HCl, 0.1 g BSA/100 mL Milli-Q water.
   - 50 mM Tris buffer (pH 8.5, 100 μg/mL BSA): 1.51 g Tris/250 mL Milli-Q water, add 2.5 mL BSA 10 mg/mL.
   - 50 mM Tris buffer (pH 2, 100 μg/mL BSA): 2 mL of 50 mM Tris buffer (pH 8.5, 100 μg/mL BSA), add 70 μL 1 M HCl.
2. Lab equipment: Magnetic Particle Concentrator for Microcentrifuge Tubes (Dynal MPC-S), Dynal Biotech, Oslo, Norway.

2.3 Trypsin Digestion

1. Buffers and reagents:
   - Basic dithiothreitol (10 mM in 250 mM NaOH): 1.34 mg/870 μL 250 mM NaOH.
   - Sequencing grade modified trypsin (Promega): Resuspend 20 μg lyophilized trypsin in 1 mL Milli-Q water.
2. Lab equipment: Water bath 40 and 60°C.

2.4 Internal Standard

Buffers and reagents:

- Mouse monoclonal antibody against human EGFR (R&D Systems, Minneapolis, USA). 500 μg resuspended in 1 mL PBS buffer (pH 7.4).
- PBS buffer (pH 7.4): 0.15 g NaH_2PO_4, 1.45 g Na_2HPO_4, 4.39 g NaCl/500 mL Milli-Q water.

2.5 LC-MS/MS

1. Buffers and reagents:
   - Mobile phase A: Water/butanol (99/1, v/v) containing 0.1% (v/v) formic acid.
   - Mobile phase B: Acetonitrile containing 0.1% (v/v) formic acid.
2. Lab equipment:
   - A triple quadrupole mass spectrometer coupled to an HPLC system running at 0.3 mL/min is necessary for quantification of the peptides derived from cetuximab. The equipment referred to in this article is a Quattro LCZ (Waters, Manchester, UK) coupled to an HPLC Alliance. Another triple quadrupole mass spectrometer can be used for this application. Assay sensitivity will be dependent on the equipment used.
   - Chromatography was performed on a C18 guard column followed by a Zorbax SB-C18 column (150×2.1 mm i.d., 5 μm particle size, 80 Å porosity) (Agilent Technologies, Massy, France).

3 Methods

The mass spectrometry methods for absolute quantification of mAbs are similar to those for other large therapeutic proteins or biomarkers, which are now well described in the literature. The strategy involved in a first step the selection of tryptic peptides representative of and specific to the mAb. The variable regions of both the light and heavy chains must be targeted for detection specificity. Generally, high-resolution instruments such as Q-TOF or Orbitrap are used for peptide identification. For quantification, these selected peptides are monitored on a triple quadrupole instrument in the SRM mode, which requires fine tuning with synthetic peptides.

Selective extraction of the mAb from serum is a crucial step due to its similarity to endogenous immunoglobulins, which are among the most abundant proteins in plasma, with concentrations around 10 mg/mL. This is 1,000–10,000-fold greater than that of circulating therapeutic antibodies. In the example of cetuximab, we implemented an immunocapture of the antibody form that binds its pharmacological target. For ease of assay development and as described previously (5), we used magnetic beads that were covalently linked to sEGFR, the target of cetuximab. We demonstrated that this selective purification of the active mAb from other IgGs and plasma proteins allows its efficient concentration from a 500 μL serum sample and reduction of the matrix effect.

Some other teams have published LC-MS/MS assays for mAbs (2–4), using less specific approaches for plasma/serum extraction, such as solid phase extraction or major plasma protein depletion (which will not be described here). These protocols are more convenient during routine bioanalytical work and also do not need the pharmacological target protein. However, sensitivity is limited to the μg/mL range (ng/mL range with immunocapture) and the mAb active form cannot be discriminated.

3.1 Proteotypic Peptide Identification by LC-MS/MS

The first step of assay development is to select specific peptides of cetuximab, from among those produced by digestion with trypsin, with the most intense signal in mass spectrometry (see Note 1). Their molecular weights are determined by both nanoLC-ESI-TOF-MS and MALDI-TOF-MS techniques and are compared with predicted peptides from theoretical digestion (using Mass-Lynx V4.0). Peaks are assigned to tryptic peptides by matching measured to expected molecular weights, taking into account the mass accuracy of the instrument. Details of this work are not described here since this is a common proteomic task that can be achieved with other lab equipment (Obitrap, FT-ICR) and protocols.

Specifically for cetuximab, we searched for peptides from the variable regions located in the Fab fragments and involved in the binding to sEGFR, which must have unique peptide sequences, allowing differentiation from human plasma IgG1. Among these potential targets for quantification, those without miscleavage and showing higher desorption and ionization efficiency were selected. This led to the choice of one peptide from the light chain (LT3) and another from the heavy chain (HT4). In addition, to estimate nonspecific binding of endogenous human IgG to magnetic beads during the immunocapture step, a third tryptic peptide (HT 23) of the Fc region was monitored. Peptide sequences are indicated in Table 1.

Table 1
Selected proteotypic peptides and main mass spectrometer settings for selected reaction monitoring detection

Peptide	Sequence	Precursor ion (*m/z*)	Daughter ion (*m/z*)	Collision energy (eV)
LT3 (25–39)	ASQSIGTNIHWYQQR	597.3 (3+)	652.1	15
		895.1 (2+)	1,303.1	34
HT4 (44–66)	GLEWLGVIWSGGNTDYNTPFTSR	857.5 (3+)	908.7	15
		1,268.2 (2+)	1,703.1	34
HT23 (304–319)	VVSVLTVLHQDWLNGK	603.8 (3+)	805.8	15
		904.8 (2+)	997.8	32

3.2 Preparation of Magnetic Beads Coated with sEGFR

In this step, the pharmacological target of cetuximab, sEGFR, was covalently linked to functionalized M-280 tosylactivated magnetic beads. The free amino groups of sEGFR react with the tosylate groups on the beads.

Another possible approach is to link an antibody against cetuximab to the beads, via protein G affinity, for instance. Anti-human antibody would be not specific enough since all IgGs from serum would be recognized, leading to bead saturation. However, an anti-idiotype, if available, that recognizes the specific binding regions of cetuximab could be an alternative.

1. Beads are stored at +4°C. Vortex the beads for 30 s and collect 300 μL (at 30 mg of beads/mL) in a 1.5 mL tube. Add 500 μL borate buffer (0.1 M, pH 9.5), place the tube on the magnet, and expel the supernatant.
2. Repeat step 1 twice.
3. Incubate the beads for 4 h at room temperature with gentle shaking with sEGFR (diluted to 100 μg/mL in borate buffer) at a ratio of 3 μg of protein per mg of beads. sEGFR covalently binds to the beads by reaction of the protein amino groups with the tosylate groups on the beads.
4. Remove non-covalently bound sEGFR by washing (as described in step 1) twice with 500 μL PBS buffer (pH 7.4, 0.1% BSA).
5. Resuspend the magnetic beads in 500 μL of Tris buffer (0.2 M, pH 8.5, 0.1% BSA) and incubate overnight at 4°C to block free tosylate groups.
6. Wash the beads once with 500 μL PBS buffer (pH 7.4, 0.1% BSA), and then resuspend in Tris–HCl (50 mM, pH 8.5, 100 μg/mL BSA) at a concentration of 90 mg of beads per mL.

3.3 Immunopurification from Serum

The method was developed using 500 μL of serum, a sample volume which allows tenfold concentration since beads are recovered, after cetuximab extraction, in 50 μL of buffer. Due to the specificity of immunocapture, extraction can be done using larger sample volumes leading to a proportional increase in sensitivity. In this case, the volume of beads should be adjusted to avoid saturation of binding sites.

An internal standard (IS) is necessary to compensate for possible fluctuation during immunopurification/digestion or LC-MS/MS quantification. It must be added at the beginning of the sample process. Here, we selected a mouse antibody against sEGFR, spiked in each serum sample to correct for immunoaffinity and digestion variability. As for cetuximab, a specific proteotypic peptide from the variable domain, co-eluting with LT3, was identified and selected for quantification.

A major difficulty with protein quantification is nonspecific binding to surfaces (Eppendorf pipettes, tubes, beads, etc.). This phenomenon was observed for cetuximab at low concentration (below 1 μg/mL) after the immunoaffinity step. Addition of BSA to the washing and elution buffers was found to avoid this nonspecific binding. Concentrations of BSA ranging from 10 to 1,000 μg/mL were tested and 100 μg/mL was selected.

1. Add the internal standard to the serum sample at a concentration of 500 ng/mL.
2. Add 0.9 mg (10 μL) of beads to 500 μL of serum.
3. Incubate the serum sample for 2 h at room temperature with gentle rotation.
4. Wash twice with 500 μL of Tris–HCl (50 mM, pH 8.5, 100 μg/mL BSA).
5. Recover the beads and discard the supernatant.
6. Elute cetuximab and internal standard under acidic conditions with 50 μL of Tris–HCl (50 mM, pH 2, 100 μg/mL BSA) for 15 min while vortexing.

3.4 Trypsin Digestion

Digestion of cetuximab and internal standard is performed with trypsin. The main precaution is to ensure that the pH is at 8.5 for maximum efficiency of trypsin. NaOH is added before trypsin to neutralize the elution buffer.

1. Transfer all the supernatant (50 μL) to an Eppendorf pipette.
2. Add 10 μL of basic dithiothreitol (10 mM in 250 mM NaOH) to ensure pH 8.5.
3. Leave the tubes for 30 min at 60°C.
4. Add 0.1 μg (in 5 μL) of trypsin.
5. Incubate the samples for 3 h at 40°C.

6. Add 10 μL of methanol containing 10% formic acid to stop the reaction.
7. Transfer the samples to an HPLC vial (40 μL injected).

3.5 Cetuximab Quantification by LC-MS/MS

Following cetuximab digestion by trypsin, quantification consists of the analysis of selected peptides. Analytical methods for peptide quantification are well described in the literature (6). The approach described below combines LC separation on a C18 reversed phase column and detection of the peptides on a triple quadrupole instrument operating in the SRM mode. The column is eluted at a flow rate of 0.3 mL/min using as mobile phase A water/butanol (99/1, v/v) and as mobile phase B acetonitrile, both containing 0.1% (v/v) formic acid. The analytical gradient profile was as follows (min/% of mobile phase B in A): 0.0/5, 4.0/25, 12.0/37, 12.1/100, 13.0/100, 13.1/5, and 17.0/5. Column temperature was 50°C.

The retention times of LT3, HT23, and HT4 were 5.9, 9.1, and 11.6 min, respectively. Mass spectrometry consists of electrospray ionization in the positive ion mode followed by the acquisition of two SRM transitions for each peptide (see Table 22.1). It was checked that the ratio of the two SRM transitions for each peptide was constant in each run to ensure detection specificity. The two specific peptides from the variable part of cetuximab were monitored (LT3 and HT4) and LT3 was used for quantification. HT4 was used as a confirmation peptide and HT4 over LT3 area ratio was measured. The peptide from the Fc part (peptide HT23) was also monitored to estimate the level of nonspecific IgG binding to beads.

A calibration curve was prepared and analyzed in each run. Human serum was spiked with cetuximab at concentrations ranging from 20 to 2,000 ng/mL. Sample treatment was as above. Cetuximab concentration in the unknown samples was then determined by comparing the measured area ratio LT3/IS with the established calibration curve (see Note 2).

4 Notes

1. Assay variability below 20% is achieved by paying particular attention during the incubation to the washing and elution steps used to immunopurify the antibody from serum. Also, selection of proteotypic peptide is a crucial step in ensuring assay specificity and robustness. Peptides containing miscleavage sites or residues likely to undergo chemical modification during sample handling (i.e., Cys, Trp, Met) should be avoided.
2. Also important is the rigorous monitoring of the whole analytical process using an appropriate internal standard. In this study, we selected a commercial mouse monoclonal antibody

specific to human sEGFR which is co-extracted with cetuximab. An alternative is the addition of tryptic peptides labeled with stable isotopes before the LC-MS/MS analysis, although this does not allow monitoring of the sample preparation steps, such as immunocapture and enzymatic digestion (7). The ideal alternative would be the introduction of stable isotope residues in the whole antibody backbone and the further monitoring of some labeled tryptic peptides (3–6).

References

1. Dubois M, Fenaille F, Clement G, Lechmann M, Tabet JC, Ezan E et al (2008) Immunopurification and mass spectrometric quantification of the active form of a chimeric therapeutic antibody in human serum. Anal Chem 80:1737–1745
2. Hagman C, Ricke D, Ewert S, Bek S, Falchetto R, Bitsch F (2008) Absolute quantification of monoclonal antibodies in biofluids by liquid chromatography-tandem mass spectrometry. Anal Chem 80:1290–1296
3. Heudi O, Barteau S, Zimmer D, Schmidt J, Bill K, Lehmann N et al (2008) Towards absolute quantification of therapeutic monoclonal antibody in serum by LC-MS/MS using isotope-labeled antibody standard and protein cleavage isotope dilution mass spectrometry. Anal Chem 80:4200–4207
4. Yang Z, Hayes M, Fang X, Daley MP, Ettenberg S, Tse FL (2007) LC-MS/MS approach for quantification of therapeutic proteins in plasma using a protein internal standard and 2D-solid-phase extraction cleanup. Anal Chem 79:9294–9301
5. Becher F, Duriez E, Volland H, Tabet JC, Ezan E (2007) Detection of functional ricin by immunoaffinity and liquid chromatography-tandem mass spectrometry. Anal Chem 79: 659–665
6. Van den Broek I, Sparidans RW, Huitema ADR, Schellens JHM, Beijnen JH (2006) Development and validation of a quantitative assay for the measurement of two HIV-fusion inhibitors, enfuvirtide and tifuvirtide, and one metabolite of enfuvirtide (M-20) in human plasma by liquid chromatography-tandem mass spectrometry. J Chromatogr B Biomed Sci Appl 837:49–58
7. Brun V, Dupuis A, Adrait A, Marcellin M, Thomas D, Court M et al (2007) Isotope-labeled protein standards: toward absolute quantitative proteomics. Mol Cell Proteomics 6:2139–2149

Index

Alain Beck (ed.), *Glycosylation Engineering of Biopharmaceuticals: Methods and Protocols*, Methods in Molecular Biology, vol. 988, DOI 10.1007/978-1-62703-327-5, © Springer Science+Business Media New York 2013

P

Q

R

S

T

U

V

Y

MIX
Papier aus verantwortungsvollen Quellen
Paper from responsible sources
FSC® C105338

If you have any concerns about our products,
you can contact us on
ProductSafety@springernature.com

In case Publisher is established outside the EU,
the EU authorized representative is:
Springer Nature Customer Service Center GmbH
Europaplatz 3, 69115 Heidelberg, Germany

Printed by Libri Plureos GmbH
in Hamburg, Germany